MATHEMATICAL SURVEYS
AND MONOGRAPHS SERIES LIST

MATHEMATICAL SURVEYS AND MONOGRAPHS SERIES LIST

ALGEBRAIC GEOMETRY FOR SCIENTISTS AND ENGINEERS

MATHEMATICAL SURVEYS
AND MONOGRAPHS

NUMBER 35

ALGEBRAIC GEOMETRY
FOR SCIENTISTS
AND ENGINEERS

SHREERAM S. ABHYANKAR

American Mathematical Society
Providence, Rhode Island

1980 *Mathematics Subject Classification* (1985 *Revision*). Primary 14-XX.

Library of Congress Cataloging-in-Publication Data

Abhyankar, Shreeram Shankar.
   Algebraic geometry for scientists and engineers/Shreeram S. Abhyankar.
      p.   cm.—(Mathematical surveys and monographs, ISSN 0076-5376; no. 35)
   Includes bibliographical references and index.
   ISBN 0-8218-1535-0 (alk. paper)
   1. Geometry, Algebraic.   I. Title.   II. Series.
QA564.A22   1990   90-815
516.3′5—dc20   CIP

Dedicated to
my father, Professor S. K. Abhyankar,
who taught me algebra and geometry
and to
my master, Professor O. Zariski,
who made it into algebraic geometry

Contents

Preface

What is algebraic geometry, and what is the need for a new book on it? First, we take up the question of what is Algebraic Geometry. Long ago, to a major extent in my father's time, and to a lesser extent in my own time, in high-school and college we learned the two subjects of analytic geometry and theory of equations. Analytic geometry consists of studying geometric figures by means of algebraic equations. Theory of equations, or high school algebra, was manipulative in nature and dealt with simplifying expressions, factoring polynomials, making substitutions, and solving equations. These two subjects were later synthesized into and started being collectively called algebraic geometry. Thus, algebraic geometry, at least in its classical form, is an amalgamation of analytic geometry and the theory of equations.

But, in the last fifty years, algebraic geometry, as such, became more and more abstract, and its original two incarnations, mentioned above, gradually vanished from the curriculum. Indeed, analytic geometry first became a chapter, and then a paragraph, and finally only a footnote in books on calculus. Likewise, its sister discipline of trigonometry, with all the proving of identities, began to be downplayed. Doing all these manipulations was certainly helpful in enhancing the skills needed for solving intricate problems. Similarly, studying subjects like analytic geometry and trigonometry was very useful in developing geometric intuition.

Now, during the last ten years or so, with the advent of the high-speed computer, the need for the manipulative aspects of algebra and algebraic geometry is suddenly being felt in the scientific and engineering community. The growing and dominating abstractions of algebraic geometry notwithstanding, my approach to it remained elementary, manipulative, and algorithmic. In my 1970 poem, "Polynomials and Power Series," and my 1976 article on "Historical Ramblings," I lamented the passing of the concrete attitude and made a plea for its rejuvenation. Thus, it is with great pleasure that I see the recent rise of the algorithmic trend, albeit at the hands of the engineers, and I am happy for the company of their kindred souls.

In this book on algebraic geometry, which is based on my recent lectures to an engineering audience, I am simply resurrecting the concrete and ancient methods of Shreedharacharya (500 A.D.), Bhaskaracharya (1150 A.D.), Newton (1660), Sylvester (1840), Salmon (1852), Max Noether (1870), Kronecker (1882), Cayley (1887), and so on.

In writing this book, I found it extremely helpful to have at my disposal the original notes of these lectures which were taken down by C. Bajaj, professor of computer science at Purdue, V. Chandru, professor of industrial engineering at Purdue, and S. Ghorpade, at one time a mathematics student at Purdue and currently a professor of mathematics at IIT Bombay. My heartfelt thanks to these note-takers, especially to Ghorpade who also helped with the TEXing.

Many people, of course, have aided me in my study of mathematics, but I am particularly grateful to the two persons to whom this book is dedicated: my father, Shankar Keshav Abhyankar, who was a mathematics professor in India and who imparted geometric intuition and manipulative skills to me, and my major professor at Harvard, Oscar Zariski, who provided ample scope to use these in solving interesting problems.

Having said that, in the lectures on which this book is based I was simply resurrecting some ancient concrete material, I should correct myself by noting that this was completely true only of the three short courses, each of which was of a month's duration and which I gave during the academic years 1986–1988. In the semester course, on which this book is mainly based and which I gave during fall 1988, in addition to presenting the concrete old stuff, I also kept motivating and explaining its links to more modern algebraic geometry based on abstract algebra. I did this partly because, for all the praise of the algorithmic ancient methods, the modern abstractions do sometimes seem to be necessary for solving, or at least clarifying, interesting problems. Moreover, even when modern abstractions are neither necessary nor better, it may be advisable to become familiar with them simply because many people choose to write in that language.

So this book is primarily meant as a textbook for a one- or two-semester course on algebraic geometry for engineers. It can, of course, also be used for independent study.

I have retained the original format of the lectures, and I have made an effort to organize the thirty lectures in such a manner that they can more or less be read in any order. This is certainly untrue of most modern writing of mathematics, including my own, in which to make any sense of what is on page 500, you must first carefully read all the previous 499 pages. At any rate, in this book I have followed the mathematical writing style that was prevalent before, say 1930.

Although mainly meant for the engineers, this book may even be found useful by those students of mathematics who are having a difficulty understanding modern algebraic geometry because the writing of it frequently lacks sufficient motivation. Such a student may find that after browsing through this book, he is in a better position to approach the modern stuff.

Now presentation of mathematics is frequently logical, but rarely is the creation of mathematics logical. Likewise, application of mathematics to science and industry is based more on heuristical understanding rather than immediate formal precision. Following this thought, the aim of the course on which this book is based was not to give formal proofs, but rather to give heuristic ideas and suggestive arguments. In other words, the aim was not to make a legal presentation, but to help people learn. This should prepare the students to read up, or better still, make up, formal proofs if and when desired. So readability is a primary goal of this book. Preference is given to motivation over formality. Thus, this book is not meant to prepare the student for formal examinations, but to really learn the subject; not qualifiers, but original investigations.

I have tried to tell the story of algebraic geometry and to bring out the poetry in it. I shall be glad if this helps the reader to enjoy the subject while learning it.

This work was partly supported by NSF grant DMS88-16286, ONR grant N00014-88-K-0402, and ARO contract DAAG29-85-C-0018 under Cornell MSI at Purdue University. I am grateful for this support. My thanks are also due to P. Keskar, W. Li, and I. Yie for help in proofreading, and to Y. Abhyankar for everything.

Shreeram S. Abhyankar
West Lafayette
18 January 1990

immediate formal precision. Following this thought, the aim of the course
on which this book is based was not to give formal proofs, but rather to give
heuristic ideas and suggestive arguments. In other words, the aim was not
to make a legal presentation, but to help people learn. This should prepare
the students to read up, or better still, make up, formal proofs if and when
desired. So readability is a primary goal of this book. Preference is given
to motivation over formality. Thus, this book is not meant to prepare the
student for formal examinations, but to really learn the subject; not qualifiers,
but original investigations.

I have tried to tell the story of algebraic geometry and to bring out the
poetry in it. I shall be glad if this helps the reader to enjoy the subject while
learning it.

This work was partly supported by NSF grant DMS88-16286, ONR grant
N00014-88-K-0402, and ARO contract DAAG29-85-C-0018 under Cornell
MSI at Purdue University. I am grateful for this support. My thanks are
also due to P. Keskar, W. Li, and I. Yie for help in proofreading, and to Y.
Abhyankar for everything.

Shreeram S. Abhyankar
West Lafayette
18 January 1990

Rational and Polynomial Parametrizations

Let us begin with some simple examples of plane curves.

Conic Sections. These include the circle, ellipse, hyperbola, and parabola defined by the equations $X^2 + Y^2 = 1$, $X^2/a^2 + Y^2/b^2 = 1$, $XY = 1$, and $Y^2 = X$ respectively. We are quite familiar with their graphs as depicted in Figure 1.1.

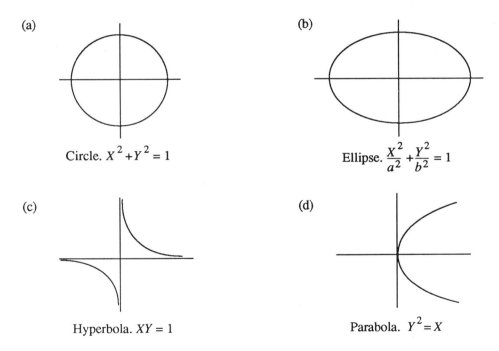

(a)

Circle. $X^2 + Y^2 = 1$

(b)

Ellipse. $\dfrac{X^2}{a^2} + \dfrac{Y^2}{b^2} = 1$

(c)

Hyperbola. $XY = 1$

(d)

Parabola. $Y^2 = X$

FIGURE 1.1

Further, we can similarly consider
Surfaces in 3-space. Some examples are shown in Figure 1.2 on page 2.

Now the examples of the plane curves that we considered previously are plane curves of *degree* 2. We can also consider curves of higher degree such as those shown in Figures 1.3–1.5 on pages 2 and 3.

Notice that near the origin we may think of the tacnodal quartic as the

(a)

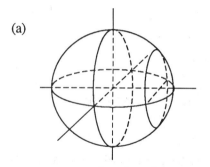

Sphere. $X^2 + Y^2 + Z^2 = 1$

(b)

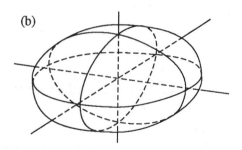

Ellipsoid. $\dfrac{X^2}{a^2} + \dfrac{Y^2}{b^2} + \dfrac{Z^2}{c^2} = 1$

FIGURE 1.2

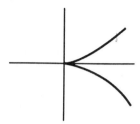

FIGURE 1.3. Cuspidal Cubic. This is defined by $Y^2 - X^3 = 0$.
Cusp = double point with only one tangent.

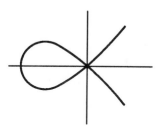

FIGURE 1.4. Nodal Cubic. This is defined by $Y^2 - X^2 - X^3 = 0$.
Node = double point with distinct tangents.

union of the two parabolas $Y = \pm X^2$ having the combined equation $Y^2 - X^4 = 0$ with the extra term Y^3 added so that the curve becomes "irreducible," that is, its defining polynomial $Y^3 + Y^2 - X^4$ doesn't factor.

As another example of a quartic (that is, a plane curve of degree 4) we could consider $X^4 + Y^4 = 1$ whose graph would look like a flat circle. We could also consider flatter and flatter circles such as $X^6 + Y^6 = 1$, $X^8 + Y^8 = 1$, and so on. Note that as the degree becomes larger and larger, these flat

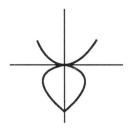

FIGURE 1.5. Tacnodal Quartic. This is defined by $Y^3 + Y^2 - X^4 = 0$.
Tacnode = double point with only one tangent but two "branches."

circles tend to become a square in some sense. Incidentally, the graphs of the corresponding odd power equations, such as $X^3 + Y^3 = 1$ or $X^5 + Y^5 = 1$, look quite different. (Draw them!)

Parametrization. Now to draw a curve such as the circle with the "implicit" equation given by $X^2 + Y^2 = 1$, we can try to express one variable in terms of the other to give us points on the curve. Another method that is less tedious is to represent the circle by the parametric equations $X = \cos\theta$ and $Y = \sin\theta$. See Figure 1.6. But for computations it may be desirable that X and Y be represented in terms of polynomials or rational functions, because then we can lessen the round off errors.

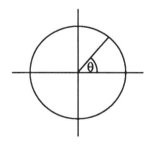

FIGURE 1.6

The above was a quick introduction to what we call *analytic geometry.* Another subject we study in high school is *calculus.* Let us now give a brief introduction to calculus.

Calculus deals with the processes of differentiation and integration. Differentiation is a direct and simple process. For example, we have $\frac{d}{dX}X^3 = 3X^2$. Integration is the reverse process, where calculating $\int r(X)dX$ amounts to finding a function $s(X)$ whose derivative is $r(X)$. This would give $\int r(X)dX = s(X) + C$ where C is an arbitrary constant.

The problem of integrating any given function can be achieved in the following cases:

1. When r is a rational function, i.e., $r(X) = p(X)/q(X)$ where $p(X)$ and $q(X)$ are polynomials, and we want to calculate $\int r(X)dX$. This uses

the method of decomposing the rational function $r(X)$ into *partial fractions*. In other words, we first factor the denominator $q(X)$ as

$$q(X) = b^* \prod_{i=1}^{h} (X - b_i)^{d_i} \quad \text{with } b^* \neq 0$$

and then we find a polynomial $p^*(X)$ and constants a_{ij} such that

$$r(X) = p^*(X) + \sum_{i=1}^{h} \sum_{j=1}^{d_i} \frac{a_{ij}}{(X - b_i)^j}.$$

If the original polynomials p and q have real coefficients and if we wish to avoid complex numbers in the partial fraction expansion, then instead of powers of linear terms in the denominator, we may also get powers of quadratic terms.

2. When r is a rational function $r(\theta) = \frac{f(\sin\theta, \cos\theta, \tan\theta, \dots)}{g(\sin\theta, \cos\theta, \tan\theta, \dots)}$ of the trigonometric functions $\sin\theta$, $\cos\theta$, $\tan\theta$, $\dots$, where f and g are polynomials, and we want to find $\int r(\theta)d\theta$. Although some cases can be solved by the use of identities, we see that a key substitution reduces the problem of integrating $r(\theta)$ to a problem of the above case 1. The substitution is $t = \tan\frac{\theta}{2}$. Using this substitution, we can write $\sin\theta = \frac{2t}{1+t^2}$ and $\cos\theta = \frac{1-t^2}{1+t^2}$, which brings us back to the parametric equations of the circle. This is the connection we wish to highlight.

We thus have the *rational parametrization* of the circle in Figure 1.7.

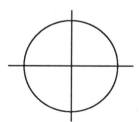

FIGURE 1.7.   Circle. $X^2 + Y^2 = 1$ $\begin{cases} X = (1 - t^2)/(1 + t^2) \\ Y = (2t)/(1 + t^2) \end{cases}$

We can try to do the same for the other conics. See Figures 1.8–1.10. For the parabola we may note that we have a polynomial parametrization. Although it is desirable to have a polynomial representation for any curve, we may not always get it. In general for any plane curve, we consider the following problem:

Given an implicit equation, when can we obtain an explicit rational parametrization? Moreover, when can we even obtain a polynomial parametrization?

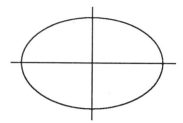

FIGURE 1.8. Ellipse. $\frac{X^2}{a^2} + \frac{Y^2}{b^2} = 1$ $\begin{cases} X = (a(1 - t^2))/(1 + t^2) \\ Y = (2bt)/(1 + t^2) \end{cases}$

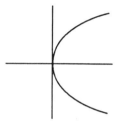

FIGURE 1.9. Parabola. $Y^2 = X$ $\begin{cases} X = t^2 \\ Y = t \end{cases}$

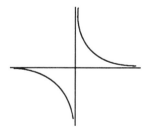

FIGURE 1.10. Hyperbola. $XY = 1$ $\begin{cases} X = t \\ Y = 1/t \end{cases}$

The question may also be asked: Is there a geometric way of realizing the above parametrization of the circle?

Answer: Certainly, as we can consider the following. See Figure 1.11 on page 6. Fix a point $P = (-1, 0)$ on the circle. Consider all lines passing through this point P. These lines intersect the circle at exactly one other point. The equation of the line of slope t passing through the point P is $Y = t(X + 1)$. Obtaining the points at which this line intersects the circle gives us the rational parametrization. To find these points of intersection, we

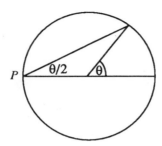

FIGURE 1.11

substitute the equation of the line into the equation of the circle to get

$$X^2 + t^2(X+1)^2 - 1 = 0$$

and simplifying we have

$$(t^2 + 1)X^2 + 2t^2 X + (t^2 - 1) = 0.$$

Solving this equation and substituting in the equation of the line and disregarding the fixed solution $(-1, 0)$ that corresponds to the point P, we get the parametrization

$$\begin{cases} X = \frac{1-t^2}{1+t^2} \\ Y = \frac{2t}{1+t^2}. \end{cases}$$

Thus, we have found the rational parametrization of the circle directly without the intervention of trigonometric functions!

The parametrization of the ellipse may be obtained in a similar manner by taking lines through the point $(-a, 0)$. Although the above parametrizations of the parabola and the hyperbola may be obtained in a similar manner, actually they can be obtained much more simply.

The parabola $Y^2 = X$ has no term in X^2, so we get its parametrization simply by putting $Y = t$ and solving for X. The parabola $X^2 = Y$ has no term in Y^2, so we get its parametrization simply by putting $X = t$ and solving for Y. The hyperbola $XY = 1$ has neither a term in Y^2 nor a term in X^2. Hence, we can parametrize it by putting $X = t$ and solving for Y, or alternatively by putting $Y = t$ and solving for X. In other words, the equations of hyperbola and parabola are linear in one of the variables. So we can easily parametrize them by putting t for the other variable and solving for the linear variable.

To use a similar method for the general conic

$$G(X, Y) = aX^2 + bY^2 + cXY + dX + eY + f$$

we seek to linearize it in one of the variables. To this effect we could make a linear transformation

$$X \to \alpha'X + \beta'Y + \gamma' \quad \text{and} \quad Y \to \alpha''X + \beta''Y + \gamma''$$

and then the equation would become

$$a(\alpha'X + \beta'Y + \gamma')^2 + b(\alpha''X + \beta''Y + \gamma'')^2$$
$$+ c(\alpha'X + \beta'Y + \gamma')(\alpha''X + \beta''Y + \gamma'') + \text{terms of degree} < 2$$
$$= (a\beta'^2 + b\beta''^2 + c\beta'\beta'')Y^2 + \cdots .$$

To get rid of the Y^2 term, put $\beta'' = 1$. Then the coefficient of Y^2 is $a\beta'^2 + c\beta' + b = 0$, and this quadratic can be solved for β'.

To express the above idea in greater generality, given any polynomial $F(X, Y)$ of degree n in X and Y, we can write

$$F(X, Y) = F_n(X, Y) + F_{n-1}(X, Y) + \cdots + F_j(X, Y) + \cdots$$

where $F_j(X, Y)$ is a homogeneous polynomial of degree j. That is, it contains only terms of degree j. By making the above linear transformation, for the transformed equation

$$F^*(X, Y) = F(\alpha'X + \beta'Y + \gamma', \alpha''X + \beta''Y + \gamma'')$$
$$= F_n(\alpha'X + \beta'Y, \alpha''X + \beta''Y) + \text{terms of degree} < n$$

we write

$$F^*(X, Y) = F_n^*(X, Y) + F_{n-1}^*(X, Y) + \cdots + F_j^*(X, Y) + \cdots$$

where $F_j^*(X, Y)$ is a homogeneous polynomial of degree j. Then we have

$$F_n^*(X, Y) = F_n(\alpha'X + \beta'Y, \alpha''X + \beta''Y) = a_0^*Y^n + a_1^*Y^{n-1}X + \cdots + a_n^*X^n$$

with $a_0^* = F_n(\beta', \beta'')$. Thus,

$$F^* \text{ is devoid of } Y^n \iff F_n^* \text{ is devoid of } Y^n \iff F_n(\beta', \beta'') = 0.$$

How shall we find the values of (β', β'') for which $F_n(\beta', \beta'') = 0$? To this end, we note that a polynomial in one variable $a_0Z^n + a_1Z^{n-1} + \cdots + a_n$ can be factored into linear factors as $a_0 \prod_{i=1}^n (Z - \alpha_i)$. If we don't allow the roots to be complex, then we could get quadratic factors etc., but let us suppress this point right now. The homogeneous polynomial

$$F_n(X, Y) = a_0Y^n + a_1Y^{n-1}X + \cdots + a_nX^n$$

of degree n in two variables, X and Y, is equivalent to the polynomial in one variable

$$\frac{F_n}{X^n} = a_0 \left(\frac{Y}{X}\right)^n + a_1 \left(\frac{Y}{X}\right)^{n-1} + \cdots + a_n = a_0 \prod_{i=1}^n \left(\frac{Y}{X} - \alpha_i\right) = \frac{a_0}{X^n} \prod_{i=1}^n (Y - \alpha_i X)$$

assuming $a_0 \neq 0$. Thus the homogeneous polynomial can always (without assuming $a_0 \neq 0$) be written as a product of n linear factors:

$$F_n(X, Y) = a_0 Y^n + a_1 Y^{n-1} X + \cdots + a_n X^n = \prod_{i=1}^{n}(\mu_i X - \lambda_i Y).$$

Geometrically this corresponds to n lines passing through the origin. Note that the factors $(\mu_i X - \lambda_i Y)$ are determined only up to proportionality. For example, in case of $n = 2$ we can write $(\mu_1 X - \lambda_1 Y)(\mu_2 X - \lambda_2 Y) = (2\mu_1 X - 2\lambda_1 Y)(\frac{1}{2}\mu_2 X - \frac{1}{2}\lambda_2 Y)$.

Now $F_n(\beta', \beta'') = 0$ iff (= if and only if) (β', β'') is proportional to (λ_i, μ_i) for some i. Assuming $a_0 \neq 0$, we can find such a (β', β'') by taking $\beta' = 1$ and solving the one variable equation $F_n(1, \beta'') = 0$, which incidentally has at most n roots.

We observe that if Y^n is present in $F(X, Y)$, then a few linear transformations would convert it to a polynomial in which the coefficient of Y^n is equal to zero. On the other hand, most linear transformations would convert it to a polynomial in which Y^n is present, irrespective of whether Y^n is present in the original polynomial. (Here, by *few* and *most* we are not counting the number but the dimensionality in the space of all linear transformations.)

So we make the following definition: If Y^n is present, $F(X, Y)$ is said to be *regular* in Y. Similarly one defines *regular* in X. What is the geometric meaning? Let us consider some examples in Figures 1.12–1.15:

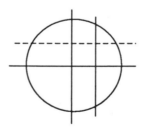

FIGURE 1.12. Circle. $X^2 + Y^2 - 1 = 0$. This is regular in both X and Y. Geometrically, every vertical line meets the circle in two points, and also every horizontal line meets the circle in two points.

A better explanation for the parabola $X^2 - Y$ not being regular in Y is that a vertical line meets it at a "point at infinity." A similar explanation can also be given for the hyperbola. (As it turns out, the above explanation can be made even more precise.)

That discussion was a sort of digression from our original topic of param-

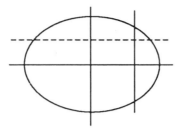

FIGURE 1.13. Ellipse. $X^2/a^2 + Y^2/b^2 - 1 = 0$. This is regular in both X and Y. Once again, every vertical line as well as every horizontal line meets the ellipse in two points.

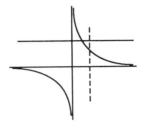

FIGURE 1.14. Hyperbola. $XY - 1 = 0$. This is not regular in X or Y. A vertical line as well as a horizontal line meets the hyperbola in one point.

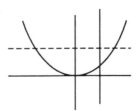

FIGURE 1.15. Parabola. $X^2 - Y = 0$. This is regular in X but not in Y. A horizontal line meets the parabola $X^2 - Y$ in two points whereas a vertical line meets it in one point.

etrization. At any rate, if from the equation of the conic

$$aX^2 + bY^2 + cXY + dX + eY + f = 0$$

we kill the term in Y^2 by a suitable linear transformation, then the equation takes the form

$$(rX + s)Y + (uX^2 + vX + w) = 0.$$

We get the rational parametrization

$$\begin{cases} X = t \\ Y = \frac{-(ut^2+vt+w)}{(rt+s)}. \end{cases}$$

Now for the conic $aX^2 + bY^2 + cXY + dX + eY + f = 0$, the "points at infinity" are given by the terms of highest degree (*degree form*). The two factors of the homogeneous polynomial $\overline{aX^2 + bY^2 + cXY}$ of degree two will correspond to two points at infinity.

Circle. As the degree form is $X^2 + Y^2 = (X + iY)(X - iY)$, the circle has two points at infinity, both of which are complex: $(1, i)$ and $(1, -i)$.

Ellipse. Similarly, the ellipse has two points at infinity, both of which are complex.

Hyperbola. As the degree form is XY, it has the two real points $(1, 0)$ and $(0, 1)$ at infinity.

Parabola. As the degree form Y^2 or X^2 of a parabola has only a single root, it has only one point at infinity.

It turns out that this exceptional behavior of the parabola allows it to be parametrized by polynomials. For the other conics the fact that we do not (yet!) have a polynomial parametrization can be explained by the previous facts about points at infinity.

To summarize we may state the following: a circle (similarly: ellipse, hyperbola) cannot be parametrized by polynomials because it has more than one point at infinity.

Concerning polynomial parametrization, the following precise theorem can be proved:

THEOREM. *If a curve has more than one point at infinity then it is not possible to parametrize it by polynomials.*

For curves of degree more than 2, the exact converse is not true. To state the appropriate converse we need the more subtle notion of *places* at infinity. Another word for "place" is "branch." In terms of places, we have the following characterization:

THEOREM. *A curve can be parametrized by polynomials iff it can be parametrized by rational functions and has only one place at infinity.*

NOTE. Every curve has at least one and at most a finite number of places at infinity. Moreover, the number of places at infinity is always $\geq$ the number of points at infinity.

Fractional Linear Transformations

Let us first review what we discussed last time.

Rational functions can be integrated by partial fractions. Integration of rational functions of trigonometric functions can be reduced to the above by the substitution $\tan \frac{\theta}{2} = t$. This indicates that the same substitution would convert the usual trigonometric parametrization $X = \cos \theta$ and $Y = \sin \theta$ of the (unit) circle into the rational parametrization

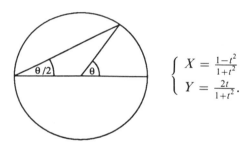

$$\begin{cases} X = \frac{1-t^2}{1+t^2} \\ Y = \frac{2t}{1+t^2}. \end{cases}$$

Geometrically, the parametrization $X = \cos \theta$ and $Y = \sin \theta$ corresponds to lines through the origin. Since lines through the origin meet the circle in two points, the parametrization must involve multivalued functions. On the other hand, since lines through the point $(-1, 0)$ meet the circle in only one other point, the corresponding parametrization is single valued. A single valued function is pretty much rational. At any rate, a single valued algebraic function is certainly rational. This justifies why lines through a point on the circle give a rational parametrization.

We also discussed the conic

$$\Gamma : G(X, Y) = aX^2 + bY^2 + cXY + dX + eY + f = 0,$$

and we stated that: *G is nonregular in Y iff the Y-axis (or a line parallel to the Y-axis) meets Γ at infinity.*

By killing the Y^2 term in G, i.e., by making it nonregular in Y, the equation of Γ takes the form

$$(rX + s)Y - (uX^2 + vX + w) = 0.$$

Then we get the rational parametrization

$$\begin{cases} X = t \\ Y = \frac{ut^2 + vt + w}{rt + s}. \end{cases}$$

To make G nonregular in Y, we find a root of the degree form, i.e., terms of degree 2, and use this to make a suitable linear transformation. However, this method does not work in the case of a circle because then the degree form has no real root. In other words, the circle has no real point at infinity.

What do we mean by a point at infinity? Let us note that any two parallel lines seem to meet at infinity (a good analogy is a pair of railroad tracks). Often we adjoin to the usual (X, Y)-plane (also called the *affine plane*) a line at infinity, converting it to the "projective $(\mathscr{X}, \mathscr{Y}, \mathscr{Z})$-plane." A point in the (X, Y)-plane is usually described by a pair (α, β). We can also describe it by all triples $(\kappa\alpha, \kappa\beta, \kappa)$ with $\kappa \neq 0$, and we call any such triple $(\kappa\alpha, \kappa\beta, \kappa)$ *homogeneous $(\mathscr{X}, \mathscr{Y}, \mathscr{Z})$-coordinates* of the point (α, β). This accounts for all triples (α, β, γ) with $\gamma \neq 0$. Moreover, there are triples for which $\gamma = 0$. We shall go ahead and say that these precisely correspond to points at infinity. Thus, the *projective plane* consists of the set of all equivalence classes of nonzero triples (i.e., triples $(\alpha, \beta, \gamma) \neq (0, 0, 0)$) where two nonzero triples are regarded as equivalent if they are proportional by a nonzero constant of proportionality. Now the affine plane consists of all points whose homogeneous $\mathscr{Z}$-coordinate is nonzero, and all points in the projective plane whose homogeneous $\mathscr{Z}$-coordinate is zero constitute the line at infinity.

Points at infinity on Γ correspond to factors of the degree form $G(X, Y)$:

$$aX^2 + bY^2 + cXY = (\mu_1 X - \lambda_1 Y)(\mu_2 X - \lambda_2 Y).$$

To reconcile the case of the circle and this method of parametrizing by making things nonregular, send $(-1, 0)$ to ∞. This is achieved by a fractional linear transformation

$$X = \frac{\alpha' X^* + \beta' Y^* + \gamma'}{\alpha^* X^* + \beta^* Y^* + \gamma^*} \quad \text{and} \quad Y = \frac{\alpha'' X^* + \beta'' Y^* + \gamma''}{\alpha^* X^* + \beta^* Y^* + \gamma^*}.$$

Now a fractional linear transformation corresponds to homogenizing the equation, making a homogeneous linear transformation, and then dehomogenizing. To homogenize the above equation of Γ, we substitute $X = \frac{\mathscr{X}}{\mathscr{Z}}$ and $Y = \frac{\mathscr{Y}}{\mathscr{Z}}$ in

$$aX^2 + bY^2 + cXY + dX + eY + f$$

and multiply throughout by $\mathscr{Z}^2$ to get

$$a\mathscr{X}^2 + b\mathscr{Y}^2 + c\mathscr{X}\mathscr{Y} + d\mathscr{X}\mathscr{Z} + e\mathscr{Y}\mathscr{Z} + f\mathscr{Z}^2.$$

Now consider the point P on the circle, where in the usual "affine" (X, Y)-coordinates P is given by $(-1, 0)$, whereas in the "homogeneous" $(\mathscr{X}, \mathscr{Y}, \mathscr{Z})$-coordinates P is given by $(-\kappa, 0, \kappa)$ for all $\kappa \neq 0$, and hence in particular by $(-1, 0, 1)$. If we wish to send the point $(-1, 0)$, i.e., the point with homogeneous coordinates $(-1, 0, 1)$, to the point at infinity along the Y-axis, we must send $(-1, 0, 1)$ to $(0, 1, 0)$.

What is the explanation:

A point on the Y-axis is like $(0, p, 1)$. Dividing it by p we get $(\frac{0}{p}, \frac{p}{p}, \frac{1}{p})$. Now letting $p \to \infty$ we get $(0, 1, 0)$ as the homogeneous coordinates of the point at infinity along the Y-axis.

We want to send $(-1, 0, 1)$ to $(0, 1, 0)$ by the transformation

$$\begin{cases} \mathscr{X} = \alpha' \mathscr{X}^* + \beta' \mathscr{Y}^* + \gamma' \mathscr{Z}^* \\ \mathscr{Y} = \alpha'' \mathscr{X}^* + \beta'' \mathscr{Y}^* + \gamma'' \mathscr{Z}^* \\ \mathscr{Z} = \alpha^* \mathscr{X}^* + \beta^* \mathscr{Y}^* + \gamma^* \mathscr{Z}^*. \end{cases}$$

Now by substituting $(\mathscr{X}, \mathscr{Y}, \mathscr{Z}) = (-1, 0, 1)$ and $(\mathscr{X}^*, \mathscr{Y}^*, \mathscr{Z}^*) = (0, 1, 0)$ we get

$$\begin{cases} -1 = \beta' \\ 0 = \beta'' \\ 1 = \beta^*, \end{cases}$$

and so we may take

$$\begin{cases} \mathscr{X} = -\mathscr{Y}^* \\ \mathscr{Y} = -\mathscr{Z}^* \\ \mathscr{Z} = \mathscr{Y}^* - \mathscr{X}^*. \end{cases}$$

This would do the job.

We transformed the circle $X^2 + Y^2 - 1 = 0$ into $\mathscr{X}^2 + \mathscr{Y}^2 - \mathscr{Z}^2 = 0$ by homogenizing, and then we changed $(\mathscr{X}, \mathscr{Y}, \mathscr{Z})$ to $(\mathscr{X}^*, \mathscr{Y}^*, \mathscr{Z}^*)$. So $\mathscr{X}^2 + \mathscr{Y}^2 - \mathscr{Z}^2 = 0$ becomes

$$\mathscr{Y}^{*2} + \mathscr{Z}^{*2} - (\mathscr{Y}^* - \mathscr{X}^*)^2 = 0.$$

How do we return to the usual affine coordinates? First we divide both sides of the previous equation by $\mathscr{Z}^{*2}$, which gives

$$\frac{\mathscr{Y}^{*2} + \mathscr{Z}^{*2} - (\mathscr{Y}^* - \mathscr{X}^*)^2}{\mathscr{Z}^{*2}} = 0.$$

By letting $X^* = \frac{\mathscr{X}^*}{\mathscr{Z}^*}$ and $Y^* = \frac{\mathscr{Y}^*}{\mathscr{Z}^*}$, the above equation converts to

$$Y^{*2} + 1 - (Y^* - X^*)^2 = 0$$

and by simplifying we get

$$2X^* Y^* = X^{*2} - 1.$$

Now the above equation has become nonregular in Y^*. Thus, we can write

$$Y^* = \frac{X^{*2} - 1}{2X^*},$$

and so we get the rational parametrization

$$\begin{cases} X^* = \dfrac{1}{t} \\[2mm] Y^* = \dfrac{1-t^2}{2t} . \end{cases}$$

The homogeneous linear transformation

$$\begin{cases} \mathscr{X} = -\mathscr{Y}^* \\ \mathscr{Y} = -\mathscr{Z}^* \\ \mathscr{Z} = \mathscr{Y}^* - \mathscr{X}^* \end{cases}$$

can be converted to a fractional linear transformation by noting that

$$X = \frac{\mathscr{X}}{\mathscr{Z}} = \frac{-\mathscr{Y}^*}{\mathscr{Y}^* - \mathscr{X}^*} = \frac{-Y^*}{Y^* - X^*} \quad \text{and} \quad Y = \frac{\mathscr{Y}}{\mathscr{Z}} = \frac{-\mathscr{Z}^*}{\mathscr{Y}^* - \mathscr{X}^*} = \frac{-1}{Y^* - X^*} .$$

By substituting the above parametrization in these equations we get the familiar rational parametrization

$$\begin{cases} X = \dfrac{-\frac{1-t^2}{2t}}{\frac{1-t^2}{2t} - \frac{1}{t}} = \dfrac{1-t^2}{1+t^2} \\[4mm] Y = \dfrac{-1}{\frac{1-t^2}{2t} - \frac{1}{t}} = \dfrac{2t}{1+t^2} . \end{cases}$$

By making the fractional linear transformation

$$X = \frac{-Y^*}{-X^* + Y^*} \quad \text{and} \quad Y = \frac{-1}{-X^* + Y^*}$$

the equation of the circle is transformed to

$$\left(\frac{-Y^*}{-X^* + Y^*} \right)^2 + \left(\frac{-1}{-X^* + Y^*} \right)^2 - 1 = 0 ,$$

which, by clearing the denominator, gives

$$Y^{*2} + 1 - (-X^* + Y^*)^2 = 0 .$$

This is, of course, the same as we had before.

In general, let us consider a curve $C : F(X, Y) = 0$ where $F(X, Y)$ is a polynomial of degree n. By collecting terms of like degree, we can write $F(X, Y)$ as

$$F(X, Y) = F_n(X, Y) + F_{n-1}(X, Y) + \cdots + F_j(X, Y) + \cdots + F_0$$

where $F_j(X, Y)$ is a homogeneous polynomial of degree j. We call $F_n(X, Y)$ the *degree form* of $F(X, Y)$. We homogenize $F(X, Y)$ by putting $X = \frac{\mathscr{X}}{\mathscr{Z}}$ and $Y = \frac{\mathscr{Y}}{\mathscr{Z}}$ to obtain

$$f(\mathscr{X}, \mathscr{Y}, \mathscr{Z}) = \mathscr{Z}^n F\left(\frac{\mathscr{X}}{\mathscr{Z}}, \frac{\mathscr{Y}}{\mathscr{Z}} \right)$$

$$= F_n(\mathscr{X}, \mathscr{Y}) + F_{n-1}(\mathscr{X}, \mathscr{Y})\mathscr{Z} + \cdots + F_j(\mathscr{X}, \mathscr{Y})\mathscr{Z}^{n-j} + \cdots + F_0 \mathscr{Z}^n ,$$

which is the defining homogeneous polynomial of the curve C in the projective plane. Substituting $\mathcal{Z} = 0$, we obtain $F_n(\mathcal{X}, \mathcal{Y}) = 0$. Now we factor the degree form of $F(X, Y)$ by writing $F_n(X, Y) = \prod_{i=1}^{n}(\mu_i X - \lambda_i Y)$. Then the points $(\lambda_i, \mu_i, 0)$ are the points of the curve C at infinity. In particular, *C has a real point at infinity iff $F_n(u, v) = 0$ for some real $(u, v) \neq (0, 0)$.* Then a real linear transformation suffices to make F nonregular in Y.

Even if F_n has no real roots, we can always make a fractional linear transformation

$$X = \frac{\alpha' X^* + \beta' Y^* + \gamma'}{\alpha^* X^* + \beta^* Y^* + \gamma^*} \quad \text{and} \quad Y = \frac{\alpha'' X^* + \beta'' Y^* + \gamma''}{\alpha^* X^* + \beta^* Y^* + \gamma^*}$$

so that

$$F\left(\frac{\alpha' X^* + \beta' Y^* + \gamma'}{\alpha^* X^* + \beta^* Y^* + \gamma^*}, \frac{\alpha'' X^* + \beta'' Y^* + \gamma''}{\alpha^* X^* + \beta^* Y^* + \gamma^*}\right)$$

is a rational function but

$$\left(\alpha^* X^* + \beta^* Y^* + \gamma^*\right)^n F\left(\frac{\alpha' X^* + \beta' Y^* + \gamma'}{\alpha^* X^* + \beta^* Y^* + \gamma^*}, \frac{\alpha'' X^* + \beta'' Y^* + \gamma''}{\alpha^* X^* + \beta^* Y^* + \gamma^*}\right)$$

is a polynomial.

Since we are going to adopt this procedure for cubics and so on, it would not hurt to reconsider the case of a circle once again. The above fractional linear transformation converts the defining polynomial $X^2 + Y^2 - 1$ of the circle into the polynomial

$$\left(\alpha^* X^* + \beta^* Y^* + \gamma^*\right)^2 \left[\left(\frac{\alpha' X^* + \beta' Y^* + \gamma'}{\alpha^* X^* + \beta^* Y^* + \gamma^*}\right)^2 + \left(\frac{\alpha'' X^* + \beta'' Y^* + \gamma''}{\alpha^* X^* + \beta^* Y^* + \gamma^*}\right)^2 - 1\right]$$

$$= \left(\alpha' X^* + \beta' Y^* + \gamma'\right)^2 + \left(\alpha'' X^* + \beta'' Y^* + \gamma''\right)^2 - \left(\alpha^* X^* + \beta^* Y^* + \gamma^*\right)^2$$

in which the coefficient of Y^{*2} is $\beta'^2 + \beta''^2 - \beta^{*2}$. For the equation $\beta'^2 + \beta''^2 - \beta^{*2} = 0$ we can easily find real roots. If the denominator of the fractional linear transformation were reduced to 1, i.e., if we were only making a linear transformation, then the coefficient of Y^{*2} would be $\beta'^2 + \beta''^2$, and the equation $\beta'^2 + \beta''^2 = 0$ has no real roots. However, for a fractional linear transformation to do the job, to begin with we must have some real points on the curve. For example, the curve $X^2 + Y^2 + 1 = 0$, which is a circle with radius $i = \sqrt{-1}$, has no real point on it, and after applying the previous fractional linear transformation the term in Y^{*2} is $\beta'^2 + \beta''^2 + \beta^{*2}$, and the equation $\beta'^2 + \beta''^2 + \beta^{*2} = 0$ has only nonreal complex roots.

To summarize: *if a curve C of degree n has a real point at infinity, then by a real linear transformation we can make the equation of the curve nonregular in Y. Otherwise, we must use a fractional linear transformation.*

As we have said, we can factor the degree form $F_n(X, Y)$ of $F(X, Y)$ by writing $F_n(X, Y) = \prod_{i=1}^{n}(\mu_i X - \lambda_i Y)$. Then the points (λ_i, μ_i) are the points of the curve C at infinity. Thus we can check if the curve has a real point at infinity, in which case a real linear transformation would suffice. Also note that a complex linear transformation would always suffice. *Examples:* (1) $X^2 + Y^2 - 1 = 0$: degree form has two complex conjugate roots. (2) $\frac{X^2}{a^2} + \frac{Y^2}{b^2} - 1 = 0$: similar. (3) $XY - 1 = 0$: $(1, 0, 0)$ and $(0, 1, 0)$ are the two real points at infinity. (4) $Y^2 - X = 0$ or $X^2 - Y = 0$: one real point at infinity.

For real polynomial parametrization we have the following

CRITERION. *A curve C has a real polynomial parametrization iff it has a rational parametrization and has only one place at infinity that is real.*

NOTE. How about a criterion for a curve to have only one (real) place at infinity?

LECTURE 3

Cubic Curves

Today, let us study the cubic curve $\Lambda: H(X, Y) = 0$ where

$$H(X, Y)$$
$$= a'X^3 + b'X^2Y + c'XY^2 + d'Y^3 + a^*X^2$$
$$+ b^*Y^2 + c^*XY + d^*X + e^*Y + f^*.$$

If it has a real point at infinity, then linear transformations are enough to make this nonregular in Y. If not, then we must use fractional linear transformations. Recall that the points at infinity are given by the degree form (i.e., terms of highest degree). For conics, we had real points at infinity only in the case of parabolas and hyperbolas. However, in the case of cubics we are lucky, because every cubic has a real point at infinity. The reason is that now the degree form always has a real root. The degree form on dehomogenizing gives

$$h(x) = a'x^3 + b'x^2 + c'x + d'.$$

Because complex roots occur in conjugates, $h(x)$ always has a real root, provided a' is nonzero. In case a' is zero, $H(X, Y)$ is already nonregular in X, and so we would be done by interchanging X and Y. Assuming that a' is nonzero, we may make a linear transformation

$$X = \alpha'X^* + \beta'Y^* \quad \text{and} \quad Y = \alpha''X^* + \beta''Y^*$$

to arrange that the cubic curve has a real point at infinity on the Y^*-axis. We are trying to make the cubic nonregular in Y^* so that the resulting degree form should be divisible by X^*. Therefore, we want the term in Y^{*3} to be zero. That is, we want $a'\beta'^3 + b'\beta'^2\beta'' + c'\beta'\beta''^2 + d'\beta''^3 = 0$. So we may take $\beta'' = 1$ and β' to be a real number such that $a'\beta'^3 + b'\beta'^2 + c'\beta' + d' = 0$. With such a choice of β' and β'', we have

$$\widehat{H}(X^*, Y^*) = H(X, Y)$$
$$= (lX^* + m)Y^{*2} + (uX^{*2} + vX^* + w)Y^* + (pX^{*3} + qX^{*2} + rX^* + s),$$

which is in the form of the usual quadratic equation.

Let us recall the old Indian method of Shreedharacharya (fifth century) for solving the quadratic equation $aZ^2 + bZ + c = 0$. This method of Shreedharacharya, which is given in verse form in Bhaskaracharya's *Beejaganit*

(1150) [**Bha**], goes something like this: multiply the equation by 4 times the coefficient of the square term to get $4a^2Z^2 + 4abZ + 4ac = 0$. Next, add and subtract the square of the coefficient of the original unknown to obtain $4a^2Z^2 + 4abZ + b^2 - (b^2 - 4ac) = 0$. Now the rest follows by noting that we clearly have $(2aZ + b)^2 - (b^2 - 4ac) = 0$.

Using the above method we get

$$
\begin{aligned}
H^*(X^*, Y^*) &= 4(lX^* + m)\widehat{H}(X^*, Y^*) \\
&= \left[2(lX^* + m)Y^* + (uX^{*2} + vX^* + w)\right]^2 \\
&\quad - \left[(uX^{*2} + vX^* + w)^2 - 4(lX^* + m)(pX^{*3} + qX^{*2} + rX^* + s)\right].
\end{aligned}
$$

Now by putting

$$
Y' = 2(lX^* + m)Y^* + (uX^{*2} + vX^* + w)
$$

we have

$$
\begin{aligned}
H'(X^*, Y') &= H^*(X^*, Y^*) \\
&= Y'^2 - g(X^*) \quad \text{with } \deg g(X^*) \le 4.
\end{aligned}
$$

We want to analyze if we can rationally parametrize the curve $Y'^2 = g(X^*)$. To do this we consider several cases as follows: $g(X^*)$ has only one distinct root, $g(X^*)$ has two distinct roots,..., and so on. In the case of multiple roots, we may use a general method to get rid of them. For any equation $Y'^2 = g(X^*)$ where $g(X^*)$ is a nonzero polynomial, we can write $g(X^*) = g'(X^*)^2 g''(X^*)$ where $g'(X^*)$ and $g''(X^*)$ are nonzero polynomials such that $g''(X^*)$ has no multiple roots. Then by putting $Y'' = Y'/g'(X^*)$ we get $Y''^2 = g''(X^*)$. In our case, because $\deg g(X^*) \le 4$, the equation reduces to

$$
Y''^2 = g''(X^*) \quad \text{with } \deg g''(X^*) \le 4.
$$

If $\deg g''(X^*) \le 2$, then this is a conic, which can always be rationally parametrized. Hence, we get a rational parametrization of our original cubic curve Λ. Otherwise, when $g''(X^*)$ has either three or four distinct roots, i.e., if $3 \le \deg g''(X^*) \le 4$, the cubic curve Λ cannot be rationally parametrized. Thus Λ can be rationally parametrized iff $\deg g''(X^*) \le 2$. It can easily be seen that $\deg g''(X^*) \le 2$ iff the cubic curve Λ is singular, i.e., it has a double point in the following sense.

Geometric viewpoint. The idea of parametrizing a circle or a conic was to fix a point on it and take lines through that point. The lines can, of course, be parametrized by a single parameter, say, the slope. Let us put down the following aphorism: *lines through a point form a line!* Indeed, this is a good definition of a *projective line*. (More generally, *lines through a point in affine n-space constitute an $(n - 1)$-dimensional projective space.*) Next, we take a cubic curve. A cubic curve is a curve which is met by most lines in three points. Indeed a curve of degree n can be defined (as is in fact done in geometry) as a curve in the plane which is met by most lines in n points. If we consider a cubic curve having a singular point, then lines through that singular point rationally parametrize it. See Figure 3.1.

FIGURE 3.1

Singular points and tangents. So what is a singular point? This leads us to a quick introduction to algebraic geometry. We know what a line is! Take a line and take a point outside it. The linear space spanned by the line and such a point is called a plane. A curve of degree n in the plane is one which is met by most lines in n points. Let $C = C_n$ be a curve of degree n. Most lines through a point P meet C outside P in $n - d$ points, where $d =$ mult $_P C =$ multiplicity of C at P. If $d = 1$ then we say that P is a *simple* point of C. If $d = 2$ then we say that P is a *double* point of C. Similarly we talk about a *d-ple* point or a *d-fold* point. If $d = 0$ then P is not on C. If $d > 1$ then we say that P is a *singular point* of C. The curve C is said to be *singular* or *nonsingular* according to whether it does or does not have a singular point. Next we come to *tangents*. Assuming $d > 0$, i.e., assuming P to be a point on C, tangents to C at P are those lines through P that meet C at P in more than d points. That is, they meet C outside P in less than $n - d$ points (counting properly). We first come across this notion in the case of a circle.

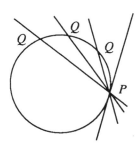

FIGURE 3.2. Usually we define a tangent as the limit of chords PQ as $Q \to P$. We may say that the tangent line meets the circle in two *coincidental* points.

Surfaces. Similarly, by intersecting a surface S in 3-space by lines in the 3-space, we can talk about the degree of S, the multiplicity of points on S, simple points and singular points on S, tangent lines to S at a point of S, and the tangent plane to S at a simple point of S. As an example, consider a sphere, or more generally, any surface S_2 of degree 2. A plane in general cuts it in a circle or a conic. What happens if the plane is a tangent plane? Well, in general, it is a degenerate conic, i.e., a pair of lines. Expanding on this observation, we can see that S_2 is the product of two projective lines.

Bezout's Theorem. Going back to curves in the plane, two circles seem to meet in two points. A line and a circle meet in two points. However, two ellipses meet in four points. An ellipse and a hyperbola meet in four points. So a natural thing is to expect two conics to meet in four points. In general we can formulate a nice theorem. A curve C_m of degree m and a curve C_n of degree n meet in mn points: $C_m \cdot C_n = mn$ points. At any rate, this is the oldest theorem in algebraic geometry and is called Bezout's Theorem [**Bez**]. (*Heuristic Proof.* Degenerate C_m into m lines and C_n into n lines; m lines and n lines meet in mn points. Hence, C_m and C_n meet in mn points.) Now a curve of degree 4 with a triple point is rational. Briefly, C_4 with P_3 is rational (as we can take lines through this point). In the same breath we can say that a curve C_e of degree e with a $(e-1)$-fold point P_{e-1} is a rational curve, and a surface S_e of degree e with an $(e-1)$-fold point P_{e-1} is a rational surface. Next, consider a curve C_4 of degree 4 having 3 double points. It is a high-school exercise to pass a conic through any five preassigned points. In addition to the three double points, fix a simple point on C_4. Now $C_4 \cdot C_2 = 8$ points. The three double points and the fixed simple point account for $2 + 2 + 2 + 1 = 7$ points. Then we can take a variable point on the curve C_4 and pass a conic through these five points, giving us a rational parametrization of the curve C_4 of degree 4 having three double points. This suggests that symbolically $P_3 = 3P_2$, or more generally $P_\nu = \frac{1}{2}\nu(\nu - 1)$ double points. That is, a ν-fold point accounts for $\frac{1}{2}\nu(\nu - 1)$ double points. See Figure 3.3.

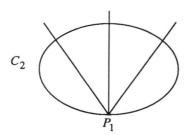
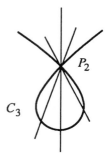

FIGURE 3.3

Genus. If we use this argument, then we can easily prove by Bezout's Theorem that a curve of degree 4 cannot have four or more double points. In general, we can show that the number of double points on a curve C_n of degree n is $\leq \frac{(n-1)(n-2)}{2}$. We can define $g = $ genus of $C_n = \frac{(n-1)(n-2)}{2} - $ the number of double points. Then we get the

THEOREM. $g = 0$ *iff* C_n *has a rational parametrization.*

In counting the number of double points, a ν-fold point is to be counted as $\frac{1}{2}\nu(\nu - 1)$ double points. Yet even this is not very precise because it works only in the case of the so called ordinary multiple points. For a multiple point, which is not ordinary, we must also take into account "infinitely near" singularities.

NOTE. As we have said, it is a natural thing to expect two conics to meet in four points, and more generally, for a curve of degree m and a curve of degree n to meet in mn points. We have called this Bezout's Theorem. But our experience doesn't support this theorem entirely. However, a viewpoint in algebraic geometry is that *a theorem is not something that is true, but is rather a nice geometric statement that you want to be true. So you adjust your definitions properly.*

Cubic Surfaces and General Hypersurfaces

We have seen that a cubic curve $H(X, Y) = 0$ can be rationally parametrized if and only if it is a singular cubic. That is, it has a double point. However, for cubic surfaces in (X, Y, Z)-space, it turns out that most of them can be rationally parametrized. To see this geometrically, first note that if we cut a cubic surface $\Theta : \phi(X, Y, Z) = 0$ by a plane then we get a cubic curve. If the plane is tangent to the surface, then something special must happen. We get a singular cubic curve, which we know can be rationally parametrized, say by a parameter t. Now consider the tangent plane to the cubic surface at the point t. We again get a singular cubic curve as the intersection, which can be parametrized by another parameter τ. (See Figure 4.1.)

FIGURE 4.1

Thus, we get a rational parametrization

$$\begin{cases} X = \frac{p(t,\tau)}{s(t,\tau)} \\ Y = \frac{q(t,\tau)}{s(t,\tau)} \\ Z = \frac{r(t,\tau)}{s(t,\tau)} \end{cases}$$

of the cubic surface Θ, where $p(t, \tau), q(t, \tau), r(t, \tau), s(t, \tau)$ are polynomials in the two parameters t and τ with $s(t, \tau) \neq 0$. However, this parametrization is not faithful; most points of Θ correspond to more than one value of (t, τ). Indeed, they correspond to six values. By modifying this construction, it is possible to obtain a faithful rational parametrization of Θ. This is also guaranteed by Castelnuovo's Theorem, which says that

23

if a surface has a rational parametrization, then it has a faithful rational parametrization. The corresponding theorem for curves is called Lüroth's Theorem.

Now to discuss the parametrization of the given cubic surface $\Theta : \phi(X, Y, Z) = 0$ algebraically, take a simple point on it. Most points are simple, so this isn't a problem. By translation bring the simple point to the origin. Now we have

$$\phi(X, Y, Z) = \hat{a}X + \hat{b}Y + \hat{c}Z + \cdots .$$

Rotating the axes, we can arrange matters so that the tangent plane $\hat{a}X + \hat{b}Y + \hat{c}Z = 0$ of Θ at the origin becomes the plane $Z = 0$. Then we get

$$\phi(X, Y, Z) = Z + \left[F_2(X, Y) + F_1(X, Y)Z + F_0 Z^2 \right]$$
$$+ \left[G_3(X, Y) + G_2(X, Y)Z + G_1(X, Y)Z^2 + G_0 Z^3 \right]$$

where F_i and G_j are homogeneous polynomials of degree i and j respectively. The intersection of Θ with the tangent plane $Z = 0$ is given by

$$F_2(X, Y) + G_3(X, Y) = 0 = Z ,$$

which is obviously a cubic curve in the (X, Y)-plane having a double point at the origin. So we can obtain a rational parametrization for it. That is, we can find rational functions $u(t)$ and $v(t)$ such that

$$F_2(u(t), v(t)) + G_3(u(t), v(t)) = 0.$$

Now bring the point $(u(t), v(t), 0)$ to the origin by the coordinate transformation

$$\begin{cases} X^* = X - u(t) \\ Y^* = Y - v(t) \\ Z^* = \phi_X(u(t), v(t), 0)(X - u(t)) \\ \qquad + \phi_Y(u(t), v(t), 0)(Y - v(t)) + \phi_Z(u(t), v(t), 0)Z. \end{cases}$$

Substituting this in $\phi(X, Y, Z)$ we get

$$\phi^*(X^*, Y^*, Z^*) = \phi(X, Y, Z)$$
$$= Z^* + \left[F_2^*(X^*, Y^*) + F_1^*(X^*, Y^*)Z^* + F_0^* Z^{*2} \right]$$
$$+ \left[G_3^*(X^*, Y^*) + G_2^*(X^*, Y^*)Z^* + G_1^*(X^*, Y^*)Z^{*2} + G_0^* Z^{*3} \right]$$

where F_i^* and G_j^* are homogeneous polynomials of degree i and j respectively. Intersecting with $Z^* = 0$ we get

$$F_2^*(X^*, Y^*) + G_3^*(X^*, Y^*) = 0 = Z^*.$$

Once again we can rationally parametrize this, say by a parameter τ. However, this time the coefficients of F_2^* and G_3^* are themselves rational functions of t. Thus, we get a rational parametrization of Θ in terms of the two parameters t and τ.

This was a parametrization of a nonsingular cubic surface. Now if the cubic surface has a double point, then it can be rationally parametrized very easily by taking lines through that double point. However, if the cubic surface has a triple point then it is a cone (homogeneous equation of degree 3). By dehomogenizing it we get a cubic curve. If this cubic curve has a singular point, then the curve and hence the surface can be rationally parametrized. If the cubic curve turns out be nonsingular, then neither the curve nor the surface can be rationally parametrized.

So far we have considered some examples of curves and surfaces. Now we could leap into higher dimensions by considering solids or 3-folds defined by equations such as the following:

$$X^2 + Y^2 + Z^2 + W^2 = 1 \qquad \text{(3-sphere in 4-space)}$$
$$X^3 + Y^3 + Z^3 + W^3 = 1 \qquad \text{(cubic 3-fold in 4-space)},$$

or we may as well consider the locus defined by the equation $X_1^3 + X_2^3 + \cdots + X_n^3 = 1$. This is then called a *cubic hypersurface* in n-space.

More generally, consider a polynomial

$$f(X_1, X_2, \ldots, X_n) = \sum_{i_1 + \cdots + i_n \le e} a_{i_1 \cdots i_n} X_1^{i_1} \cdots X_n^{i_n}$$

of some given degree e in n variables where n can be any positive integer.

A typical term in the expansion is of the form

$$\text{(coefficient)} \times \text{(product of powers of variables)}.$$

Such a term (excluding the coefficient) is called a *monomial*. We define the degree of a monomial, which is also the same as the degree of the corresponding term, as the sum of the exponents. We define the degree of f as the degree of the highest degree term present in its expansion (present = having nonzero coefficient).

A polynomial $f(X_1, \ldots, X_n)$ of degree e in n variables defines a *hypersurface* of degree e in n-space. Thus we have curves, surfaces, solids, ..., and in general hypersurfaces. For these geometric objects we would now like to discuss the notions of simple points, singular points, tangents, etc.

Geometrically, as in the case of plane curves, a tangent may be defined as a limit of chords, or equivalently, as a line which meets the hypersurface in two coincidental points. Now we may attempt to give an "algebraic" definition of this concept. Before we do this, let us digress for a moment to remark about the nature of the subject, namely, algebraic geometry. A quick definition of algebraic geometry is "geometry done algebraically." Classically, one studied subjects such as analytic geometry, the study of graphs or configurations defined by algebraic equations, and theory of equations, which is concerned with solving an equation or a system of equations. We may say that

$$\text{algebraic geometry} = \text{analytic geometry} + \text{theory of equations}.$$

Now, what do we mean by algebra? To answer this, we note that algebra may be divided into three parts as shown in Figure 4.2.

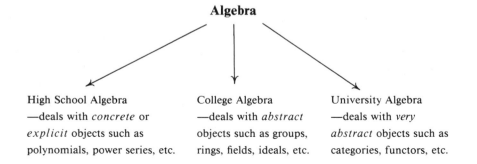

High School Algebra	College Algebra	University Algebra
—deals with *concrete* or *explicit* objects such as polynomials, power series, etc.	—deals with *abstract* objects such as groups, rings, fields, ideals, etc.	—deals with *very abstract* objects such as categories, functors, etc.

FIGURE 4.2

Let us now go back to defining tangents algebraically. Consider a curve, a surface, or more generally a hypersurface $S : f(X_1, X_2, \ldots, X_n) = 0$.

To define tangent lines at a point of S we must first understand what we mean by a line through a point in n-space. Let us just suppose that the point under consideration is the origin. Now lines through the origin in the plane are either given by implicit equations such as $Y = mX$ (where m is the slope) or by explicit parametric equations such as

$$\begin{cases} X = at \\ Y = bt. \end{cases}$$

The latter is perhaps a better description since we are not giving preference to either X or Y. Moreover we can also generalize it to *lines L in n -space* (through the origin) as being given by the parametric equations

$$L : \begin{cases} X_1 = a_1 t \\ X_2 = a_2 t \\ \vdots \\ X_n = a_n t. \end{cases}$$

[We may think of $a_1, a_2, \ldots, a_n$ as "direction cosines," at least when $n = 3$.]

Points of intersection of line L with the hypersurface $f(X_1, \ldots, X_n) = 0$ are given by the solutions of $f(a_1 t, \ldots, a_n t) = 0$. If $f(X_1, \ldots, X_n)$ is of degree e, then $f(a_1 t, \ldots, a_n t) = b_0 + b_1 t + \cdots + b_e t^e$ would be a polynomial in t whose degree is at most e. The degree would be equal to e for most values of $a_1, \ldots, a_n$. Incidentally, this leads us to a *geometric definition of the degree of a hypersurface S* as the number of points in which a line meets

S. For the above coefficients $b_0, b_1, \ldots, b_e$, we can determine the integer $\epsilon \geq 0$ such that $b_0 = b_1 = \cdots = b_{\epsilon-1} = 0$ and $b_\epsilon \neq 0$. Then we can write

$$f(a_1 t, \cdots, a_n t) = b_\epsilon t^\epsilon + b_{\epsilon+1} t^{\epsilon+1} + \cdots + b_e t^e.$$

The number ϵ is called the *intersection multiplicity* of S and the line L at the point P (which we have taken as the origin). We denote this number by $I(S, L; P)$.

We also define the *multiplicity* of a point P on S, denoted by $\mathrm{mult}_P S$, as

$$\mathrm{mult}_P S = \min_L I(S, L; P),$$

where the minimum is taken over all lines L passing through P. The integer $\mathrm{mult}_P S$ may also be called the *multiplicity* of S at P.

Having defined multiplicity and intersection multiplicity, we can define tangent lines. A line L through a point P on S is said to be a *tangent* to S at P if

$$I(S, L; P) > \mathrm{mult}_P S.$$

Heuristically, most lines through P would meet S at P in $\mathrm{mult}_P S$ coincidental points, but some may meet S at P in more coincidental points. These are precisely the tangent lines. Thus, a tangent line at a point meets the curve (or in general, the hypersurface) at that point more than other lines.

Now that we have defined the notion of multiplicity, we can readily give the definitions of simple points and singular points. A point P on S is said to be *simple* if $\mathrm{mult}_P S = 1$, and P is said to be *singular* if $\mathrm{mult}_P S > 1$. Moreover, if $\mathrm{mult}_P S = 2, 3, \ldots, d, \ldots$ then we may call P a *double* point, a *triple* point, $\ldots$, a d-*ple* point ($= d$-*fold* point), $\ldots$.

Degree. We have given a geometric definition of the degree of a hypersurface $S: f(X_1, \ldots, X_n) = 0$ as the number of points in which a line meets the hypersurface. A question is which line? We can answer this in two ways:

Answer 1: most lines.

Answer 2: maximize over all lines.

We also gave an algebraic definition of the degree of S as the degree of the defining polynomial $f(X_1, \ldots, X_n)$. Now we have a theorem that asserts that the geometric definition coincides with the algebraic definition. This can be easily verified in view of the earlier discussion.

We can also relate multiplicity with its algebraic counterpart, namely, the *order*. Recall that for a polynomial $f(X_1, \ldots, X_n)$ we defined the degree of f to be equal to e if the highest degree term present is of degree e. We can likewise consider the lowest degree term present in the expansion of f and its degree is then called the *order* of f or, more precisely, the order of f at the origin. If P is any other point then we can define the order of f at P, denoted by $\mathrm{ord}_P f$, by considering the "expansion of f at P" obtained by translation of coordinates. Thus if $\mathrm{ord}_P f = d$ and P is the origin (say) then we can write

$$f(X_1, \ldots, X_n)$$
$$= f_d(X_1, \ldots, X_n) + f_{d+1}(X_1, \ldots, X_n) + \cdots + f_j(X_1, \ldots, X_n) + \cdots$$

where each f_j is a homogeneous polynomial of degree j (i.e., a polynomial in which every term has degree j) and $f_d \neq 0$. More generally, if P is not the origin but has coordinates $(\alpha_1, \ldots, \alpha_n)$, then the expansion of f at P would be

$$f(X_1, \ldots, X_n) = \sum c_{i_1 \cdots i_n} (X_1 - \alpha_1)^{i_1} \cdots (X_n - \alpha_n)^{i_n}$$

and the order d of f at P would be the smallest integer such that $c_{i_1 \cdots i_n} \neq 0$ for some $(i_1, \ldots, i_n)$ with $i_1 + \cdots + i_n = d$.

Analogous to the theorem for degree, we can state this result that relates geometry to algebra.

THEOREM. $\operatorname{ord}_P f = \operatorname{mult}_P S$.

NOTE. The algebraic analogues which we discussed so far are high-school algebra analogues. We could also use the language of college algebra to define and study the notions discussed here. Key terms would be rings, fields, ideals, local rings, etc. Roughly speaking, considering polynomials corresponds to being in the realm of a ring, viz., the ring of all polynomials in $X_1, \ldots, X_n$. Coefficients of these polynomials vary in a field such as the rational number field $\mathbb{Q}$, the real number field $\mathbb{R}$, or the complex number field $\mathbb{C}$. Ideals relate to equations. Finally, local rings may be loosely described as power series rings. We shall make these ideas precise and discuss them in greater detail as we proceed.

Again let

$$f(X_1, \ldots, X_n) = \sum_{i_1 + \cdots + i_n \leq e} a_{i_1 \cdots i_n} X_1^{i_1} \cdots X_n^{i_n}$$

be a polynomial in n variables, such that $a_{i_1 \cdots i_n} \neq 0$ for some $(i_1, \ldots, i_n)$ with $i_1 + \cdots + i_n = e$, so that the *degree* of f is e. We write this as $\deg f = e$. We also defined the *order* of f by putting

$$\operatorname{ord} f = d \quad \text{if } a_{i_1 \cdots i_n} = 0 \quad \text{for all } i_1 + \cdots + i_n < d,$$
$$\text{and} \quad a_{i_1 \cdots i_n} \neq 0 \quad \text{for some } i_1 + \cdots + i_n = d.$$

For reasons of symmetry, we may also write the definition of $\deg f$ in an analogous manner as follows:

$$\deg f = e \quad \text{if } a_{i_1 \cdots i_n} = 0 \quad \text{for all } i_1 + \cdots + i_n > e,$$
$$\text{and} \quad a_{i_1 \cdots i_n} \neq 0 \quad \text{for some } i_1 + \cdots + i_n = e.$$

The advantage of doing this is that then the definitions of $\deg f$ and $\operatorname{ord} f$ make sense even when f is a power series (i.e., when the summation in the

above expansion of f is taken over *all* non-negative integers $i_1, i_2, \ldots i_n$),
and we have that

$$f \text{ is a polynomial} \Leftrightarrow \deg f < \infty.$$

Alternatively, we can define $\deg f$ and $\operatorname{ord} f$ by first defining the *support*
of f as $\operatorname{supp}(f) = \left\{ (i_1, \ldots, i_n) : a_{i_1 \cdots i_n} \neq 0 \right\}$, i.e., the set of all $(i_1, \ldots, i_n)$
such that $a_{i_1 \cdots i_n} \neq 0$, and for an n-tuple $i = (i_1, \ldots, i_n)$, denoting by $|i|$
the sum $i_1 + \cdots + i_n$, we define

$$\deg f = \max \{|i| : i \in \operatorname{supp}(f)\},$$
$$\operatorname{ord} f = \min \{|i| : i \in \operatorname{supp}(f)\},$$

where " $\in$ " is the usual symbol for "in" or "is an element of."

Let us also remark that, by convention, the degree of the zero polynomial
is taken to be $-\infty$, whereas the order of the zero polynomial is taken to be
∞. As two basic properties of degree and order, we note that obviously

$$\deg fg = \deg f + \deg g,$$
$$\operatorname{ord} fg = \operatorname{ord} f + \operatorname{ord} g$$

for any two polynomials f and g, where the equation for order remains
valid when f and g are power series.

Given any polynomial (or power series) $f(X_1, \ldots, X_n)$, we can "collect
terms of like degree" to write

$$f = f_d + f_{d+1} + \cdots + f_j + \cdots$$

as a sum of homogeneous polynomials. Here each f_j is a homogeneous
polynomial of degree j, and we have $f_d \neq 0$ where $d = \operatorname{ord} f$.

NOTE. The notion of order can also be defined in the language of college
algebra. So let k denote the field of coefficients (for example, k could be
the rationals $\mathbb{Q}$, the reals $\mathbb{R}$, or the complex numbers $\mathbb{C}$, or even a finite
field) and let $k[[X_1, \ldots, X_n]]$ be the set of all power series in $X_1, \ldots, X_n$
with coefficients in k. We mostly consider formal power series, disregarding
convergence. Let M be the unique maximal ideal in $k[[X_1, \ldots, X_n]]$. Now
for any power series f in $X_1, \ldots, X_n$ with coefficients in k we may define

$$\operatorname{ord}_R f = \max d \text{ such that } f \in M^d.$$

Thus if $d = \operatorname{ord}_R f$, then $f \in M^d$ and $f \notin M^{d+1}$. We shall explain the
terms used above in later lectures.

Now let us go back to our geometric discussion of hypersurfaces. Consider
a hypersurface $S : f(X_1, \cdots, X_n) = 0$ of degree e in n-space so that
$\deg f = e$. Note that *conic sections* correspond to the case of $n = 2$ and
$e = 2$, the *cubics* correspond to the case of $n = 2$ and $e = 3$, and the sphere
is an example of the case of $n = 3$ and $e = 2$.

Let $\widehat{L}$ be any line in n-space given by

$$\begin{cases} X_1 = \alpha_1 + a_1 t \\ \vdots \\ X_n = \alpha_n + a_n t. \end{cases}$$

Note that $\widehat{L}$ need not pass through the origin, i.e., we may have $\alpha_i \neq 0$ for some i. To find where S intersects $\widehat{L}$, we consider

$$A(t) = f(\alpha_1 + a_1 t, \ldots, \alpha_n + a_n t)$$

and we let $e' = \deg A(t)$. Note that

$$A(t) = f_e(a_1, \ldots, a_n) t^e + \text{ terms of degree } < e.$$

where f_e is the *degree form* of f, i.e., the sum of the terms of highest degree in the expansion of f. Now for most values of $a_1, \ldots, a_n$ we have $f_e(a_1, \ldots, a_n) \neq 0$. Thus we always have $e' \leq e$ with equality for most lines $\widehat{L}$. We can factor $A(t)$ into linear factors as

$$A(t) = \gamma^* \prod_{i=1}^{h} (t - \gamma_i)^{e_i} \quad \text{with } \gamma^* \neq 0.$$

Now the points $P_i = (\alpha_1 + a_1 \gamma_i, \ldots, \alpha_n + a_n \gamma_i)$, for $i = 1, \ldots, h$, are exactly the points of intersection of S and $\widehat{L}$, and $I(S, \widehat{L}; P_i) = e_i$. That is, the intersection multiplicity of S and $\widehat{L}$ at P_i is e_i. Note that the sum of these intersection multiplicities is e', i.e., $\sum_{i=1}^{h} e_i = e_1 + \cdots + e_h = e'$. In case $e' < e$, there are some "additional points of intersection" of S and $\widehat{L}$ which have "gone to infinity." These points are obtained by looking at the degree form f_e of f.

Question: What if we cannot factor over a given coefficient field k?

Answer: Indeed, this can be the case (for example, when the coefficient field is $k = \mathbb{R}$ or $\mathbb{Q}$), and if this happens then we simply factor as much as we can, i.e., we write

$$A(t) = \gamma^* \prod_{i=1}^{h} A_i(t)^{e_i} \quad \text{with } \gamma^* \neq 0$$

where $A_1(t), A_2(t), \ldots, A_h(t)$ are distinct *monic* irreducible polynomials in t with coefficients in k (monic means the coefficient of the highest degree term is 1). In this case, the equation $\sum_{i=1}^{h} e_i = e'$ is not quite true, but is to be replaced by the corrected equation $\sum_{i=1}^{h} e_i d_i = e'$ where $d_i = \deg A_i$ for $1 \leq i \leq h$.

EXAMPLE. Let $k = \mathbb{Q}$ or $\mathbb{R}$ and consider the curve $C : F(X, Y) = 0$ in the rational plane or the real plane, where

$$F(X, Y) = 4Y^5 - X^2 Y^2 - 4Y + 4,$$

and let us intersect C by the line (through the origin) given by

$$\begin{cases} X = 2t \\ Y = t. \end{cases}$$

Now we get $A(t) = F(2t, t) = 4(t^5 - t^4 - t + 1)$. We can factor it into powers of irreducible polynomials (with coefficients in k) as

$$A(t) = 4 \left(t^2 + 1 \right) (t - 1)^2 (t + 1).$$

Thus, in this example we have

$$e' = e = 5,$$

and

$$\gamma^* = 4, \ h = 3, \ e_1 = 1, \ d_1 = 2, \ e_2 = 2, \ d_2 = 1, \ e_3 = 1, \ d_3 = 1.$$

The equation $e_1 d_1 + \cdots + e_h d_h = e'$ is easily verified, whereas the equation $e_1 + \cdots + e_h = e'$ is no longer true.

REMARK. Most problems in algebraic geometry ultimately reduce to (or have to be reduced to) questions about polynomials in one variable since this is the only case that we know well enough. Indeed, many theorems in algebraic geometry come down to the following

Basic Fact: A polynomial in one variable of degree n has n roots.

Now what do we mean by this? Unless we interpret this statement correctly, it is false. First of all, we have to count properly. That is, we have to take the multiplicity of each root into consideration; moreover, we must also multiply it by the "degree" of the corresponding root. Thus, if

$$A(t) = \left(c_0 t^n + c_1 t^{n-1} + \cdots + c_n \right) = c_0 A_1(t)^{e_1} A_2(t)^{e_2} \cdots A_h(t)^{e_h} \quad \text{with } c_0 \neq 0$$

is the factorization of a polynomial in one variable t into powers of distinct monic irreducible polynomials $A_1(t), A_2(t), \ldots, A_h(t)$, and if $f_i = \deg A_i$, then we have an obvious

THEOREM. $\sum_{i=1}^{h} e_i f_i = n$.

Here, of course, n is not the number of variables, and f_i's are not the homogeneous components. Indeed, the reason that we used n to denote the degree of $A(t)$ instead of e', and f_i to denote the degree of $A_i(t)$ instead of d_i, is that the college algebra analogue of this formula reads exactly the same as above. For this we refer to Chapter V of Zariski-Samuel's book [ZSa], or to the 1882 paper of Dedekind-Weber [DWe] where it originated. The said paper of Dedekind-Weber appeared in the *Crelle* journal and was written in German. One aim of this paper was to unify the theory of algebraic numbers and algebraic curves (or functions). At this point we may give a

Guide to the literature:

I. *College Algebra*

C$_1$ Zariski-Samuel, *Commutative algebra* [ZSa].

C$_2$ van der Waerden, *Modern algebra* [Wa2].

II. *High school algebra*

H_1 Chrystal, *Textbook of algebra* (1886) [**Chy**].

H_2 Böcher, *Higher algebra* (1907) [**Boc**].

At any rate, it would be useful to study some *old* book on algebra. Here by old we mean anything before 1930, since, in 1930, the first book on college algebra appeared (van der Waerden's). In this connection, the above book of Chrystal would be ideal. On the other hand, the 1882 paper of Dedekind and Weber, although chronologically old, is a precursor of college algebra, which we highly recommend as an introduction to that. We should also add to the list some books on

III. *Analysis*

A_1 Edwards, *Differential calculus* (1899) [**Ed1**].

A_2 Forsyth, *Theory of functions of a complex variable* (1893) [**For**].

NOTE. A careful study of the above books by Chrystal, Edwards, Forsyth, and Zariski-Samuel should reveal that they are essentially talking about the same thing!

LECTURE 5

Outline of the Theory of Plane Curves

Today let us give an outline of the theory of plane curves.

Let $F(X, Y) = \sum_{i+j \leq n} a_{ij} X^i Y^j$ be a polynomial of degree n and let $C_n : F(X, Y) = 0$ be the plane curve of degree n defined by $F(X, Y)$. We may plot the curve in the (X, Y)-plane to get something like Figure 5.1.

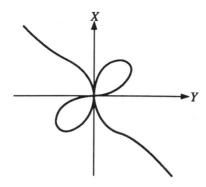

FIGURE 5.1

We now state some facts about plane curves. We shall fill in the details in later lectures.

1. *Bezout's Theorem.* $C_m \cdot C_n = mn$ points. By this we mean that a curve of degree m and a curve of degree n meet in exactly mn points, provided they do not have a common component. However, for this theorem to be true in general, the points of intersection must be "counted properly."

2. *Singularity bounds.* If C_n is devoid of multiple components (that is, if $F(X, Y)$ is not divisible by the square of any nonconstant polynomial), then C_n has at most $\frac{n(n-1)}{2}$ double points. Moreover if C_n is an irreducible curve (that is, if $F(X, Y)$ is an irreducible polynomial), then C_n has at most $\frac{(n-1)(n-2)}{2}$ double points. Again the double points have to be counted properly.

We can make a table (see 5.1) of these bounds for the first few values of n.

TABLE 5.1

Degree	1	2	3	4	5	6
Bound in the square-free case	0	1	3	6	10	15
Bound in the irreducible case	0	0	1	3	6	10

Thus, in case C_n is irreducible, if we define

$$g = \frac{(n-1)(n-2)}{2} - \text{the number of double points}$$

where the double points are counted properly, then we always have that $g \geq 0$. Moreover, we can show that $g = 0 \iff C_n$ *can be rationally parametrized.*

To explain what we mean by a curve being rationally parametrizable let us look at the following

Example. Consider the circle $X^2 + Y^2 = 1$. See Figure 5.2.

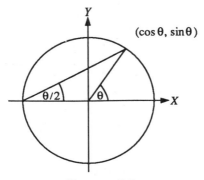

FIGURE 5.2

This may be described by the usual parametric equations $X = \cos \theta$ and $Y = \sin \theta$, which use trigonometric functions. By making the substitution $\tan \theta/2 = t$, these equations are converted to $X = (1 - t^2)/(1 + t^2)$ and $Y = (2t)/(1 + t^2)$ so that X and Y are expressed as rational functions (quotients of polynomial functions) of a single parameter t. We thus say that the circle can be rationally parametrized. (This way of converting trigonometric parametrization to rational parametrization by means of the substitution $\tan \frac{\theta}{2} = t$ is the key to integrating rational functions of trigonometric functions by a general method instead of by hundred-and-one separate reduction formulae, cf. the book of Edwards on integral calculus [**Ed2**].)

More generally, we say that a plane curve C_n can be rationally parametrized if we can find rational functions $\phi(t)$ and $\psi(t)$ of a single variable t such that $F(\phi(t), \psi(t)) = 0$. To avoid $(\phi(t), \psi(t))$ giving only one point of the curve or tracing only one of its components, we also require $(\phi(t), \psi(t))$ to yield most (i.e., all except finitely many) points of the curve as t takes all possible values. Equivalently, this can be achieved by requiring F to be irreducible and at least one of $\phi(t)$ and $\psi(t)$ to be nonconstant.

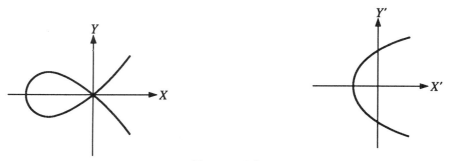

FIGURE 5.3

3. *Birational invariance of g*. For any irreducible plane curve C, we have defined the number $g = g(C)$. This number is called the *genus* of C. It has the basic property that for any other irreducible plane curve C': *C is birationally equivalent to $C' \Rightarrow g(C) = g(C')$.*

To explain what we mean by a curve being birationally equivalent to another, let us discuss the following

EXAMPLE. Consider the nodal cubic $Y^2 - X^2 - X^3 = 0$. If we make the transformation $X' = X$ and $Y' = Y/X$ so that $X = X'$ and $Y = X'Y'$, then we get $(Y^2 - X^2 - X^3) \longrightarrow (X'^2Y'^2 - X'^2 - X'^3) = X'^2(Y'^2 - 1 - X')$. See Figure 5.3. Throwing away the extraneous factor X'^2 we get $Y'^2 - 1 - X' = 0$, which is a parabola. Thus, the nodal cubic is transformed into a parabola by the transformation

$$\begin{cases} X' = X \\ Y' = Y/X \end{cases} \quad \text{or} \quad \begin{cases} X = X' \\ Y = X'Y'. \end{cases}$$

Now the equations defining this transformation are rational both ways. Such a transformation is called a birational transformation. Thus, in this example, we may say that the nodal cubic and the parabola are birationally equivalent. Now, we could also consider the same transformation applied to the cuspidal cubic $Y^2 - X^3 = 0$. See Figure 5.4.

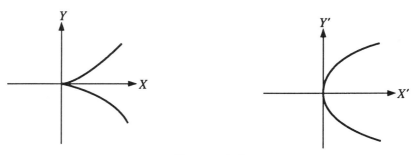

FIGURE 5.4

It yields $X'^2(Y'^2 - X') = 0$ and, throwing away the extraneous factor X'^2, we get $Y'^2 - X' = 0$, which is again a parabola. Thus, we may conclude that the cuspidal cubic and the parabola are birationally equivalent. Therefore, the nodal cubic and the cuspidal cubic are also birationally equivalent. It can be shown that any conic section is birationally equivalent to a line. So it follows that any two conic sections are birationally equivalent. In particular, we see that all the curves considered in this example have the same genus, namely zero.

Now one way to prove birational invariance of genus is to give an alternate definition of genus that easily implies birational invariance because by its very nature it is invariantive. In order to do this we come to the notion of

Differentials. We shall discuss the notion of a differential on a plane curve C, which is assumed irreducible. Then it will turn out that

$$2g - 2 = \text{the number of zeros of a differential on } C,$$

where $g = g(C)$ is the genus of C as defined. Thus the above equation can give an alternate definition of the genus of C. To prepare the groundwork we shall also consider functions on C and their zeros.

We do come across the idea of a differential in studying calculus where we deal with objects such as dx, dy, dv, etc. If perhaps we don't study these on their own, we certainly encounter expressions such as $\int u\,dx$ or $\int u\,dx + v\,dy$ or $\int u\,dv$. Thus, we do have some idea of what a differential is (namely, integrand = integral − the integral sign) or what it should be (namely, something that obeys the rules of transformation of integrands). What should we mean by functions on a curve C? We are familiar with functions as such, e.g., $X^2 + 3$, $\sin X$, etc. Suppose we are given a curve C. Say, for example, C is the circle $X^2 + Y^2 - 1 = 0$, or in general C may be given by an equation $F(X, Y) = 0$. Then we consider functions of x and y, where x and y are assumed to be related by the constraint $x^2 + y^2 - 1 = 0$, or more generally by the constraint $F(x, y) = 0$. In some cases we can solve for x or y in terms of the other. For example, if C is the circle $X^2 + Y^2 - 1 = 0$, then $x^2 + y^2 - 1 = 0$ gives $y = \sqrt{1 - x^2}$. The function $x^2 + x^3 y$ on C may be explicitly expressed as $x^2 + x^3 \sqrt{1 - x^2}$. In general on a plane curve C, we can consider polynomial functions $G(x, y)$, with polynomial expressions $G(X, Y)$, as well as rational functions $u = G(x, y)/H(x, y)$, with polynomial expressions $G(X, Y)$ and $H(X, Y)$ such that $H(x, y) \neq 0$, where x and y are related by $F(x, y) = 0$. In fact, birational equivalence can be alternately defined by saying that C is birationally equivalent to C' if and only if functions on C are the same as functions on C'. By this we mean that the field of rational functions on C is isomorphic to the field of rational functions on C'. Now we consider

4. *Zeros of a function on a curve C*. Here the basic result is that for any (nonzero) rational function on a curve C,

$$\text{number of zeros} = \text{number of poles}.$$

For example, if C is the X-axis, and $r(X) = p(X)/q(X)$ is a rational function on C where $p(X)$ and $q(X)$ are polynomials having no common factor, then the zeros of $r(X)$ are the zeros of the numerator $p(X)$, and the poles of $r(X)$ are the zeros of the denominator $q(X)$. The difference in the degrees of $p(X)$ and $q(X)$ is the order of zero (or pole) of $r(X)$ at infinity. Hence, we see that, for $r(X)$, the number of zeros is the same as the number of poles. In greater detail, let

$$p(X) = b_0 X^d + b_1 X^{d-1} + \cdots + b_d = b_0 \prod_{i=1}^{\mu} (X - \beta_i)^{d_i} \quad \text{with } b_0 \neq 0,$$

and

$$q(X) = c_0 X^e + c_1 X^{e-1} + \cdots + c_e = c_0 \prod_{i=1}^{\nu} (X - \gamma_i)^{e_i} \quad \text{with } c_0 \neq 0.$$

Now $r(X)$ has a zero of order d_i at the (finite) value β_i of X. $r(X)$ has a pole of order e_i at the (finite) value γ_i of X. If $d > e$ then $r(X)$ has a pole of order $d - e$ at ∞. If $d < e$ then $r(X)$ has a zero of order $e - d$ at ∞. Finally, if $d = e$ then $r(X)$ has neither a pole nor a zero at ∞. Thus counting the zeros and poles of $r(X)$ with their multiplicities, we have that

if $d > e$ then: number of zeros

$$= d_1 + \cdots + d_\mu = d = e_1 + \cdots + e_\nu + (d - e) = \text{number of poles},$$

whereas

if $d < e$ then: number of zeros

$$= d_1 + \cdots + d_\mu + (e - d) = e = e_1 + \cdots + e_\nu = \text{number of poles}.$$

Finally

if $d = e$ then: number of zeros

$$= d_1 + \cdots + d_\mu = d = e = e_1 + \cdots + e_\nu = \text{number of poles}.$$

Now let us also consider (nonzero) differentials on a curve C which are expressions of the form $w = u\,dv$ where u and v are rational functions on C (such that u is nonzero and v is nonconstant).

5. *Zeros of a differential on a curve C*. Given a curve C and a (nonzero) differential $w = u\,dv$ on C we have that

$$2g - 2 = \text{ number of zeros of } w - \text{ number of poles of } w,$$

where g is the genus of C.

By the number of zeros (resp. poles) of $w = u\,dv$ we mean the sum of the orders of zeros (resp. poles) of w at points of C, where the sum is taken over those points of C where w does have a zero (resp. pole). To find the order of zero (or pole) of a differential at a point of C we proceed as follows.

Let C again be defined by $F(X, Y) = 0$, and let x and y be the corresponding rational functions on C. Given any point P of C, we choose some t such that t is a rational function of x and y, x and y can be expressed as nice functions of t (viz., power series), and the value of t at P is zero. We then call t a *uniformizing parameter* at the point P. Now, the order of zero (or pole) of a differential $w = u\,dv$ at P is, by definition, the order of zero (or pole) of the function $u\frac{dv}{dt}$ at P. If t' is any other uniformizing parameter at P, then $\frac{dv}{dt} = \frac{dv}{dt'} \cdot \frac{dt'}{dt}$, and $\frac{dt'}{dt}$ has neither a zero nor a pole at P because the expansion of t' as a power series in t has a nonzero linear term. Therefore, the order of zero (or pole) of $w = u\,dv$ at P is independent of the choice of the uniformizing parameter.

If we don't insist on t being zero at P, then we can actually use *uniformizing coordinates*. Thus, if at $P = (\alpha, \beta)$ we don't have a vertical tangent, i.e., if $F_Y(\alpha, \beta) \neq 0$ (where, as usual, F_Y denotes the partial derivative of F with respect to Y), then, by the Implicit Function Theorem we can expand y in terms of x near P. Now x is a uniformizing coordinate at P, which means that $x - \alpha$ is a uniformizing parameter at P. Note that

$$\frac{dv}{d(x - \alpha)} = \frac{dv}{dx} \cdot \frac{dx}{d(x - \alpha)} = \frac{dv}{dx}.$$

Hence the order of zero (or pole) of $w = u\,dv$ at P is the order of zero (or pole) of $u\frac{dv}{dx}$ at P. Likewise, if at $P = (\alpha, \beta)$ we don't have a horizontal tangent, then y is a uniformizing coordinate at P, and the order of zero (or pole) of $w = u\,dv$ at P is the order of zero (or pole) of $u\frac{dv}{dy}$ at P. Thus, for calculating the order of zero (or pole) of a differential, at nonvertical tangent points we can use x, whereas at nonhorizontal tangent points we can use y. To illustrate these notions let us again consider the example of a circle.

EXAMPLE. Let C be the circle $X^2 + Y^2 - 1 = 0$ and let $P = (\alpha, \beta)$ be a point on C. We can choose either x or y as a uniformizing coordinate at points P having neither vertical nor horizontal tangents, i.e., at points P other than $(\pm 1, 0)$ and $(0, \pm 1)$. See Figure 5.5. At points $P = (\alpha, \beta)$ with $\alpha = \pm 1$, y can be taken as a uniformizing coordinate. Likewise at points $P = (\alpha, \beta)$ with $\beta = \pm 1$, x can be taken as a uniformizing coordinate.

Let us verify that the order of a zero (or pole) of a differential $w = u\,dv$ at points $P = (\alpha, \beta)$ other than $(\pm 1, 0)$ and $(0, \pm 1)$ is independent of whether we choose x or y as a uniformizing coordinate at P. Now $\frac{u(dv/dx)}{u(dv/dy)} = \frac{dy}{dx}$. Therefore, it suffices to show that $\frac{dy}{dx}$ has neither a zero nor a pole at P. From $x^2 + y^2 - 1 = 0$, by implicit differentiation, we obtain $2x\,dx + 2y\,dy = 0$, and hence $\frac{dy}{dx} = \frac{-x}{y}$. Obviously, $\frac{-x}{y}$ has no zeros or poles at points $P = (\alpha, \beta)$ with $\alpha \neq 0$ and $\beta \neq 0$, i.e., at points $P = (\alpha, \beta)$ other than $(\pm 1, 0)$ and $(0, \pm 1)$.

Let us also note that the difference (number of zeros $-$ number of poles) is independent of what differential we choose. Namely, if $w = u\,dv$ and $w' = u'\,dv'$ are any two differentials on a curve C, then $w' = u'\frac{dv'}{dv}\,dv$ and $\frac{u'}{u} \cdot \frac{dv'}{dv}$ is a function on C. Hence, our claim follows from the theorem on

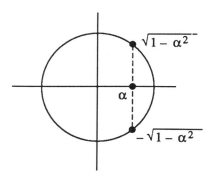

FIGURE 5.5

zeros of functions on C. (Informally, we could simply say that $\frac{w'}{w} = \frac{u'dv'}{udv} = \frac{u'}{u} \cdot \left(\frac{dv'}{dv}\right) = $ a function on C.)

Now let us try to work out a more substantial example.

EXAMPLE. Let C be defined by

$$Y^2 = s(X) \quad \text{with } s(X) = \alpha^* \prod_{i=1}^{h}(X - \alpha_i)$$

where $\alpha^* \neq 0$ and $\alpha_i \neq \alpha_j$ for $i \neq j$. Assuming h to be even, the genus of the curve is going to turn out to be $g = \frac{1}{2}(h - 2)$. Thus we let $h = 2g + 2$ and then we expect that the difference (number of zeros − number of poles) for a differential w on C should come out to be $2g - 2$. (If h is odd then by writing $h = 2g + 1$, this difference would still come out to be $2g - 2$. So again the genus would be g. The cases of $2g + 1$ and $2g + 2$ can be transformed into each other by a "fractional linear transformation.")

Since it doesn't matter which differential w we choose, let us take $w = dx$. If P is any point on C and α is its X-coordinate, then $t = x - \alpha$ serves as a uniformizing parameter at P provided α is different from $\alpha_1, \ldots, \alpha_h$. So in this case $dx/dt = 1$, and thus at these points dx has no zeros or poles. If $\alpha = \alpha_i$ for some i, then y can serve as a uniformizing parameter, and by implicit differentiation we have that $2ydy = \dot{s}(x)dx$, where $\dot{s}$ is the derivative of s. Therefore, $dx/dy = 2y/\dot{s}(x)$, since α_i is a simple root of $s(X)$, we must have $\dot{s}(\alpha_i) \neq 0$. Hence, at such a point dx has a zero of order 1 (and therefore no pole!). Thus $\alpha_1, \ldots, \alpha_h$ account for $1 + 1 + \cdots + 1 = h = 2g + 2$ zeros. Now to get the desired number $2g - 2$, we must consider "points at infinity." It turns out that this curve has two "branches" at infinity. At each of them, dx has a pole of order 2. Therefore, the difference (number of zeros of dx − number of poles of dx) equals $2g + 2 - 2(2) = 2g - 2$ as asserted before.

NOTE. Most of the calculations for differentials would involve techniques of implicit differentiation as in our informal computation in the examples.

Affine Plane and Projective Plane

In the previous lecture we discussed zeros and poles of functions on a curve mainly in the case when the curve is a line. We shall now enlarge this discussion to include the case of any irreducible plane curve $C : F(X, Y) = 0$.

Let us consider a rational function $u = G(x, y)/H(x, y)$ on C, with polynomials $G(X, Y)$ and $H(X, Y)$ that are not multiples of $F(X, Y)$ (so that $G(x, y) \neq 0 \neq H(x, y)$), where x and y are related by the equation $F(x, y) = 0$. By the zeros of u we mean the points at which the curve $G(X, Y) = 0$ meets C, and by the poles of u we mean the points at which the curve $H(X, Y) = 0$ meets C. However, we warn that this is only a first approximation since we could have a case in which both G and H meet C at P. See Figure 6.1(a).

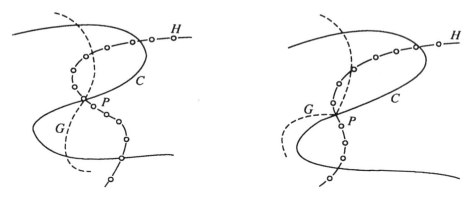

FIGURE 6.1

Worse still, we could have a case in which G and H both meet C at P, and G is tangent to C at P but H is not. See Figure 6.1(b). In this case, P is to be counted as a "double zero" of G, whereas P may be a "simple zero" of H. Then we say P is a zero of u of order $(2 - 1) = 1$. Likewise, if at P, both G and H meet C, and G is not tangent to C at P but H is, then P is a zero of order $(1 - 2) = -1$, i.e., a pole of order 1, and so on. To include all the cases, we define the intersection multiplicity $I(C, u; P)$

of C and u at a point P by putting

$$I(C, u; P) = I(C, G; P) - I(C, H; P)$$

where $I(C, G; P)$ and $I(C, H; P)$ respectively denote the intersection multiplicities of C with the curves $G(X, Y) = 0$ and $H(X, Y) = 0$ at the point P. We define a point P to be a *zero* of u if $I(C, u; P) > 0$. The number $I(C, u; P)$ is then called the order of this zero. On the other hand, we define P to be a *pole* of u if $I(C, u; P) < 0$ and the number $-I(C, u; P)$ is then called the order of this pole.

So the result that for a rational function u on C,

$$\text{number of zeros} = \text{number of poles}$$

can be stated more precisely as

$$\sum_P I(C, u; P) = 0.$$

The above summation must be taken over all points P of C including "those at ∞." So now let us discuss

Points at ∞: To find the points at ∞ of $C : F(X, Y) = 0$, where

$$F(X, Y) = \sum_{i+j \leq n} a_{ij} X^i Y^j = F_0(X, Y) + F_1(X, Y) + \cdots + F_n(X, Y)$$

with $\deg F = n$ and with homogeneous polynomials $F_0, F_1, \ldots, F_n$ of respective degrees $0, 1, \ldots, n$, we first *homogenize* $F(X, Y)$ to obtain

$$f(\mathscr{X}, \mathscr{Y}, \mathscr{Z}) = \sum_{i+j \leq n} a_{ij} \mathscr{X}^i \mathscr{Y}^j \mathscr{Z}^{n-i-j}$$

$$= F_0(\mathscr{X}, \mathscr{Y}) \mathscr{Z}^n + F_1(\mathscr{X}, \mathscr{Y}) \mathscr{Z}^{n-1} + \cdots + F_n(\mathscr{X}, \mathscr{Y}).$$

The points at ∞ of C are obtained by putting $\mathscr{Z} = 0$ in $f(\mathscr{X}, \mathscr{Y}, \mathscr{Z})$. Thus, briefly speaking we "homogenize and put $\mathscr{Z} = 0$." But this simply amounts to looking at the "degree form" $F_n(X, Y)$, which by definition consists of the highest degree terms in $F(X, Y)$. We factor the degree form by writing

$$F_n(X, Y) = \prod_{i=1}^{n} (\mu_i X - \lambda_i Y).$$

Thereby we obtain the *homogeneous coordinates* $(\lambda_i, \mu_i, 0)$ of the points at infinity of $C : F(X, Y) = 0$. These factors are unique only up to proportionality. That is, $Y^2 - X^2 = (Y - X)(Y + X) = (2Y - 2X)(Y/2 + X/2)$. Thus, $(\lambda_i, \mu_i, 0)$ is to be considered equivalent to $(\kappa \lambda_i, \kappa \mu_i, 0)$ where κ can be any nonzero constant. In other words, we are considering points in the projective plane. So let us now discuss what that means.

Affine plane and projective plane. In classical (Euclidean) geometry we have a theorem asserting that any two lines in the same plane meet at exactly one point except when they are parallel. See Figure 6.2. To take care of

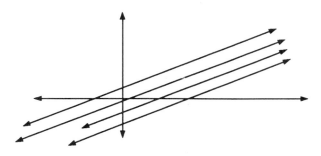

FIGURE 6.2

the exceptional case we postulate a point at infinity in each direction and declare that the corresponding parallel lines meet at that point. If we further postulate that these extra points (at ∞) form a line, called the line at ∞, then we obtain an extension of the usual plane, which is called the *projective plane*. It is denoted by $\mathbb{P}^2$. The usual plane is then called the *affine plane*, and it is denoted by $\mathbb{A}^2$.

In greater detail, a point in the *affine* (X, Y)-*plane* is given by a pair (α, β) where α is the X-coordinate and β is the Y-coordinate. In the projective plane, such a point (α, β) is represented by all triples $(\kappa\alpha, \kappa\beta, \kappa)$ with $\kappa \neq 0$. We call any such triple $(\kappa\alpha, \kappa\beta, \kappa)$ *homogeneous* $(\mathscr{X}, \mathscr{Y}, \mathscr{Z})$-*coordinates* of the point (α, β). This creates room for "points" whose homogeneous $\mathscr{Z}$-coordinate is zero; we call these the *points at infinity*. The original points are sometimes called *points at finite distance*.

More directly, the projective $(\mathscr{X}, \mathscr{Y}, \mathscr{Z})$-plane is obtained by considering all triples (α, β, γ) and identifying proportional triples. In other words, (α, β, γ) and $(\hat{\alpha}, \hat{\beta}, \hat{\gamma})$ represent the same point if and only if $(\hat{\alpha}, \hat{\beta}, \hat{\gamma}) = (\kappa\alpha, \kappa\beta, \kappa\gamma)$ for some $\kappa \neq 0$. Here we exclude the zero triple $(0, 0, 0)$ from consideration. The line at infinity is now given by $\mathscr{Z} = 0$; it is sometimes denoted by L_∞. To a point (α, β, γ) with $\gamma \neq 0$, i.e., to a point not on the line at infinity, there corresponds the point $(\alpha/\gamma, \beta/\gamma)$ in the affine plane. In other words, if $(\mathscr{X}, \mathscr{Y}, \mathscr{Z})$ are homogeneous coordinates of a point in the projective plane which is not on the line at infinity, then the (X, Y)-coordinates of the corresponding point in the affine plane are given by $X = \mathscr{X}/\mathscr{Z}$ and $Y = \mathscr{Y}/\mathscr{Z}$. Thus, by setting $X = \mathscr{X}/\mathscr{Z}$ and $y = \mathscr{Y}/\mathscr{Z}$, we get a one-to-one correspondence

$$\mathbb{P}^2 - \{\mathscr{Z} = 0\} \approx \mathbb{A}^2.$$

Another way of thinking of the projective plane $\mathbb{P}^2$ is to think of it as consisting of all lines through the origin in affine 3-space (or, equivalently, as consisting of all one-dimensional subspaces of a 3-dimensional vector space). If we do think of the projective plane $\mathbb{P}^2$ as consisting of all lines through the origin in the affine $(\mathscr{X}, \mathscr{Y}, \mathscr{Z})$-space, then planes through the origin in the affine $(\mathscr{X}, \mathscr{Y}, \mathscr{Z})$-space correspond to lines in $\mathbb{P}^2$. Now planes through the origin in the affine $(\mathscr{X}, \mathscr{Y}, \mathscr{Z})$-space are given by $a\mathscr{X} + b\mathscr{Y} + c\mathscr{Z} = 0$. If $(a, b) \neq (0, 0)$ then these correspond to $a(\mathscr{X}/\mathscr{Z}) + b(\mathscr{Y}/\mathscr{Z}) + c = 0$, i.e., to $aX + bY + c = 0$, thus giving us lines in the affine plane. On the other hand, if $(a, b) = (0, 0)$ then the plane reduces to $c\mathscr{Z} = 0$, and it corresponds to the line at ∞ in the projective plane $\mathbb{P}^2$. If we want to realize $\mathbb{P}^2$ without going to affine 3-space, then we have to take the drastic step of regarding $c = 0$ to be the equation of a line in the plane.

Incidentally, one does find such bold steps taken in some old books. For example, we may encounter something like $\partial F/\partial 1$ meaning, thereby you homogenize, differentiate, and put $\mathscr{Z} = 1$. In this connection we may cite the charming book, *Algebraic plane curves*, by Coolidge (1928) **[Coo]**.

As we said, we get a one-to-one correspondence $\mathbb{P}^2 - \{\mathscr{Z} = 0\} \approx A^2$. Now because the situation is symmetric we also have the analogous one-to-one correspondences

$$\mathbb{P}^2 - \{\mathscr{X} = 0\} \approx A^2_{\mathscr{X}} \quad \text{and} \quad \mathbb{P}^2 - \{\mathscr{Y} = 0\} \approx A^2_{\mathscr{Y}}.$$

Thus, we may say that the *projective plane is covered by three affine planes.* Symbolically, we may write $\mathbb{P}^2 = A^2 \cup A^2_{\mathscr{X}} \cup A^2_{\mathscr{Y}}$. (Note that "$\cup$" is the usual symbol for "*union*", and so $A^2 \cup A^2_{\mathscr{X}} \cup A^2_{\mathscr{Y}}$ is the set of all points that are either in A^2 or in $A^2_{\mathscr{X}}$ or in $A^2_{\mathscr{Y}}$.) Each of the affine planes A^2, $A^2_{\mathscr{X}}$, $A^2_{\mathscr{Y}}$ have a different coordinate system, namely,

$$(X, Y) : X = \mathscr{X}/\mathscr{Z}, \qquad Y = \mathscr{Y}/\mathscr{Z}$$
$$(Y', Z') : Y' = \mathscr{Y}/\mathscr{X}, \qquad Z' = \mathscr{Z}/\mathscr{X}$$
$$(X^*, Z^*) : X^* = \mathscr{X}/\mathscr{Y}, \qquad Z^* = \mathscr{Z}/\mathscr{Y}.$$

The projective plane $\mathbb{P}^2$ may be drawn as shown is Figure 6.3. The triangle in the figure is called the *fundamental triangle.*

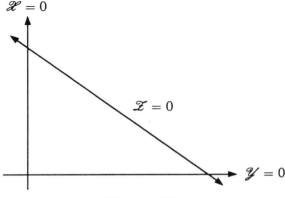

FIGURE 6.3

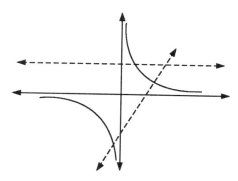

FIGURE 6.4

An algebraic curve $C : F(X, Y) = 0$ in the affine plane corresponds
to the curve in the projective plane defined by $f(\mathscr{X}, \mathscr{Y}, \mathscr{Z}) = 0$ where
$f(\mathscr{X}, \mathscr{Y}, \mathscr{Z})$ is obtained by homogenizing $F(X, Y)$, or equivalently, by
putting $X = \mathscr{X}/\mathscr{Z}$ and $Y = \mathscr{Y}/\mathscr{Z}$ and then multiplying throughout by
$\mathscr{Z}^n$. Thus $F(X, Y) = \sum a_{ij} X^i Y^j$ and $\deg f = n \Rightarrow f(\mathscr{X}, \mathscr{Y}, \mathscr{Z}) =
\sum a_{ij} \mathscr{X}^i \mathscr{Y}^j \mathscr{Z}^{n-i-j}$. To get back $F(X, Y)$, we may either divide by $\mathscr{Z}^n$
to obtain $f = F/\mathscr{Z}^n$, or we may simply put $\mathscr{Z} = 1$ (and straighten out the
calligraphic ex and why!).

Many results in algebraic geometry are true only when we pass to the
projective plane. Thus, for example, two lines in the affine plane may not
always intersect in one point. Algebraically, a system of linear equations

$$\begin{cases} aX + bY + c = 0 \\ a'X + b'Y + c' = 0 \end{cases}$$

may not always have a solution. But such is not the case when we consider
a system of homogeneous linear equations. That is, a system such as

$$\begin{cases} a\mathscr{X} + b\mathscr{Y} + c\mathscr{Z} = 0 \\ a'\mathscr{X} + b'\mathscr{Y} + c'\mathscr{Z} = 0 \end{cases}$$

always has a solution. In fact, there always exists a "solution ray."

Bezout's theorem, which says that $C_m \cdot C_n = mn$ points, (that is, curves
C_m and C_n of respective degrees m and n, having no common component,
meet in exactly mn points), is true only if we take the points at ∞ common
to C_m and C_n into account, provided we consider C_m and C_n as curves in
the projective plane. As an example, consider the hyperbola $C : XY - 1 = 0$.
Oblique lines meet it in two points, whereas lines parallel to the X-axis or
the Y-axis seem to meet it in only one point. See Figure 6.4. For example,
consider the vertical line $L : Y = 1$; it meets C at $(1, 1)$. To obtain the
other point of intersection we homogenize the equations of C and L to get

$$\mathscr{X}\mathscr{Y} - \mathscr{Z}^2 = 0 \quad \text{and} \quad \mathscr{Y} = \mathscr{Z}.$$

Solving these, we find the two solutions $\mathcal{Y} = \mathcal{Z} = \mathcal{X}$ and $\mathcal{Y} = \mathcal{Z} = 0$ giving $(1, 1, 1)$ and $(1, 0, 0)$ as the two points of intersection of C and L. (The triples are unique only up to proportionality.)

The lines $X = 0$ and $Y = 0$ that do not meet C in the affine plane are the factors of the highest degree term and can be seen to be tangents at the points at ∞ of C. Traditionally these lines are called the *asymptotes* to the hyperbola C. So we may say that

$$asymptotes \ \leftrightarrow \ tangents \ at \ \infty.$$

To see this, we make a *change of coordinates* which brings a point at ∞ of C to the origin. In the affine plane, a point $P = (\alpha, \beta)$ (at finite distance) can be brought to the origin by making the translation

$$\begin{cases} \xi = X - \alpha \\ \eta = Y - \beta. \end{cases}$$

Now to bring a point P at ∞, say $P = (\lambda, \mu, 0)$ with $\lambda \neq 0$, to the origin, we first bring it to finite distance by putting $\mathcal{X} = 1$, i.e., by passing to the coordinates

$$(Y', Z') : Y' = \mathcal{Y}/\mathcal{X}, \quad Z' = \mathcal{Z}/\mathcal{X}.$$

This gives us $(1, \mu/\lambda, 0)$, which can now be brought to the origin by the translation

$$\begin{cases} \eta' = Y' - \mu/\lambda \\ \zeta' = Z'. \end{cases}$$

NOTE. To examine the behavior at a point (at finite distance or at infinity), bring it to the origin. Having done this, you may use power series or other tools of analysis—either in their algebraic incarnation or in the original analysis-form.

Sphere with Handles

For any (irreducible) plane curve $C : F(X, Y) = 0$ we have defined the genus g of C in two ways.

(1) In terms of a differential on C, i.e, by the formula

number of zeros of a differential $-$ number of its poles $= 2g - 2$

(2) In terms of lack of singularities or "the defect," i.e., by the formula

$$g = \frac{(n-1)(n-2)}{2} - \text{number of double points.}$$

In (1), the zeros and poles have to be counted with their multiplicities. Likewise in (2), the double points have to be counted properly. Now yet another way to define the genus is

(3) In terms of the number of handles.

Let us now discuss this description of the genus. So $C : F(X, Y) = 0$ is an (irreducible) plane curve. We may draw it in the (X, Y)-plane. See Figure 7.1.

FIGURE 7.1

But the usual picture is in the real plane $\mathbb{R}^2$. What about the picture of the curve in the complex plane $\mathbb{C}^2$? Here X and Y are complex variables. That is, $X = u + iv$ and $Y = z + iw$ so that $\mathbb{C}^2$ becomes the real 4-space $\mathbb{R}^4$ of the four real variables (u, v, z, w). For the original polynomial we have $F(X, Y) = \sum a_{ij} X^i Y^j$ with complex coefficients a_{ij}. By separating into real and imaginary parts, this becomes

$$F(u + iv, z + iw) = h(u, v, z, w) + i\ell(u, v, z, w)$$

where h and ℓ are polynomials with real coefficients. The given equation $F = 0$ is equivalent to the pair of equations $h = 0$ and $\ell = 0$. (The reader may wonder why suddenly h is paired with ℓ rather than with the usual g. This is because the letter g has been used for genus. In old books on Riemann surfaces p was the genus. In modern times p became the characteristic, so the genus had to become g!)

Thus, C is given by two equations in (real) 4-space, each of which represents a solid. Now, just as two surfaces in 3-space intersect along a curve, two solids in 4-space would intersect along a surface. So the two equations would define a surface. Thus the bivariate (two-variable) complex equation $F = 0$, being equivalent to the two quartivariate (four-variable) real equations $h = 0$ and $\ell = 0$, defines a surface. It can be shown that, except for a finite number of points, the resulting surface is a sphere with a certain number of handles. Of course, we know what a sphere looks like.

FIGURE 7.2

A sphere with one handle can be thought of either as in Figure 7.3(a) or Figure 7.3(b).

(a)   (b)

teakettle   doughnut

FIGURE 7.3

Figure 7.4 shows a picture of a sphere with several handles.

Now these are examples of closed surfaces. ("Closed" means that you don't fall off the cliff!) It can be shown that any closed surface "looks like" a sphere with a certain number of handles. Moreover, spheres with different numbers of handles are "distinct" surfaces. (The precise statement is given a little later.)

Thus, to the given (irreducible) plane curve $C : F(X, Y) = 0$, we have associated a closed surface which is called its *Riemann surface*. Let us denote this surface by R. Now R looks like a sphere with g handles for some nonnegative integer g. We define g to be the genus of C. It can be shown

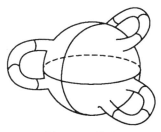

FIGURE 7.4

that this definition of genus is equivalent to the other definitions discussed before. (Unlike the definition in terms of differentials, this definition in terms of handles makes it evident that the genus is a non-negative integer.)

Let us again consider the special example

$$C' : Y^2 - s(X) = 0 \quad \text{with } s(X) = \alpha^* \prod_{i=1}^{2g'+2} (X - \alpha_i)$$

where $\alpha^* \neq 0$ and $\alpha_1, \ldots, \alpha_{2g'+2}$ are distinct. Here we are temporarily making use of the letter g' instead of g, because g is now used to denote the number of handles. Also, we are using the letter C' to denote the above special curve so as not to confuse it with the general curve C. Let R' denote the Riemann surface of C'. We shall soon find that g' equals the number of handles of the surface R'.

Now, for all values of X other than the α_i's, Y has two values, whereas for $X = \alpha_i$, Y has only one value. The points $\alpha_1, \ldots, \alpha_{2g'+2}$ are called the *branch points*.

We cut up the (u, v)-plane (i.e., the complex X-axis) into triangles such that the finite number of branch points $\alpha_1, \alpha_2, \ldots, \alpha_{2g'+2}$ are among the vertices. The (u, v)-plane may be thought of as a sphere minus a point. Notice that a sphere could be projected onto a plane from a fixed point, say the north pole N. Adding an extra point at ∞ (to correspond to N), converts the (real) plane into a sphere, i.e., the complex affine line into the complex projective line. See Figure 7.5 on page 50.

This sphere is called the *Riemann sphere*; let us denote it by S. Triangulating a sphere should not be difficult because a sphere is "like" a cube.

At any rate, we have a triangulation of the sphere S such that the branch points $\alpha_1, \cdots, \alpha_{2g'+2}$ are among the vertices. The triangulation on S can be lifted to a triangulation on R'. Let d_0, d_1, d_2 (resp. D_0, D_1, D_2) denote the number of vertices, edges, and faces of the triangulation of R' (resp: S). Let λ and Λ denote the alternating sums $d_0 - d_1 + d_2$ and $D_0 - D_1 + D_2$ respectively.

Now above every point of S except the α_i's there lie two points of R',

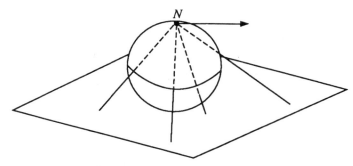

FIGURE 7.5

whereas there is exactly one point of R' lying above each α_i. Hence,

$$d_0 = 2D_0 - (2g' + 2), \qquad d_1 = 2D_1, \qquad d_2 = 2D_2.$$

Therefore,

$$\lambda = 2\Lambda - (2g' + 2).$$

Before we proceed, let us recall one of the oldest theorems in topology, namely,

EULER'S THEOREM (special case). *For a convex polyhedron*
(number of vertices) − (number of edges) + (number of faces) = 2.

As an illustration, we can calculate this sum for a tetrahedron and a cube. See Figure 7.6.

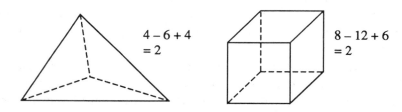

$$4 - 6 + 4 = 2 \qquad\qquad 8 - 12 + 6 = 2$$

FIGURE 7.6

Using this result, from the equation $\Lambda = 2\lambda - (2g' + 2)$ we obtain that $d_0 - d_1 + d_2 = \lambda = 2 - 2g'$ since a (triangulated) sphere is certainly a convex polyhedron so that $\Lambda = 2$. Thus, we have shown that for the surface R' obtained from C' we have $d_0 - d_1 + d_2 = 2 - 2g'$.

Euler's Theorem [**Eul**] extends to a sphere with any number of handles as follows.

EULER'S THEOREM (general case). *If δ_0, δ_1, and δ_2 are the number of vertices, edges, and faces of any triangulation of a sphere with g handles, then*

$$\delta_0 - \delta_1 + \delta_2 = 2 - 2g.$$

Before we outline a proof of this theorem, we note that it is a basic theorem in topology that the alternating sum $\delta_0 - \delta_1 + \delta_2$ depends only on the surface and not on a specific triangulation. This is usually proved by showing that $\delta_0 - \delta_1 + \delta_2 = B_0 - B_1 + B_2$ where the integer B_i is a topological invariant of the surface. The alternating sum $B_0 - B_1 + B_2$ is called the *Euler characteristic* of the surface. The integer B_i is called its i-th Betti number; Betti was a friend of Riemann!

To sketch a proof of Euler's general theorem, let us consider a cube with a smaller cube punctured from it, i.e., a cube with a cubical hole. This is "like" a sphere with one handle or a doughnut. Now if we consider g such cubes with a hole inside each of them, then we obtain a model of a sphere with g handles. See Figure 7.7.

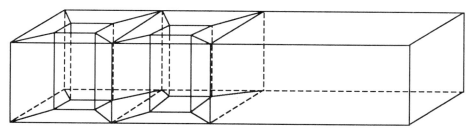

FIGURE 7.7

First of all, let us note that the sum $\delta_0 - \delta_1 + \delta_2$ is unchanged if we make a polygonal division instead of a triangular division. To see this, convert a polygonal division into a triangular division by adding edges joining vertices of a polygon and observe that addition of each edge increases δ_1 and δ_2 both by 1. With this in mind, and by (polygonally) dividing the g-punctured cubes as in Figure 7.7, we find that

$$\delta_0 = 2[2g + 2 + 4g] = 12g + 4$$
$$\delta_1 = 2[g + 1 + 2g + 8g + g + 1 + 2g] = 28g + 4$$
$$\delta_2 = 2[4g + g + 1 + 2g] = 14g + 2,$$

and so

$$\delta_0 - \delta_1 + \delta_2 = 2 - 2g.$$

For the Riemann surface R' of the curve $C' : Y^2 - s(X) = 0$ we have shown that $d_0 - d_1 + d_2 = 2 - 2g'$. Supposing R' to be equivalent to a sphere with g handles, by Euler's general theorem we have $d_0 - d_1 + d_2 = 2 - 2g$. Therefore $g' = g$, and thus the genus of the plane curve C', calculated in terms of differentials, equals the number of handles of its Riemann surface R'.

Reverting to the general case of any (irreducible) plane curve $C : F(X, Y) = 0$, let n be the Y-degree of F, i.e., let

$$F(X, Y) = a_0(X)Y^n + a_1(X)Y^{n-1} + \cdots + a_n(X)$$

with $a_0(X) \neq 0$, where $a_0(X), a_1(X), \ldots, a_n(X)$ are polynomials in X. Recall that by S we are denoting the Riemann sphere, that is, the complex X-axis plus a point at ∞. By R we are denoting the Riemann surface of C. Now to most points of S there are n corresponding points of R. However, for a finite number of points of S there are less than n corresponding points of R. These are called the *branch points*. (Likewise, for most values of X there are n corresponding values Y. However, for a finite number of values of X there are less than n corresponding values of Y. These are called the *discriminant points*. It can be shown that every branch point is a discriminant point, but the converse need not be true.)

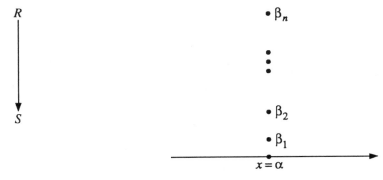

FIGURE 7.8

Again we triangulate the sphere S so that the branch points are among the vertices, and we lift this triangulation to R. Let δ_0, δ_1, and δ_2 (resp. Δ_0, Δ_1, and Δ_2) denote the number of vertices, edges, and faces of this triangulation of R (resp. S). Now

$$\delta_0 = n\Delta_0 - W, \quad \delta_1 = n\Delta_1, \quad \delta_2 = n\Delta_2$$

where

$$W = \text{the number of branch points (counted properly).}$$

Therefore,

$$2 - 2g = \delta_0 - \delta_1 + \delta_2 = n(\Delta_0 - \Delta_1 + \Delta_2) - W = 2n - W,$$

where the first and the last equalities follow by Euler's general and special theorems respectively. Thus, we get the following formula for the genus:

$$2 - 2g = 2n - W \quad \text{or equivalently} \quad g = 1 - n + \frac{1}{2}W.$$

To count the number of branch points properly, a branch point with "branching exponent" equal to e is to be counted as $(e-1)$ ordinary branch points. Thus

$$W = \sum(e - 1)$$

where the summation is over all points of R. The branching exponent e at a point P of R equals the number of "sheets" of R which "come together" at P. In greater detail, near P and Q, where Q is the point of S that corresponds to P, the situation is like the e-th root map. That is, we can find (analytic) uniformizing parameters t and τ at P and Q, respectively, such that $t = \tau^{1/e}$. The number e is also called the "ramification index" at P. The notation "W" is due to Riemann [**Rie**]. Dedekind [**DWe**], trying to interpret Riemann's results algebraically, introduced the notion of *different*, which is denoted by the German letter D, i.e., by $\mathfrak{D}$. The different $\mathfrak{D}$ is defined as the "divisor" on R (a finite set of points of R with assigned multiplicities) given by the equation

$$\mathfrak{D} = \sum \mathfrak{D}(P) P,$$

where the summation is over all points P of R and where

$$\mathfrak{D}(P) = \text{order of zero at } P \text{ of the function } \frac{d\tau}{dt}.$$

Here, as before, t and τ are uniformizing parameters at P and Q respectively. Now $\tau = t^e$ where e is the "ramification index" at P. By differentiating we get

$$\frac{d\tau}{dt} = et^{e-1} \quad \text{and hence} \quad \mathfrak{D}(P) = e - 1.$$

This being so, at every point of R, we get $W = \deg \mathfrak{D}$, where by the degree of a divisor we mean the number of points in it counted with multiplicities. In view of this interpretation of W, the above formula becomes

$$2g - 2 = -2n + \deg \mathfrak{D} \quad \text{or equivalently} \quad g = 1 - n + \frac{1}{2}\deg \mathfrak{D}.$$

Now let us discuss the "topological" result that the closed surfaces are "like" spheres with a certain number of handles in precise terms. We find that besides closedness we also need another condition, namely *orientability*. Roughly speaking, this means that if we take an oriented circle on the surface and move it around, no matter how, then it always comes back with the same orientation. A typical example of a nonorientable surface is a Möbius strip. See Figure 7.9 on page 56.

To construct a Möbius strip, identify the indicated sides of the rectangle with a twist as shown in Figure 7.9. If on the so-constructed strip we walk the depicted oriented circle around, then it reverses its direction.

The technical name for the closedness property is *compactness*. The topological result referred to previously is the following

THEOREM. *A compact orientable 2-dimensional real manifold is homeomorphic to a sphere with a certain number of handles.*

Question. How do we describe orientability precisely?

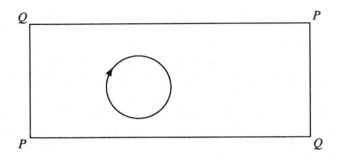

FIGURE 7.9

Answer. Let us consider a differentiable manifold M of dimension q. First of all, M is a "topological Hausdorff space" in which every point P has a local coordinate system, i.e., a homeomorphism of a neighborhood N_P of P in M onto the unit disc $\{(x_1, \ldots, x_q) \in \mathbb{R}^q : |x_1|^2 + \cdots + |x_q|^2 < 1\}$. Let us "identify" N_P with the said unit disc. (This only makes M into a "topological manifold.") Given another point P' in M, we likewise identify a neighborhood $N_{P'}$ of P' in M with the unit disc $\{(x'_1, \ldots, x'_q) \in \mathbb{R}^q : |x'_1|^2 + \cdots + |x'_q|^2 < 1\}$. On $N_P \cap N_{P'}$, we can now express $x'_1, \ldots, x'_q$ as functions of $x_1, \ldots, x_q$. On a "differentiable manifold," it is assumed that these functions have continuous partial derivatives. Hence, in particular, we can talk about the Jacobian determinant $\det (\partial x'_a / \partial x_b)_{1 \le b \le q}^{1 \le a \le q}$. Now orientability can be defined by requiring that, for every pair of points P and P' of M, the value of the said determinant is positive on $N_P \cap N_{P'}$. (Note that "$\cap$" is the usual symbol for *intersection*, and so $N_P \cap N_{P'}$ is the set of all points common to N_P and $N_{P'}$.)

By changing $\mathbb{R}$ to $\mathbb{C}$ in the definition of a differentiable manifold, we get the definition of a complex manifold M of (complex) dimension q. By splitting the complex local coordinates $x_1, \ldots, x_q$ (resp. $x'_1, \ldots, x'_q$) into real and imaginary parts $x_a = u_a + iv_a$ (resp. $x'_a = u'_a + iv'_a$), M becomes a differentiable manifold of dimension $2q$. In case of $q = 1$, by Cauchy-Riemann differential equations, the (continuous) existence of dx'_1 / dx_1 on $N_P \cap N_{P'}$ yields the positivity of the Jacobian determinant

$$\det \begin{pmatrix} \partial u'_1 / \partial u_1 & \partial u'_1 / \partial v_1 \\ \partial v'_1 / \partial u_1 & \partial v'_1 / \partial v_1 \end{pmatrix}.$$

This is how one can show that the Riemann surface R is orientable. By the same argument we can see that a complex manifold, of any dimension, is orientable.

NOTE. When a surface is orientable you don't go out of your mind walking on it, i.e., you retain a certain definite "sense of direction," which in technical jargon is sometimes called the *indicatrix*. See the third page of Lefschetz's famous 1921 monograph "L' Analysis Situs . . . " [Le1]. For a typical example of a nonorientable surface you may look up the picture of a Möbius strip in any topology book. Remember Walter Pigeon playing the role of mysterious Dr. Möbius in the classic science fiction movie, *Forbidden planet*!

Seifert and Threlfall [STh], now even translated into English, is certainly a wonderful book for all the topology involved.

What about an "old" book on Riemann surfaces? Well, in my personal diary I have an entry about the book of Stahl [Sta] saying that it is "the book I enjoyed most."

Functions and Differentials on a Curve

Before we discuss differentials, let us give a

Summary of the lecture on sphere with handles. We have a sphere S and a surface R above it, or geometrically we have the X-axis (together with a point at ∞) and a plane curve C above it. (The symbols S and R are used here to indicate the Riemann sphere and Riemann surface respectively.) See Figure 8.1.

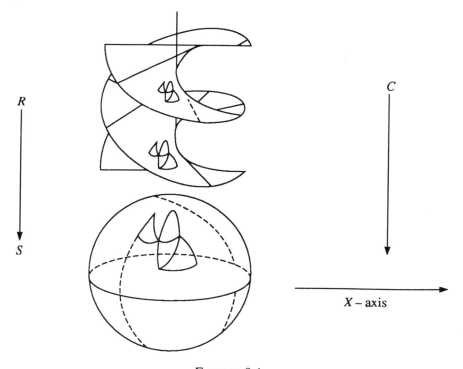

FIGURE 8.1

Let Δ_0, Δ_1, Δ_2 denote the number of vertices, edges, and faces corresponding to a suitable triangulation of S that includes the branch points as vertices. Let δ_0, δ_1, δ_2 be similarly defined numbers for R when we lift this triangulation to R. Let n^* be the covering degree of R over S. That is, let n^* be the Y-degree of the defining polynomial $F(X, Y)$ of C. Finally, let W

be the number of branch points counted properly, i.e.,

$$W = \sum_P (e_P - 1)$$

where the summation is taken over all points P of R, and where e_P is the ramification index at P. Now clearly,

$$\delta_0 = n^* \Delta_0 - W, \quad \delta_1 = n^* \Delta_1, \quad \delta_2 = n^* \Delta_2.$$

By Euler's theorem we have

$$2 - 2g = \delta_0 - \delta_1 + \delta_2 \quad \text{and} \quad 2 - 2g^* = \Delta_0 - \Delta_1 + \Delta_2$$

where g and g^* are the number of handles of R and S respectively. Thus we get the topological genus formula

$$\boxed{2 - 2g = n^*(2 - 2g^*) - W},$$

which is variously ascribed to Riemann, Hurwitz, and Zeuthen (in chronological order).

Note that we haven't substituted $g^* = 0$ (which is the case for a sphere). This is to indicate that the above considerations also apply in the case when S is not necessarily a sphere but another Riemann surface, i.e., when we are projecting the curve C of genus g onto another curve C^* of genus g^*, and we are denoting the "covering degree" of C over C^* by n^*.

$$C$$
$$\downarrow$$
$$C^*$$

Again, following Dedekind, we define the different $\mathfrak{D}^*$ of C over C^* as the divisor on C given by the equation

$$\mathfrak{D}^* = \sum \mathfrak{D}^*(P)\, P,$$

where the summation is over all points P of C. Here,

$$\mathfrak{D}^*(P) = \text{ the order of zero at } P \text{ of the function } \frac{dt^*}{dt},$$

where t and t^* are uniformizing parameters at P and P^* respectively, with P^* being the point of C^* which corresponds to P. Now

$$W = \sum \mathfrak{D}^*(P) = \deg \mathfrak{D}^*,$$

where the summation is over all points P of C. Therefore, the topological genus formula yields the algebraic genus formula

$$2g - 2 = n^*(2g^* - 2) + \deg \mathfrak{D}^*,$$

where

$$n^* = \text{ the "covering degree" of } C \text{ over } C^*.$$

To explain the etymology of the term "different," let w^* be a differential on C^*. We could regard w^* to be a differential on C. Then we may denote it by w_C^*. The different $\mathfrak{D}^*$ indicates how different w^* and w_C^* are from each other. In greater detail, by writing $w^* = u^* dt^*$ at P^*, we get $w_C^* = u^* \frac{dt^*}{dt} dt$ at P. Thus w^* and w_C^* differ by $\frac{dt^*}{dt}$ at the point pair (P^*, P). Actually, continuing in this manner, and by using the formula $\sum e_i f_i = n^*$ (see Chapter V of [ZSa]), and by replacing Euler's theorem by the equations

$$2g - 2 = \text{(number of zeros of } w_C^*) - \text{(number of poles of } w_C^*),$$
$$2g^* - 2 = \text{(number of zeros of } w^*) - \text{(number of poles of } w^*),$$

we can directly prove the algebraic genus formula instead of deducing it from the topological genus formula.

For further discussion of differentials and their use in computing the genus, let us revert to the case when C^* is the X-axis.

Example: Let us consider the curve $C : Y^p - X^q = 0$ where p and q are coprime positive integers, and let x and y be the corresponding rational functions on C. In this case, we know the answer, namely that the genus $g = 0$. This is true because we can parametrize the curve as

$$\begin{cases} X = t^p \\ Y = t^q \end{cases}$$

so that the curve is rational. Hence the genus is zero. When $X \neq 0$, Y is a function of X which takes p values. Thus, when $X \neq 0$ (and hence $Y \neq 0$), both x and y are uniformizing coordinates, and dx (say) has neither a zero nor a pole at such a point. At $X = 0$ (and hence $Y = 0$), both x and y are "bad." In this case t is a uniformizing parameter, and we have $dx = pt^{p-1} dt$. Thus dx has a zero of order $p - 1$ at the origin. It only remains to consider the behavior when $X = \infty$. (Note that our answer should be $2(0) - 2 = -2$, and so dx should have a pole of order $p + 1$ at ∞.) To do this put

$$x = \frac{1}{\xi} \quad \text{and} \quad y = \frac{1}{\eta}$$

and note that

$$\xi^q = \eta^p.$$

We are considering the origin $P' = (0, 0)$ in the (ξ, η)-plane. Now $\xi^q = \eta^p$ can be parametrized as

$$\begin{cases} \xi = \tau^p \\ \eta = \tau^q, \end{cases}$$

and τ is a uniformizing parameter at P'. We have $x = \tau^{-p}$. Therefore,

$$dx = -p\tau^{-(p+1)} d\tau.$$

So sure enough, dx has a pole of order $p+1$ at P', and thus for a differential on the above curve C we have

(number of zeros $-$ number of poles) $= [(p-1) - (p+1)] = -2$.

Hence, $g = 0$.

Note that the calculations in the above example were similar to the ones made for the old example

$$Y^2 = s(X) \quad \text{with } s(X) = \alpha^* \prod_{i=1}^{2g+2} (X - \alpha_i)$$

that we considered before. In this old example, we didn't however complete the calculations at ∞. This is because they are not as simple as in this example. This is so because the curve $Y^2 = s(X)$ has two "branches" above $X = \infty$. At each of them dx has a pole of order 2. To explain matters, let us consider a special case of this old example, namely the curve defined by $Y^2 = X^h - 1$ where h is even. Let x and y be the corresponding rational functions on the curve. We know that dx has h zeros at finite distance. For calculations at ∞ we again put

$$x = \frac{1}{\xi} \quad \text{and} \quad y = \frac{1}{\eta},$$

and we note that then

$$\eta^{-2} = \xi^{-h} - 1 = \frac{1}{\xi^h} - 1.$$

Hence,

$$\eta^2 = \frac{\xi^h}{1 - \xi^h} = \xi^h \left(1 - \xi^h\right)^{-1}.$$

By putting $h = 2\ell$ in the equation we get

$$\eta^2 = \xi^{2\ell} \left(1 - \xi^{2\ell}\right)^{-1}.$$

Now by the Binomial Theorem (for fractional exponent) we can write

$$\eta = \xi^\ell \left(1 - \xi^{2\ell}\right)^{-1/2}$$
$$= \pm \xi^\ell \left(1 + \frac{1}{2}\xi^{2\ell} + \frac{(1/2)(3/2)}{2}\xi^{4\ell} + \cdots\right)$$
$$= \pm \left[\xi^\ell + \frac{1}{2}\xi^{2\ell+\ell} + \cdots\right].$$

Thus ξ does work as a uniformizing parameter, but there are two expansions of η in terms of ξ, i.e., there are two "branches" at ∞. The curve has a *high tacnode* at the origin in the (ξ, η)-plane. See Figure 8.2. Now since $x = \xi^{-1}$, we have $dx = -\xi^{-2}d\xi$. So dx has a pole of order 2. But this is

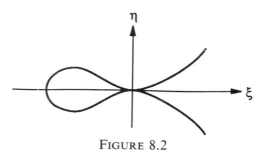

FIGURE 8.2

the case for each of the two branches at ∞. Thus we may conclude that for a differential on $Y^2 = X^h - 1$, we have

$$(\text{number of zeros} - \text{number of poles}) = (2g + 2) - (2 + 2) = 2g - 2,$$

where we have put $h = 2g + 2$. This is possible since h is assumed to be even. Thus, we see that the difference is as desired.

So far, we have described heuristically the ideas of zeros and poles of differentials on a curve and shown calculations for these in an operational manner. Such a situation is comforting as well as discomforting. It is comforting because we understand what is going on. However, it is also discomforting because we may not know what always happens (in the general situation). The heuristic discussion may help us to understand the "why" behind the general theorems to be presented later.

Recall that we are considering a plane (irreducible) curve $C : F(X, Y) = 0$, and x and y are the corresponding rational functions on C. Also recall that for a rational function $u = \frac{G(x,y)}{H(x,y)}$ on C, where $G(X, Y)$ and $H(X, Y)$ are polynomials, we first defined the intersection multiplicity of C and u at a point P by putting

$$I(C, u; P) = I(C, G; P) - I(C, H; P).$$

Then, in case $I(C, u, P) > 0$ (resp. $I(C, u, P) < 0$) we defined $I(C, u, P)$ (resp. $-I(C, u, P)$) as the order of zero (resp. pole) at P.

What if the curve C has two branches B and B' at P? (For example, if $F(X, Y) = Y^2 - X^2 - X^3$ then

$$F(X, Y) = Y^2 - X^2 - X^3 = Y^2 - X^2(1 + X).$$

By the binomial theorem we have

$$(1 + X)^{1/2} = \pm \left(1 + \frac{1}{2}X - \frac{1}{8}X^2 + \cdots \right).$$

Hence, the curve C has two branches at the origin.) See Figure 8.3 on page 62. Well, first we suitably extend the notion of intersection multiplicity of two curves to include the case of a curve and a branch. Now we put

$$I(B, u; P) = I(B, G; P) - I(B, H; P)$$

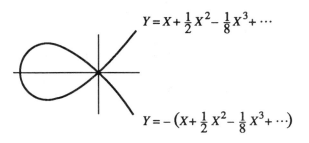

$$Y = X + \tfrac{1}{2}X^2 - \tfrac{1}{8}X^3 + \cdots$$

$$Y = -\left(X + \tfrac{1}{2}X^2 - \tfrac{1}{8}X^3 + \cdots\right)$$

FIGURE 8.3

and

$$I(B', u; P) = I(B', G; P) - I(B', H; P)$$

and then in case ... and so on. Thus, for calculating the number of zeros or poles of a function, it doesn't matter whether we consider zeros or poles at a point P or at each of the branches at P since the numbers add up so as to cause no harm to the calculations involving number of zeros or poles of a function. But such is not the case when we are considering differentials. Then we must take the branches into consideration. We already noted this when doing computations at ∞ for the curve $Y^2 = X^h - 1$ $(h = 2g + 2)$.

To discuss branches at a point in greater detail, let us suppose that

$$F(X, Y) = Y^n + a_1(X)Y^{n-1} + \cdots + a_n(X),$$

where $a_1(X), \ldots, a_n(X)$ are polynomials in X. Now we have *Newton's Theorem*, which says that we can factor $F(X, Y)$ into linear factors involving fractional power series in X. That is,

$$F(X, Y) = \prod_{i=1}^{n} \left[Y - \eta_i(X^{1/m})\right],$$

where $\eta_i(X^{1/m})$ are power series in $X^{1/m}$ for some positive integer m. In other words, there exists a positive integer m such that if we put $X = T^m$ then

$$F(T^m, Y) = \prod_{i=1}^{n} \left[Y - \eta_i(T)\right],$$

where $\eta_i(T)$ are power series in T. To see that we may indeed need fractional power series, take $F(X, Y) = Y^2 - X^3 - X^4$ and note that then

$$F(X, Y) = Y^2 - X^3 - X^4 = Y^2 - X^3(1 + X).$$

By the Binomial Theorem we have

$$[X^3(1 + X)]^{1/2} = \pm X^{3/2}\left(1 + \tfrac{1}{2}X - \tfrac{1}{8}X^2 + \cdots\right).$$

Hence,

$F(X, Y) =$

$$\left[Y - X^{3/2}\left(1 + \frac{1}{2}X - \frac{1}{8}X^2 + \cdots \right) \right]\left[Y + X^{3/2}\left(1 + \frac{1}{2}X - \frac{1}{8}X^2 + \cdots \right) \right].$$

Therefore, the "roots" η_1, η_2 need not be power series in X. (Here $m = 2$, and $X = T^2$.) This gives us a (local) parametrization

$$\begin{cases} X = T^m \\ Y = \eta_i(T) \end{cases}$$

at the point $(0, \eta_i(0))$ of C. Note that $(0, \eta_1(0)), \ldots, (0, \eta_n(0))$ are precisely the points of C whose X-coordinate is 0. (These n points need not be distinct.)

More generally, if α is any value of X, then by applying Newton's Theorem to $F^*(X, Y) = F(X + \alpha, Y)$ we get

$$F(T^{m^*}, Y) = \prod_{i=1}^{n}\left[Y - \eta_i^*(T) \right],$$

where m^* is a positive integer and $\eta_1^*(T), \ldots, \eta_n^*(T)$ are power series in T. This gives us a parametrization

$$\begin{cases} X = \alpha + T^{m^*} \\ Y = \eta_i(T) \end{cases}$$

of C at the point $(\alpha, \eta_i^*(0))$. Again we note that $(\alpha, \eta_1^*(0)), \ldots, (\alpha, \eta_n^*(0))$ are precisely the points of C whose X-coordinate is α. (Once again, these n points need not be distinct.)

Thus, given any point (α, β) of C we have a *parametrization*

$$\begin{cases} X = \lambda(T) \\ Y = \mu(T) \end{cases}$$

of C at (α, β), where $\lambda(T)$, $\mu(T)$ are power series in T such that

$$(\lambda(0), \mu(0)) = (\alpha, \beta) \quad \text{and} \quad F(\lambda(T), \mu(T)) = 0.$$

Such a parametrization corresponds to a *branch* of C at (α, β). Now if we have any similar parametrization, we can substitute say $T + T^2$ for T to obtain another parametrization, which is really not very different from the previous parametrization. Likewise, if we substitute $T^2 + T^3$ for T in the parametrization then we get a parametrization that is somewhat silly.

We define a parametrization $(\lambda^*(T), \mu^*(T))$ to be *redundant* if it can be obtained from another parametrization $(\lambda(T), \mu(T))$ by substituting for T some power series in T of order > 1. A parametrization is called *irredundant* if it is not redundant.

We define the notion of a branch more precisely as follows: A *branch* at a point is an equivalence class of irredundant parametrizations at that point where two parametrizations are considered equivalent if one can be obtained from the other by substituting a power series of order 1.

At any point of C there is at least one and at most a finite number of branches. Now we can define zeros and poles of functions and differentials at a branch as follows:

First, if B is a branch at a point of C and $u = \frac{G(x,y)}{H(x,y)}$ is any rational function on C where $G(X, Y)$ and $H(X, Y)$ are polynomials, then we define the *order of u at the branch B* by putting

$$\text{ord}_B u = \text{ord}\,(G(\lambda(T), \mu(T))) - \text{ord}\,(H(\lambda(T), \mu(T)))\,,$$

where $(\lambda(T), \mu(T))$ is some representative of the branch B. In case $\text{ord}_B u > 0$ (resp. $\text{ord}_B u < 0$), we say that u has a zero (resp. pole) of order $\text{ord}_B u$ (resp. $-\text{ord}_B u$) at B.

Note that the order of a rational function at a branch clearly depends only on the branch and not on the representative we choose. Also note that ord_B satisfies the two properties saying that for any two rational functions u and u' on C we have

$$\text{ord}_B(uu') = \text{ord}_B u + \text{ord}_B u' \quad \text{and} \quad \text{ord}_B(u + u') \geq \min(\text{ord}_B u, \text{ord}_B u').$$

So we may abstract the idea of an order function as a function having these two properties. The language used for such an abstraction is that of valuations on the set of all rational functions on C. One can show that every such valuation is "centered" at some point of C and corresponds to a branch of C at that point. We shall discuss this in more detail later.

The theorem on zeros of functions can now be stated more precisely as follows:

THEOREM. *If u is any rational function on C then*

$$\sum \text{ord}_B u = 0\,,$$

where the sum is taken over all the branches B at each point of C.

We refine the notion of a divisor on C by declaring that it is a finite set of branches of C with assigned multiplicities. We define the degree of such a divisor to be the number of branches in it counted with multiplicities. Moreover, we define the divisor (u) of a rational function u on C by putting

$$(u) = \sum \text{ord}_B(u)B\,,$$

where the sum is taken over all the branches B of C and where the parenthesis on the right side is ordinary! The above theorem on zeros of functions may now be restated by saying that *for any rational function u on C we have*

$$\deg(u) = 0.$$

Next, for a differential $u\,dv$ where u and v are rational functions on C, we define its *order at the branch B* as the order of the function $u\frac{dv}{dT}$. That is,

$$\operatorname{ord}_B(udv) = \operatorname{ord}_B\left(u\frac{dv}{dT}\right).$$

Again, in case $\operatorname{ord}_B(udv) > 0$ (resp. $\operatorname{ord}_B(udv) < 0$), we say that udv has a zero (resp. pole) of order $\operatorname{ord}_B(udv)$ (resp. $-\operatorname{ord}_B(udv)$) at B. The theorem on zeros of differentials may now be stated more precisely.

THEOREM. *If w is any differential on C then*

$$\sum \operatorname{ord}_B w = 2g - 2,$$

where g is the genus of C and the sum is again taken over all the branches B at each point of C.

Once again, we define the divisor (udv) of a differential udv on C by putting

$$(udv) = \sum \operatorname{ord}_B(udv)B,$$

where the sum is taken over all the branches B of C and where, needless to say, the parenthesis on the right side is ordinary. Now the theorem on zeros of differentials may be restated by saying that *for any differential udv on C we have*

$$\deg(udv) = 2g - 2,$$

where g is the genus of C.

REMARK. In the discussion of the theorems on zeros of functions and differentials we should have included branches at points at infinity. This can be achieved by bringing them to finite distance and then applying Newton's Theorem.

NOTE. Let $\mathfrak{D}$ be the different when we project the curve $C : F(X, Y) = 0$ onto the X-axis, and let x and y be the corresponding rational functions on C. The part of $\mathfrak{D}$ at finite distance may be thought of as an "ideal" $\mathfrak{D}'$ in the "affine coordinate ring" A' of the "desingularization" of C at finite distance. Along with the the ideal $\mathfrak{D}'$, Dedekind introduced another ideal in A' called the *conductor* and denoted it by the German letter C, ($\mathfrak{C}$). He then related the ideals $\mathfrak{C}$ and $\mathfrak{D}'$ in A' by connecting them with the partial Y-derivative F_Y of F by means of the equation

$$F_Y(x, y)A' = \mathfrak{C}\mathfrak{D}',$$

where $F_Y(x, y)A'$ denotes the ideal in A' generated by $F_Y(x, y)$. In effect, the conductor $\mathfrak{C}$ gives the singularities of C counted properly. So the equation can be paraphrased by saying that "the discriminant locus equals the branch locus plus the projection of the singular locus." For this and other relevant aphorisms, see *Historical ramblings* [A25].

Polynomials and Power Series

In discussing Newton's Theorem last time, we assumed that

$$F(X, Y) = Y^n + a_1(X)Y^{n-1} + \cdots + a_n(X),$$

where the coefficients $a_i(X)$ are polynomials in X. Now, sometimes we may have a leading coefficient $a_0(X)$ which is not necessarily equal to 1. (That is, F is not necessarily "monic in Y".) To include this case, we can divide by the leading coefficient and allow $a_i(X)$ to be rational functions of X.

For example, consider

$$(1 - X)Y^2 + XY + 2X^3,$$

and let us divide by $(1 - X)$ to obtain

$$Y^2 + \frac{X}{1 - X}Y + \frac{2X^3}{1 - X}.$$

Now we can carry out the division:

$$
\begin{array}{r}
X + X^2 + X^3 + \cdots \\
1 - X \overline{)\ X } \\
\underline{X - X^2} \\
X^2 \\
\underline{X^2 - X^3} \\
X^3 \\
\underline{X^3 - X^4} \\
X^4 \cdots
\end{array}
$$

and in this way, we have the coefficients as *power series in* X, where a power series may be defined (illogically) as a polynomial that doesn't (necessarily) stop!

Let us consider another example:

$$\left(X^2 - X^3\right)Y^2 + XY + 2, \quad \text{which gives} \quad Y^2 + \frac{X}{X^2 - X^3}Y + \frac{2}{X^2 - X^3}.$$

In this case, by carrying out the long-division, we get

$$
X^2 - X^3 \overline{) \begin{array}{l} X^{-1} + 1 + X + \cdots \\ X \\ \underline{X - X^2} \\ X^2 \\ \underline{X^2 - X^3} \\ X^3 \cdots \end{array}} \quad .
$$

Here the coefficients turn out to be *meromorphic series* in X. A meromorphic series (in one variable) may (again illogically) be defined as a power series with finitely many negative terms. By negative terms we mean terms with negative exponents.

We extend the notion of order of a power series to *order of a meromorphic series* in an obvious manner, namely, the degree of the smallest degree term present.

Actually, when we talked about the order of a differential at a branch, we have tacitly used the notions of a meromorphic series and its order. With these notions at our disposal, we can now easily show that, given any differential $u dv$ on the curve $C : F(X, Y) = 0$, the order of $u dv$ at a branch B of C depends only on the branch B and not on choice of its representative parametrization $(\lambda(T), \mu(T))$. To see this, let $(\lambda'(T), \mu'(T))$ be any other parametrization belonging to B. By the definition of equivalence of parametrizations, we must have

$$
(\lambda'(T), \mu'(T)) = (\lambda(\sigma(T)), \mu(\sigma(T)))
$$

for some power series $\sigma(T)$ of order 1. Let

$$
\theta(T) = u(\lambda(T), \mu(T)) \frac{dv(\lambda(T), \mu(T))}{dT}
$$

and

$$
\theta'(T) = u(\lambda'(T), \mu'(T)) \frac{dv(\lambda'(T), \mu'(T))}{dT}.
$$

The definitions of $\operatorname{ord}_B(u dv)$ according to parametrizations $(\lambda(T), \mu(T))$ and $(\lambda'(T), \mu'(T))$ are respectively given by $\operatorname{ord} \theta(T)$ and $\operatorname{ord} \theta'(T)$. However, by the chain rule

$$
\theta'(T) = \theta(\sigma(T)) \frac{d\sigma(T)}{dT}.
$$

Clearly,

$$
\operatorname{ord} \theta(\sigma(T)) = \operatorname{ord} \theta(T) \quad \text{and} \quad \operatorname{ord} \frac{d\sigma(T)}{dT} = 0.
$$

Therefore,

$$
\operatorname{ord} \theta(T) = \operatorname{ord} \theta'(T).
$$

As we have noticed, in discussing Newton's Theorem it is better to deal with a monic polynomial in Y whose coefficients are meromorphic series in X. In turn, the coefficients of these meromorphic series vary in some field

k, where by a *field* we mean a set where we can do addition, subtraction, multiplication, and division. If we do not require the ability to divide, then we call it a (commutative) *ring*. Thus, for example, $\mathbb{Q}, \mathbb{R}, \mathbb{C}$ are fields, whereas the set of integers $\mathbb{Z}$ is an example of a ring which is not a field. In the ring $\mathbb{Z}$ the product of any two nonzero elements is nonzero. A ring with this property is called a *domain*. Just as we go from $\mathbb{Z}$ to $\mathbb{Q}$, we can pass from a domain to its *quotient field*.

Now having fixed a field k of coefficients, we can set up the following notation:

$k((X))$ = field of all meromorphic series in X with coefficients in k.

$k[[X]]$ = ring of all power series in X with coefficients in k.

$k(X)$ = field of all rational functions in X with coefficients in k.

$k[X]$ = ring of all polynomials in X with coefficients in k.

To see that $k((X))$ is a field, in fact the quotient field of $k[[X]]$, we consider the geometric series

$$\frac{1}{1-Z} = 1 + Z + Z^2 + \cdots .$$

Then we substitute any positive order power series $\phi(X)$ for Z to get

$$\frac{1}{1-\phi(X)} = 1 + \phi(X) + \phi(X)^2 + \cdots ,$$

thereby obtaining a recipe for taking the reciprocal of any nonzero meromorphic series $\psi(X)$ by writing $\psi(X) = \alpha^* X^\nu (1 - \phi(X))$ with $\nu = \operatorname{ord} \psi(X)$ and $\operatorname{ord} \phi(X) > 0$ and $0 \neq \alpha^* \in k$.

For the four rings we have the following inclusion diagram

$$\begin{array}{ccc} k[X] & \subset & k[[X]] \\ \cap & & \cap \\ k(X) & \subset & k((X)). \end{array}$$

As usual, "$\subset$" is the symbol for "is a subset of." The diagram excludes any relationship between $k(X)$ and $k[[X]]$. Indeed, for this pair we neither have $k(X) \subset k[[X]]$ nor $k[[X]] \subset k(X)$. Now the intersection $k(X) \cap k[[X]]$ obviously consists of those rational functions that are power series in X. That is, they are well behaved at the origin. In other words

$k(X) \cap k[[X]]$ = the set of all rational functions

that do not have a pole at the origin

= the set of all rational functions that are regular

(or analytic or holomorphic) at the origin.

We call $k(X) \cap k[[X]]$ the *local ring* of the origin on the X-axis.

The ring $k(X) \cap k[[X]]$ may also be denoted by $k[X]_{M_0}$, where M_0 is the prime ideal in $k[X]$ consisting of all polynomials $D(X)$ such that $D(0) = 0$.

To explain this notation, let N be an *ideal* in a ring A. That is, N is a nonempty subset of A, such that for all $c \in N$ and $c' \in N$ we have $c - c' \in N$, and for all $c \in N$ and $a \in A$ we have $ac \in N$. Let us assume that the ideal N is a *prime ideal* in A, i.e., for all $b \in A \setminus N$, and $b' \in A \setminus N$ we have $bb' \in A \setminus N$ where, as usual, $A \setminus N$ denotes the set of all elements of A which are not in N. Now, assuming A to be a domain, in the quotient field of A we form the ring

$$A_N = \{ \frac{a}{b} : a \in A, \, b \in A \setminus N \}.$$

We call A_N the *localization* of A at N, or the *quotient ring* of A with respect to the prime ideal N.

Question. Why are we just considering the local ring of the origin?

Answer. Of course, we can consider other points $X = \alpha$. The local ring at $X = \alpha$ would consist of those rational functions $r(X)$ that do not have a pole at α, i.e., those rational functions $r(X)$ that can be expressed in the form $\frac{p(X)}{q(X)}$, where $p(X)$ and $q(X)$ are polynomials in X with $q(\alpha) \neq 0$. In the notation, this local ring equals $k[X]_{M_\alpha}$, where $M_\alpha = \{D(X) \in k[X] : D(\alpha) = 0\}$, which is obviously a prime ideal in $k[X]$. The prime ideals M_α can also be described as the set of all polynomials divisible by $(X - \alpha)$. This follows from what we study in high school under the name of *remainder theorem* or *factor theorem*. It says that

For any polynomial $D(X)$ we have $D(X) = (X - \alpha)E(X) + D(\alpha)$ for some polynomial $E(X)$. So, in particular, for any polynomial $D(X)$ we have $D(\alpha) = 0$ if and only if $D(X)$ is divisible by $X - \alpha$.

The zero ideal (the set consisting only of 0) is obviously a prime ideal in any domain. If $k = \mathbb{C}$, or more generally if k is "algebraically closed," then as α varies over k, M_α varies over all nonzero prime ideals in $k[X]$. Now a rational function $r(X)$ has poles at only finitely many points. Namely, by writing $r(X)$ in reduced form, that is, by writing $r(X) = \frac{p(X)}{q(X)}$ where $p(X)$ and $q(X)$ are polynomials in X having no nonconstant common factor, we see that the poles of $r(X)$ are at the roots of the denominator $q(X)$. If the rational function $r(X)$ is not a polynomial (if $q(X)$ is nonconstant), then $r(X)$ is not in some local ring. Namely, for each α with $q(\alpha) = 0$ we have $r(X) \notin k[X]_{M_\alpha}$. Thus, for any domain A we have motivated the following important equation

$$A = \bigcap_{\substack{\text{over all prime} \\ \text{ideals } N \text{ in } A}} A_N,$$

because we have just verified it when $A = k[X]$ with $k = \mathbb{C}$ or k algebraically closed.

Now among (nonzero) prime ideals, we can consider maximal and minimal elements, and we may consider similar equations. Indeed, it turns out that for any domain A we *always* have

$$A = \bigcap_{\substack{\text{over all prime} \\ \text{ideals } N \text{ in } A}} A_N = \bigcap_{\substack{\text{over all maximal} \\ \text{ideals } M \text{ in } A}} A_M.$$

Sometimes we also have

$$A = \bigcap_{\substack{\text{over all minimal} \\ \text{prime ideals } L \text{ in } A}} A_L.$$

Roughly speaking, this "sometimes" is when we have *"no singularities in codimension one,"* i.e., when the singularities of a d-dimensional object are contained in a $(d-2)$-dimensional subobject. In our example A corresponded to the X-axis, but A could represent any curve. In that case every prime ideal is maximal and hence also minimal. Consequently, the distinction between maximal and minimal prime ideals becomes nontrivial only when we consider surfaces (or solids, etc.). So let us briefly consider the situation on a surface.

Consider the surface $S : f(X, Y, Z) = X^2 + Y^2 - Z^2 = 0$, which is a cone with vertex at the origin. Note that the vertex of the cone (the origin) is the only singular point of S. See Figure 9.1.

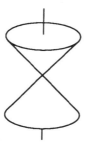

FIGURE 9.1

Let $R = k[X, Y, Z]$ where, as usual, $k[X, Y, Z]$ denotes the ring of all polynomials $g(X, Y, Z)$ with coefficients in k. Now the points (α, β, γ) in the (X, Y, Z)-space give rise to the maximal ideals $M_{(\alpha, \beta, \gamma)}$ in R given by

$$M_{(\alpha, \beta, \gamma)} = \{g(X, Y, Z) \in R : g(\alpha, \beta, \gamma) = 0\}.$$

Let $(X - \alpha, Y - \beta, Z - \gamma)R$ be the "ideal in R generated by $(X - \alpha)$, $(Y - \beta)$, and $(Z - \gamma)$," that is, the set of all linear combinations of $(X - \alpha)$, $(Y - \beta)$, $(Z - \gamma)$ with coefficients in R. Then, as in the one-variable case, it turns out that $M_{(\alpha, \beta, \gamma)} = (X - \alpha, Y - \beta, Z - \gamma)R$. This follows from the following 3-variable incarnation of the 1-variable remainder theorem.

THEOREM. *Given any* $g(X, Y, Z)$ *in* R *and* α, β, γ *in* k, *we can write*

$g(X, Y, Z) - g(\alpha, \beta, \gamma)$

$\quad = h_\alpha(X, Y, Z)(X - \alpha) + h_\beta(X, Y, Z)(Y - \beta) + h_\gamma(X, Y, Z)(Z - \gamma)$

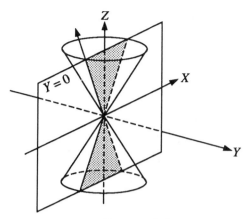

FIGURE 9.2

for some $h_\alpha(X, Y, Z)$, $h_\beta(X, Y, Z)$, *and* $h_\gamma(X, Y, Z)$ *in* R.

To prove this is easy. So let us do it now. To establish the remainder theorem for one-variable, one uses the identity

$$X^m - \alpha^m = (X - \alpha)\left(X^{m-1} + \alpha X^{m-2} + \cdots\right).$$

For dealing with many variables, we can use the calculus trick. Add and subtract suitable quantities so that we can then use the previous identity. Thus, a typical term on the LHS of the equation can be written as

$$X^i Y^j Z^\ell - \alpha^i \beta^j \gamma^\ell$$
$$= X^i Y^j Z^\ell - \alpha^i Y^j Z^\ell + \alpha^i Y^j Z^\ell - \alpha^i \beta^j Z^\ell + \alpha^i \beta^j Z^\ell - \alpha^i \beta^j \gamma^\ell$$
$$= \left(X^i - \alpha^i\right) Y^i Z^\ell + \left(Y^j - \beta^j\right) \alpha^i Z^\ell + \left(Z^\ell - \gamma^\ell\right) \alpha^i \beta^j.$$

This, by our one-variable considerations, is in $(X - \alpha, Y - \beta, Z - \gamma)R$.

Let A be the ring $k[x, y, z]$ consisting of all expressions $g(x, y, z)$ with $g(X, Y, Z) \in R$, where x, y, z are related by the equation $f(x, y, z) = 0$. We call A the *affine coordinate ring* of S. Now maximal ideals in A correspond to points on S. Ideals in A are easily seen to be in one-to-one correspondence with those ideals in R that contain the polynomial $f(X, Y, Z)$. Thus

points (α, β, γ) on S

$\leftrightarrow$ maximal ideals $M_{(\alpha, \beta, \gamma)}$ in R with $f(X, Y, Z) \in M_{(\alpha, \beta, \gamma)}$

$\leftrightarrow$ maximal ideals $M^*_{(\alpha, \beta, \gamma)}$ in A

where

$$M^*_{(\alpha, \beta, \gamma)} = \{g(x, y, z) : g(X, Y, Z) \in M_{(\alpha, \beta, \gamma)}\}.$$

What about the minimal prime ideals in A? They correspond to irreducible curves on the cone S. In particular, lines on the cone S are represented by some of the minimal prime ideals in A. See Figure 9.2. An example of a line on the cone S is $X - Z = 0 = Y$. In A it is given by the ideal $(x - z, y)A$, whereas in R it is given by the ideal $P = (X - Z, Y)R$. This line is on the cone means $f(X, Y, Z) \in P$. Note that $f(X, Y, Z) = (X - Z)(X + Z) + (Y)(Y)$. The surface $f(X, Y, Z) = 0$ intersected with the plane $Y = 0$ clearly gives us the two lines $X - Z = 0 = Y$ and $X + Z = 0 = Y$. This may be expressed ideal theoretically as

$$(f(X, Y, Z), Y)R = L \cap L'$$

where $L = (X - Z, Y)R$ and $L' = (X + Z, Y)R$.

Now we may ask whether L is generated by $f(X, Y, Z)$ and $X - Z$, i.e., whether $L = (f(X, Y, Z), X - Z)R$. However, this is not true since the surface $f(X, Y, Z) = 0$ intersects the plane $X = Z$ in "one line repeated twice." This may be expressed by the equation

$$L = \sqrt{(f(X, Y, Z), X - Z)R},$$

where we note that *for any ideal J in any ring R*, by $\sqrt{J}$ we denote the *radical* of J in R, which is the ideal in R defined by

$$\sqrt{J} = \mathrm{rad}_R J = \{b \in R : b^m \in J \text{ for some positive integer } m\}.$$

The fact that the ideal L is not actually generated by $f(X, Y, Z)$ and $X - Z$ may be expressed by saying that the line $X - Z = 0 = Y$ is not a *"complete intersection"* of the surface $f(X, Y, Z) = 0$ and the plane $X - Z = 0$.

NOTE. Space for rent!

LECTURE 10

Review of Abstract Algebra

As we said in the last lecture, in Newton's Theorem we deal with a monic polynomial in Y with coefficients in $k((X))$ where k is a field. Actually, for the validity of Newton's Theorem we need to assume that the field k is algebraically closed. We also need to assume that the Y-degree of the said polynomial is nondivisible by the characteristic of k. By k being *algebraically closed* we mean that every univariate (one-variable) polynomial with coefficients in k has a root in k (for example, $k = \mathbb{C}$). By the *characteristic* of k we mean..., and so on. In fact, let us once and for all list the formal definitions of terms such as group, ring, field, ideal, etc., giving a brief review of abstract algebra.

Set. A set is a collection of objects that are called its elements. A *subset* is a subcollection. In other words, a set Ω' is a subset of a set Ω if every element of Ω' is also an element of Ω. We denote this by $\Omega' \subset \Omega$. For example, $\mathbb{Z} \subset \mathbb{Q}$. Every integer is a rational number.

Group. By a group we mean a set Γ together with a binary operation, which is associative (i.e., $a(bc) = (ab)c$ for all a, b, c in Γ), and for which there exists an identity element 1 (i.e., an element such that $a1 = 1a = a$ for all a in Γ), and with respect to which every element a in Γ has an inverse a^{-1} (i.e., an element such that $aa^{-1} = a^{-1}a = 1$). A *subgroup* of a group Γ is a subset Γ' of Γ such that Γ' is a group with respect to the same operation as in Γ.

Abelian group. A group Γ is said to be abelian or *commutative* if $ab = ba$ for all a, b in Γ. In this case the binary operation is often written as an addition instead of multiplication, and the identity element in Γ is denoted by 0 instead of 1. To make this explicit we may then say that Γ is an "additive abelian group." Some examples of groups are $\mathbb{Z}$, $\mathbb{Q}$, $\mathbb{R}$, $\mathbb{C}$ with respect to the usual operation of addition. These are all abelian groups. Moreover, for any integer n, the subset $n\mathbb{Z}$ of $\mathbb{Z}$ consisting of all integral multiples of n clearly forms a subgroup of $\mathbb{Z}$.

Ring. By a ring we mean an additive abelian group R together with an additional associative binary operation (called "multiplication") such that the two operations are connected by the distributive laws, i.e., $a(b+c) = ab+ac$ and $(b+c)a = ba + ca$ for all a, b, c in R. The *rings* we consider will tacitly be *assumed to be commutative* (i.e., $ab = ba$ for all a, b in R)

75

and to contain an element 1 (i.e., an element such that $1a = a1 = 1$ for all a in R). Examples of rings are $\mathbb{Z}$, $\mathbb{Q}$, $\mathbb{R}$, $\mathbb{C}$ with usual addition and multiplication. A *subring* of a ring R is a subset R' of R such that R' is a ring with respect to the same operations as in R. For example, $\mathbb{Z}$ is a subring of $\mathbb{R}$. Given any element a in a ring R, by a *multiplicative inverse* of a in R we mean an element a^{-1} in R such that $aa^{-1} = 1$. Note that if a^{-1} exists it is unique. By a *unit* in R we mean an element in R that has a multiplicative inverse in R. By a *nonunit* in R we mean an element in R which is not a unit in R. For example, 1 and -1 are the only units in $\mathbb{Z}$. An element a in a ring R is said to be a *zerodivisor* in R if $aa' = 0$ for some $0 \neq a' \in R$. By the *null* ring we mean the ring containing only one element (which automatically must be 0). Note that the null ring is the only ring in which $1 = 0$.

Domain. By a domain we mean a non-null ring that does not contain any nonzero zerodivisors. Equivalently, a domain is a non-null ring R in which the law of cancellation holds. That is, for any a, b, c in R we have: $ab = ac$ with $a \neq 0 \Rightarrow b = c$. For example, for any positive integer n, the polynomial ring $\mathbb{Z}[X_1, \ldots, X_n]$ in $X_1, \ldots, X_n$ with coefficients in $\mathbb{Z}$ is a domain. Likewise, the polynomial ring $k[X_1, \ldots, X_n]$ in $X_1, \ldots, X_n$ with coefficients in a field k is a domain. Finally, the power series ring $k[[X_1, \ldots, X_n]]$ in $X_1, \ldots, X_n$ with coefficients in k is also a domain. A subring of a domain may be called a *subdomain*. For example, $\mathbb{Z}$ is a subdomain of $\mathbb{Z}[X_1, \ldots, X_n]$.

Field. By a field we mean a non-null ring in which every nonzero element is a unit. Equivalently, a field is a domain in which the nonzero elements form a group with respect to multiplication. Examples of fields are $\mathbb{Q}$, $\mathbb{R}$, $\mathbb{C}$. A *subfield* of a field is a subring which is itself a field; it contains the multiplicative inverse of every nonzero element in it. For example, $\mathbb{Q}$ is a subfield of $\mathbb{R}$. A field is said to be the *quotient field* of a subring if every element in the field can be expressed as a quotient of elements from that subring. For example, $\mathbb{Q}$ is the quotient field of $\mathbb{Z}$. For any positive integer n, by $k(X_1, \ldots, X_n)$ we denote the field of all rational functions in $X_1, \ldots, X_n$ with coefficients in k. In other words $k(X_1, \ldots, X_n)$ is the quotient field of $k[X_1, \ldots, X_n]$.

Characteristic. The characteristic of a ring R (abbreviation: char R) is defined to be the smallest positive integer p (if it exists) such that in R we have

$$\underbrace{1 + 1 + \cdots + 1}_{p \text{ times}} = 0.$$

If there is no such positive integer p, then R is said to be of *characteristic zero*. If R is a domain then we easily see that the characteristic of R is either a prime number or zero. Also note that the null ring is the only ring of characteristic 1.

Example. For any positive integer n, the residue class ring $\mathbb{Z}/n\mathbb{Z}$ is a ring of characteristic n. The said ring $\mathbb{Z}/n\mathbb{Z}$ consists of residues of integers when divided by n. More concretely, $\mathbb{Z}/n\mathbb{Z}$ may be viewed as consisting of elements $0, 1, \ldots, n-1$. Addition consists of adding these elements in an ordinary manner and then taking the remainder after dividing by n, likewise for multiplication.

Exercise. If p is prime then prove that $\mathbb{Z}/p\mathbb{Z}$ is a field. (*Hint:* If you are old, use the long division process for finding GCD. If you are young, use Euclid's algorithm from computer science. After all, they are the same!)

For more details of such material, refer to the book of Birkhoff-MacLane, *Survey of Modern Algebra* [**BMa**], the book of van der Waerden, *Modern Algebra* [**Wa2**], or any one of the other abstract algebra books of which there is now no dearth! Let's continue with our list of definitions:

Ideal. By an ideal in a ring R we mean an additive subgroup J of R such that for all $a \in R$ and $c \in J$ we have $ac \in J$. For any ring R, the set $\{0\}$ (the set containing only 0), as well as the entire set R, are obvious examples of ideals in R. These are respectively called the *zero ideal* and the *unit ideal*. By a *maximal* ideal in a ring R we mean a nonunit ideal M in R such that M and R are the only ideals in R which contain M.

Vector space. A vector space over a field R is an additive abelian group Γ together with a scalar multiplication, which to every $a \in R$ and $c \in \Gamma$ associates an element $ac \in \Gamma$, such that for all $c \in \Gamma$ we have $1c = c$, for all $a, a' \in R$ and $c \in \Gamma$ we have $a(a'c) = (aa')c$ and $(a + a')c = ac + a'c$, and for all $a \in R$ and $c, c' \in \Gamma$ we have $a(c + c') = ac + ac'$. A *subspace* of a vector space Γ over a field R is an additive subgroup Γ' of Γ such that for all $a \in R$ and $c \in \Gamma'$ we have $ac \in \Gamma'$.

Module. If in the above definition of a vector space and its subspaces we let R be any ring (which need not be a field), we get the definition of a module over the ring R and its *submodules*, that is, the definition of an R-module and its R-submodules. A ring R may be regarded as a module over itself. Then an ideal in R is the same thing as a R-submodule of R.

Ideal generation. Given a ring R and any finite number of elements $c_1, \ldots, c_m$ in R, by $(c_1, \ldots, c_m)R$ we denote the set of all linear combinations $a_1 c_1 + \cdots + a_m c_m$ with $a_1, \ldots, a_m$ in R. This is clearly an ideal in R. It is called the ideal generated by $c_1, \ldots, c_m$ in R. In case of $m = 1$, we may denote $(c_1)R$ by $c_1 R$. By convention, an empty sum is taken to be equal to zero. For any subset H of R, by HR we denote the union of $(c_1, \ldots, c_m)R$ taken over all finite sequences $c_1, \ldots, c_m$ in H (of various lengths). Alternatively, HR may be characterized as the smallest ideal in R containing H. We call HR the ideal generated by H in R. An ideal is said to be *finitely generated* if it can be expressed as HR for some finite subset H of R. An ideal generated by a single element is called a *principal* ideal.

Principal ideal domain and Euclidean domain. A domain in which every ideal is principal is called a *principal ideal domain* (PID). For example, the ring $\mathbb{Z}$ of integers, as well as the ring $k[X]$ of polynomials in one variable X over a field k, is a PID. This is proved by the division algorithm which in case of $\mathbb{Z}$ (resp. $k[X]$) says that given any nonzero elements a and b in $\mathbb{Z}$ (resp. $k[X]$) we can write $a = qb + r$ with $q, r \in \mathbb{Z}$ and $0 \le r < |b|$ (resp. $q, r \in k[X]$ and $\deg r < \deg b$). By abstracting this idea of division algorithm one obtains the notion of a *Euclidean Domain* (ED). For the formal definition of a Euclidean domain, see any of the abstract algebra books.

Factorization domain. By abstracting the idea of a prime number, we define a *prime element* in a domain R to be a nonzero nonunit p in R such that $a, b \in R$, $p|ab \Rightarrow p|a$ or $p|b$. (Note that for p, a in a domain R, "$p|a$" means that "p divides a." That is, there is some element q in R such that $a = pq$. Likewise "$p \nmid a$" means that "p does not divide a.") By abstracting the idea of a prime number in a different manner, we define an *irreducible* element in a domain R to be a nonzero nonunit in R which cannot be expressed as the product of any two nonzero nonunits in R. A ring R is said to be a factorization domain if R is a domain in which every nonzero nonunit can be expressed as a finite product of irreducible factors. Moreover, if the said factorization is unique up to order and unit factors, then R is called a *unique factorization domain* (UFD). Finally, a ring R is said to be an *irreducible factorization domain* (IFD) if R is a factorization domain in which every irreducible element is prime. In the abstract algebra books it is proved that for any ring we have ED $\Rightarrow$ PID $\Rightarrow$ UFD $\Leftrightarrow$ IFD. The basic example of a UFD is the polynomial ring in any number of variables with coefficients either in a field or in the ring of integers $\mathbb{Z}$.

Prime ideal. Analogous to the definition of a prime element, we define a *prime ideal* in a ring R to be a nonunit ideal N in R such that $a, b \in R$, $ab \in N \Rightarrow a \in N$ or $b \in N$. The set of all prime ideals in a ring R is called the *spectrum* of R and is denoted by $\mathrm{Spec}\,(R)$, and the set of all maximal ideals in R is called the *maximal spectrum* of R and is denoted by $\mathrm{Maxspec}\,(R)$ or $\mathrm{Spec}_0(R)$. Note that clearly $\mathrm{Maxspec}\,(R) \subset \mathrm{Spec}\,(R)$. By a *minimal* prime ideal in a domain R we mean a nonzero prime ideal L in R such that L is the only nonzero prime ideal in R that is contained in L. Note that a ring is a domain if and only if the zero ideal in it is a prime ideal. Likewise, a ring is a field if and only if the zero ideal in it is the only prime ideal.

Linear dependence. For a vector space Γ over a field k', the *vector space dimension* of Γ over k', which we denote by $[\Gamma : k']$, is defined thus. Elements $z_1, \ldots, z_q$ in Γ are *linearly dependent* over k' if there exist elements $c_1, \ldots, c_q$ in k', at least one of which is nonzero, such that $c_1 z_1 + \cdots + c_q z_q = 0$. The elements $z_1, \ldots, z_q$ are *linearly independent* over k' if they are not linearly dependent over k'. If there is a non-negative integer q such that there exist q elements $z_1, \ldots, z_q$ in Γ that are linearly independent over k', but there do not exist any $q + 1$ elements in Γ that are linearly independent over k', then we put $[\Gamma : k'] = q$. If there is no such non-negative integer q, we put $[\Gamma : k'] = \infty$. If $[\Gamma : k'] = q < \infty$, then any q elements $z_1, \ldots, z_q$ in Γ which are linearly independent over k' are said to form a *vector space basis* of Γ over k'.

Integral dependence. A ring R is said to be *integral* over a subring R' if every $y \in R$ is *integral* over R', that is, it satisfies an *equation of integral dependence* over R' by which we mean an equation of the form $\theta(y) = 0$ where $\theta(Y)$ is a monic polynomial in Y with coefficients in R'. Note that a *monic* polynomial is a nonzero polynomial in which the coefficient of the highest degree term is 1.

Algebraic dependence. A field K' is said to be *algebraic* over a subfield k' if every $y \in K'$ is *algebraic* over k'. That is, it satisfies an equation of the form $\theta(y) = 0$ with $0 \neq \theta(Y) \in k'[Y]$. More generally, elements $z_1, \ldots, z_q$ in K' are said to be *algebraically dependent* over k' if there exists $0 \neq \Theta(Z_1, \ldots, Z_q) \in k'[Z_1, \ldots, Z_q]$ such that $\Theta(z_1, \ldots, z_q) = 0$. The elements $z_1, \ldots, z_q$ are said to be *algebraically independent* over k' if they are not algebraically dependent over k'. The *transcendence degree* of K' over k', denoted by $\mathrm{trdeg}_{k'} K'$, is defined to be the unique non-negative integer q, if it exists, such that K' contains q elements which are algebraically independent over k' but does not contain any $q + 1$ elements which are algebraically independent over k'. If no such q exists, we put $\mathrm{trdeg}_{k'} K' = \infty$. If $\mathrm{trdeg}_{k'} K' = q < \infty$, then any q elements $z_1, \ldots, z_q$ of K' which are algebraically independent over k' are said to constitute a *transcendence basis* of K' over k'. Note that then $\mathrm{trdeg}_{k'} K' = 0 \Leftrightarrow K'$ is algebraic over k'. By the *degree of the field extension* K'/k' we mean the vector space dimension $[K' : k']$. We may also call this the *field degree* of K'/k'. (The symbol K'/k' may be read as K' over k'.) Finally, by a *finite algebraic field extension* K'/k', we mean a field K' together with a subfield k' such that $[K' : k'] < \infty$.

Let us now consider some examples.

Example of line. In the ring $k[X]$ we clearly have the following one-to-one correspondences:

> nonzero ideals in $k[X]$ $\leftrightarrow$ monic polynomials,
>
> nonzero prime ideals in $k[X]$ $\leftrightarrow$ monic irreducible polynomials.

Now if $k = \mathbb{C}$ (or more generally if k is algebraically closed), then all the monic irreducible polynomials in $k[X]$ are of the form $(X - \alpha)$ with $\alpha \in \mathbb{C}$. On the other hand, if $k = \mathbb{R}$ then the monic irreducible polynomials in $k[X]$ are of the form $(X - \alpha)$ with $\alpha \in \mathbb{R}$ or of the form $(X^2 + bX + c)$ with $b, c \in \mathbb{R}$ such that $(b^2 - 4c) < 0$. So if we let k^1 denote the X-axis (the affine line) and if k is algebraically closed ($k = \mathbb{C}$) then we have the correspondence:

> $k^1 \leftrightarrow \mathrm{Maxspec}\,(k[X]) = $ the set of all nonzero prime ideals in $k[X]$.

However, if $k = \mathbb{R}$ then Maxspec $(k[X])$ doesn't correspond any more to k^1. Instead, it corresponds to $\mathbb{R}^1 \cup \{u + iv : u \in \mathbb{R}, 0 < v \in \mathbb{R}\}$ (because the two roots of the quadratic are complex conjugates and are thus determined by a single nonreal complex number).

Example of plane. The set of points in the affine plane is denoted by k^2 or $\mathbb{A}_k^2$. Thus $\mathbb{A}_k^2 = \{(\alpha, \beta) : \alpha, \beta \in k\}$. Considering the ring $k[X, Y]$, we now have the correspondences:

points in $\mathbb{A}_k^2 \leftrightarrow$ maximal ideals in $k[X, Y]$,

irreducible curves in $\mathbb{A}_k^2 \leftrightarrow$ minimal prime ideals in $k[X, Y]$

which are, strictly speaking, true provided k is algebraically closed. At any rate, a point (α, β) in $\mathbb{A}_k^2$ corresponds to the maximal ideal $M_{(\alpha, \beta)}$ in $k[X, Y]$ given by

$$M_{(\alpha, \beta)} = \{G(X, Y) \in k[X, Y] : G(\alpha, \beta) = 0\}.$$

Notice that in the last lecture, we sketched a proof of the remainder theorem for polynomials in one variable or three variables. In a similar manner, for any positive integer n, we obtain the following general *remainder theorem*, which says that for any $g(X_1, \ldots, X_n)$ in $k[X_1, \ldots, X_n]$ and any $\alpha_1, \ldots, \alpha_n$ in k we have:

$$g(X_1, \ldots, X_n) - g(\alpha_1, \ldots, \alpha_n) = \sum_{i=1}^{n} h_i(X_1, \ldots, X_n)(X_i - \alpha_i)$$

for some $h_i(X_1, \ldots, X_n)$ in $k[X_1, \ldots, X_n]$.

As a consequence, we see that

$$M_{(\alpha, \beta)} = (X - \alpha, Y - \beta)k[X, Y]$$

and likewise in the more general

Example of n-space. Considering polynomial ring $R_n = k[X_1, \ldots, X_n]$, for any point $(\alpha_1, \ldots, \alpha_n)$ in the affine n-space $\mathbb{A}_k^n$, we see that for the maximal ideal $M_{(\alpha_1, \ldots, \alpha_n)}$ in R_n given by

$$M_{(\alpha_1, \ldots, \alpha_n)} = \{g(X_1, \ldots, X_n) \in R_n : g(\alpha_1, \ldots, \alpha_n) = 0\},$$

in view of the general remainder theorem we have

$$M_{(\alpha_1, \ldots, \alpha_n)} = (X_1 - \alpha_1, \ldots, X_n - \alpha_n) R_n.$$

At any rate, to every point $(\alpha_1, \ldots \alpha_n)$ in $\mathbb{A}_k^n$ we associate the maximal ideal $M_{(\alpha_1, \ldots, \alpha_n)}$ in R_n. It is easily seen that in this correspondence, distinct points correspond to distinct maximal ideals. It also turns out that k is algebraically closed $\Leftrightarrow$ each maximal ideal in R_n corresponds to some point in $\mathbb{A}_k^n$.

Now, if k is not algebraically closed, then as we saw in the case of a line, there can be maximal ideals that do not correspond to points but instead correspond to "conjugate bunches" in the "algebraic closure" of k. (To make this precise, one has to use Galois Theory.)

So, how do we prove the converse? In other words, given a maximal ideal M in R_n, and assuming k to be algebraically closed, how do we find a point $(\alpha_1, \ldots, \alpha_n)$ in $\mathbb{A}_k^n$ such that $M = (X_1 - \alpha_1, \ldots, X_n - \alpha_n)R_n$? The converse essentially amounts to finding a common solution to several equations, namely all the polynomials in the given ideal. This, in fact, is the basic problem of theory of equations. It may be stated thus: given a system of equations, how do we find a common solution? Of course, we cannot do this if the ideal M is a unit ideal. In that case we can find $g_1, \ldots, g_m$ in the ideal M and $h_1, \ldots, h_m$ in the ring R_n such that

$$h_1(X_1, \ldots, X_n)\, g_1(X_1, \ldots, X_n) + \cdots + h_m(X_1, \ldots, X_n)\, g_m(X_1, \ldots, X_n)$$
$$= 1.$$

Substituting the common solution $(X_1, \ldots, X_n) = (\alpha_1, \ldots, \alpha_n)$ would give $0 = 1$, which would be a contradiction.

The result that says the question has an affirmative answer if k is algebraically closed is called the *Hilbert Nullstellensatz* or the *Hilbert Zero-Theorem*.

To approach the question, it would certainly be nice to show that any system of (polynomial) equations is equivalent to a finite system of equations. This actually turns out to be another result proved by Hilbert, called the *Hilbert Basis Theorem*. In the language of ideals, Hilbert Basis Theorem can be stated as follows.

THEOREM. *Every ideal in* $k[X_1, \ldots, X_n]$ *is finitely generated.*

Hilbert proved these theorems around 1890. Rings in which every ideal is finitely generated were later (after 1930) called *Noetherian rings* in honor of Emmy Noether. Hilbert Basis Theorem is sometimes stated in a "stronger form."

THEOREM. *For any ring* A *we have* A *Noetherian* $\Rightarrow A[X]$ *Noetherian.*

Now as a consequence, by induction it would follow that $k[X_1, \ldots, X_n]$ is Noetherian.

A third theorem of Hilbert also deserves to be mentioned in this regard. Roughly speaking, it says that

Hilbert function $=$ Hilbert polynomial   for large d.

By the Hilbert function of any ideal J in $k[X_1, \ldots, X_n]$ we mean the integer valued function of the integer variable d, which Hilbert associated with J.

This theorem (which, indeed, defines the Hilbert polynomial) plays a very important role in algebraic geometry. For example, we gave three definitions of the genus of a curve: using differentials, in terms of singularities, and as the number of handles of a sphere. One can also define the genus as the constant coefficient of the Hilbert polynomial of the "ideal of the curve." Moreover, the degree of the Hilbert polynomial corresponding to an "algebraic variety" gives the dimension of the variety. The leading coefficient gives the degree of that variety. A local version of the Hilbert polynomial can also be used to define the intersection multiplicity of several hypersurfaces meeting at a point. We shall define these terms and give more explanations later.

NOTE. The three theorems, very basic to algebraic geometry, arose out of one or two papers of Hilbert [Hi1] [Hi2] on invariant theory, which he wrote around 1890 while trying to solve the so-called Gordan's Problem in invariant theory. Thus, one may be tempted to raise questions. Is algebraic geometry a part of invariant theory, or is it the other way around? At any rate, a question such as "What is algebraic geometry?" doesn't seem to have a precise answer.

Some Commutative Algebra

Once again let k be a field. Recall that the *affine n-space* over k is defined by

$$\mathbb{A}_k^n = k^n = \underbrace{k \times k \times \cdots \times k}_{n \text{ times}} = \{(\alpha_1, \ldots, \alpha_n) : \alpha_i \in k \text{ for } 1 \leq i \leq n\}.$$

Let the polynomial ring $k[X_1, \ldots, X_n]$ be denoted by R_n. In the last lecture, to any point $(\alpha_1, \ldots, \alpha_n)$ in $\mathbb{A}_k^n$ we associated the maximal ideal $M_{(\alpha_1, \ldots, \alpha_n)}$ in R_n given by

$$M_{(\alpha_1, \ldots, \alpha_n)} = (X_1 - \alpha_1, \ldots, X_n - \alpha_n) R_n.$$

We noted that, in this correspondence, different points of $\mathbb{A}_k^n$ correspond to different maximal ideals in R_n. Moreover, we also noted the

HILBERT NULLSTELLENSATZ. *The field k is algebraically closed if and only if every maximal ideal in R_n is of the form $M_{(\alpha_1, \ldots, \alpha_n)}$ for some $(\alpha_1, \ldots, \alpha_n)$ in $\mathbb{A}_k^n$.*

Thus, roughly speaking,

$$\text{Maximal ideals} \leftrightarrow \text{points.}$$

Likewise, we would like to say that

$$\text{minimal prime ideals} \leftrightarrow \begin{cases} \text{curves (case of } n = 2 \text{), or} \\ \text{surfaces (case of } n = 3 \text{), or} \\ \text{hypersurfaces (case of general } n \text{),} \end{cases}$$

where the objects on the right side are assumed irreducible, and where by a *hypersurface* we mean something given by one equation. In a ring A that is a UFD we have the correspondence

$$\text{principal prime} \leftrightarrow \text{irreducible element} \leftrightarrow \text{minimal prime.}$$

A minimal prime ideal can be characterized as a prime ideal of height 1 according to the following definition.

Given a ring A, for any prime ideal N in A we define

$$\text{ht}'_A N = \max h \text{ such that there exist prime ideals}$$
$$N_0, N_1, \ldots, N_h \quad \text{in } A \text{ with} \quad N_0 \subsetneq N_1 \subsetneq \cdots \subsetneq N_h = N.$$

Now for any ideal J in A we define

$$\mathrm{ht}_A J = \min_{J \subset N} \mathrm{ht}'_A N,$$

where the minimum is taken over all prime ideals N in A that contain J.
By $\mathrm{ht}_A J$ we mean the *height* of J in A. Note that for any prime ideal N
in a ring A we obviously have $\mathrm{ht}'_A N = \mathrm{ht}_A N$.

We also define the analogous notions of *depth* of an ideal J in a ring A
by saying that

$$\mathrm{dep}_A J = \text{depth of } J \text{ in } A$$
$$= \max d \text{ such that there exist prime ideals } N'_0, \dots, N'_d$$
$$\text{in } A \text{ with } J \subset N'_0 \subsetneq N'_1 \subsetneq \cdots \subsetneq N'_d.$$

Finally, we define the *dimension* of a ring A by putting

$$\dim A = \text{dimension of } A = \max \text{ length of chains of prime ideals in } A,$$

or equivalently,

$$\dim A = \mathrm{dep}_A(0)A.$$

It would be reasonable to expect that for any ideal J in a ring A we have
$\mathrm{dep}_A J + \mathrm{ht}_A J = \dim A$. Indeed, this turns out to be true in nice rings such as
polynomial rings (in finitely many variables) over a field, or more generally,
in "finitely generated ring extensions of a field." For a proof of this refer to
my Princeton book [A04].

Motivation for the idea of the dimension of a ring is as follows:

The n-dimensional affine space $\mathbb{A}_k^n$ corresponds to the n-variable polynomial ring $R_n = k[X_1, X_2, \dots, X_n]$. Thus the definition of dimension
should be such that $\dim R_n = n$. This is indeed the case as we have the
chain of prime ideals

$(0)R_n \subset (X_1)R_n \subset \qquad \cdots \subset (X_1, \dots, X_{n-1})R_n \subset (X_1, X_2, \dots, X_n)R_n$

$\updownarrow \qquad\qquad \updownarrow \qquad\qquad\qquad\qquad \updownarrow \qquad\qquad\qquad\qquad \updownarrow$

$\mathbb{A}_k^n \quad (X_2, \dots, X_n)\text{-space} \cdots \quad X_n\text{-axis} \qquad\qquad \text{origin } (0, \dots, 0).$

Exercise. Show that $\dim k[X_1, \dots, X_n] = n$. That is, prove that the above
chain is a chain of distinct prime ideals of length n and that no chains exist
of prime ideals of length $> n$.

The definition of the dimension of a ring is due to W. Krull (active years:
1925–65). This dimension is sometimes called as the *Krull dimension* to
distinguish it from other possible definitions of dimension.

REMARK. One way to learn college algebra would be to read all papers written by Krull, including [Kr1] [Kr2] [Kr3]. However, knowledge of German
would be necessary.

We can easily see that for any ideal J in a ring A we have $\mathrm{dep}_A J = \dim A/J$. To understand this, we must explain what we mean by the residue
class ring A/J (read as A mod J).

Example. Consider groups first. That is, let $A = \mathbb{Z}$ and $J = 12\mathbb{Z}$. Now $\mathbb{Z}/12\mathbb{Z}$ gives the model of a clock.

Formally, we consider boxes (or equivalence classes) of elements of an additive abelian group A such that elements a and b in A are in the same box (or equivalent) if and only if their difference $a - b$ belongs to a given subgroup J of A. See Figure 11.1.

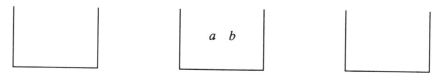

FIGURE 11.1

These boxes can be added in an obvious manner. Choose some representatives, add them, and look at the box that contains the sum. It can be easily checked that this addition is well defined; it depends only on the boxes and not on the particular representatives. In this way we obtain an additive abelian group called the *factor group*, which is denoted by A/J. Moreover, if A is a ring and J is an ideal in A, then we can multiply these boxes as well. Then the set of all these boxes forms a ring called the *residue class ring*, which is again denoted by A/J.

Now by sending each element a in A to the box containing it, we get the residue class map $\Phi : A \rightarrow A/J$, which is a surjective ring homomorphism. To explain these terms, let us add a review on maps and homomorphisms.

A *map* $\Phi : \Omega \rightarrow \Omega^*$ of a set Ω into a set Ω^* is a "rule" which to every element $a \in \Omega$ associates a unique element $\Phi(a) \in \Omega^*$. Sometimes the "solid arrow" $a \mapsto a^*$ is used to indicate that $\Phi(a) = a^*$. For any $\Omega' \subset \Omega$, by $\Phi(\Omega')$ we denote the set of all $a^* \in \Omega^*$ such that $\Phi(a') = a^*$ for some $a' \in \Omega'$. We put $\mathrm{im}\,\Phi = \Phi(\Omega)$ where "im" stands for image, and we say that Φ is *surjective* to mean that $\mathrm{im}\,\Phi = \Omega^*$. Likewise, we say that Φ is *injective* to mean that for all $a \neq b$ in Ω we have $\Phi(a) \neq \Phi(b)$. Finally, we say that Φ is *bijective* to mean that Φ is injective and surjective.

A (group) *homomorphism* $\Phi : \Gamma \rightarrow \Gamma^*$ of a group Γ into a group Γ^* is a "set–theoretic" map such that $\Phi(ab) = \Phi(a)\Phi(b)$ for all $a, b \in \Gamma$. The *kernel* of Φ, denoted by $\ker\Phi$, is the set of all elements $a \in \Gamma$ such that $\Phi(a) = 1$ or 0 according to whether we write the group operations in Γ multiplicatively or additively; $\ker\Phi$ is clearly a subgroup of Γ.

A (ring) *homomorphism* $\Phi : R \rightarrow R^*$ of a ring R into a ring R^* is a group homomorphism of the "underlying additive groups" such that $\Phi(ab) = \Phi(a)\Phi(b)$ for all $a, b \in R$. This time $\ker\Phi$ is actually an ideal in R.

As an example of a group homomorphism (which is also a ring homomorphism) we can look at the "clock" $\mathbb{Z} \rightarrow \mathbb{Z}/12\mathbb{Z}$ or the "weekly calendar" $\mathbb{Z} \rightarrow \mathbb{Z}/7\mathbb{Z}$. As a classical example of a group homomorphism, we have the De Moivre map $E : \mathbb{R} \rightarrow \mathbb{C}^*$ given by $E(\theta) = \cos 2\pi\theta + i \sin 2\pi\theta$.

Here $\mathbb{R}$ is the additive group of reals, and $\mathbb{C}^*$ is the multiplicative group of nonzero complex numbers. Moreover, $\ker E = \mathbb{Z}$, and $\operatorname{im} E = $ the unit circle $\{z \in \mathbb{C} : |z| = 1\}$.

It should now be clear that for any subgroup J of an additive abelian group A, the residue class map $\Phi : A \to A/J$ is a surjective group homomorphism.

Likewise, for any ideal J in a ring A, the residue class map $\Phi : A \to A/J$ is a surjective ring homomorphism. Moreover, in this case, $J' \mapsto \Phi(J')$ gives a one-to-one inclusion preserving correspondence between overideals J' of J in A and all ideals in A/J. (Note that "over" is the reverse of "sub." For instance, if A' is a subring of A, we may say that A is an overring of A'.) This correspondence preserves prime ideals as well as maximal ideals. Hence,

$$J \text{ is maximal} \Leftrightarrow A/J \text{ is a field},$$

$$J \text{ is prime} \Leftrightarrow A/J \text{ is a domain}.$$

The notion of a prime ideal is motivated by that of a prime number. Likewise the idea of powers of prime numbers (e.g., $27 = 3^3$) motivates the notion of primary ideals which is defined as follows:

An ideal J in a ring A is *primary* if $J \neq A$ and

$$a \in A, \ b \in A \setminus J, \ ab \in J \Rightarrow a^m \in J \quad \text{for some integer } m > 0.$$

Equivalently, J is primary $\Leftrightarrow A/J$ is a non-null ring in which every zerodivisor is nilpotent ($a \in A$ is said to be *nilpotent* if $a^m = 0$ for some $m > 0$).

Now from $27 = 3^3$ we can get back the prime number 3 by taking the appropriate root. For primary ideals, we can obtain the *associated prime ideal* by taking its radical. Recall that for an ideal J in a ring A, the *radical* of J in A is defined by

$$\operatorname{rad} J = \sqrt{J} = \{a \in A : a^m \in J \text{ for some integer } m > 0\} = \sqrt[\infty]{J}.$$

Exercise. If J is primary then show that $\sqrt{J}$ is a prime ideal.

If J is primary and $N = \sqrt{J}$ then we say that J is *primary for* N. Equivalently, for nonunit ideals J and N in a ring A we can say

$$J \text{ is primary for } N \Leftrightarrow (a \in A, \ b \in A \setminus N, \ ab \in J \Rightarrow a \in J).$$

The depth of an ideal J in a ring A can be expressed in terms of dimension by the equation

$$\operatorname{dep}_A J = \dim A/J.$$

Can we give a similar characterization of height as well? This can be done, at least for a prime ideal N in a domain A, by noting that

$$\operatorname{ht}_A N = \dim A_N,$$

where A_N is the subring of the quotient field of A given by

$$A_N = \text{the } \textit{localization} \text{ of } A \text{ at } N$$
$$= \text{the } \textit{quotient ring} \text{ of } A \text{ with respect to } N$$
$$= \{\frac{a}{b} : a \in A, \, b \in A \setminus N\}.$$

Now a *quasilocal ring* is a ring B having exactly one maximal ideal, which we denote by $M(B)$. A *local ring* is a Noetherian quasilocal ring. Thus for a prime ideal N in a domain A, the ring A_N is a quasilocal domain. If A is Noetherian, then A_N is local domain.

Question. When are the chains in the definitions of depth and height unique, or more generally, when are the ideals in a domain linearly ordered?

Answer. In valuation rings.

Valuation rings may be defined roughly as rings in which the ideals are linearly ordered. Another description would be

$$\text{Valuation rings } \leftrightarrow \text{ L'Hospital's rule holds.}$$

The familiar L'Hospital's rule in calculus enables us to determine forms of indeterminacy for functions of one variable under certain conditions. For example,

$$\lim_{x \to 0} \frac{\sin x}{x} = 1, \quad \text{by L'Hospital's rule.}$$

If we wish to say this without the gadgets of analysis, then we have valuations or valuation rings. In fact, for functions of more variables, there are indeterminate forms from which indeterminacies cannot be removed. For example,

$$\frac{x^2 + y^2}{x^2 - y^2} \text{ has no unique limit as } (x, y) \to (0, 0).$$

Valuations would give us all possible limits by allowing us to apply one-variable considerations for studying functions of several variables. More about valuations follows later.

Let us now get back to the geometric interpretation of minimal prime ideals in the domain $R_n = k[X_1, \ldots, X_n]$. Since R_n is a UFD, minimal prime ideals are principal. Namely, let L be a minimal prime in R_n. Since L is a nonzero prime, it contains a nonzero irreducible element f. Now fR_n is a nonzero prime contained in the minimal prime L. Hence, we must have $L = fR_n$.

Thus in $k[X, Y]$ we have

minimal primes	maximal ideals
$\updownarrow$	$\updownarrow$
(plane) curves	points.

In $k[X, Y, Z]$ we have

minimal primes	height 2 primes	maximal ideals
$\updownarrow$	$\updownarrow$	$\updownarrow$
surfaces	(space) curves	points,

where the curves and surfaces are assumed irreducible.

NOTE. We have noted that in the 3-variable polynomial ring $R = k[X, Y, Z]$, every minimal prime ideal is of the form fR with irreducible $f \in R$, and, assuming k to be algebraically closed, the maximal ideals are of the form $(X - \alpha, Y - \beta, Z - \gamma)R$ with α, β, γ in k. However, no such easy recipe exists for irreducible space curves or height 2 prime ideals. For instance, Macaulay [**Mac**] has given examples of irreducible space curves whose prime ideals require arbitrarily large number of generators. For an elaboration of Macaulay's examples see [**A23**]. We can at least ask whether every irreducible space curve is a set-theoretic complete intersection. In other words, is every height 2 prime ideal in R the radical of an ideal generated by two polynomials? That is, is it of the form $\sqrt{(f, g)R}$ for some f and g in R? Recently, Cowsik and Nori [**CNo**] have shown this to be true when k is a field of nonzero characteristic. Still, it is not known what happens for zero characteristic. For a general discussion about space curves, refer to the Montreal Notes [**A21**] or the Springer Notes [**ASa**]. Here is a brief history of the number of equations required to specify an irreducible space curve, to define it set-theoretically:

Kronecker	4 enough	1880 [**Kro**]
Vahlen	3 not enough	1905 [**Vah**]
Perron	Vahlen is wrong	1940 [**Per**]
Severi	Vahlen is right	1950
Kneser	3 always enough	1960 [**Kne**]
	2 enough?	open

Hensel's Lemma and Newton's Theorem

Now let us finally state and prove

NEWTON'S THEOREM. *Let* k *be an algebraically closed field, and let*

$$F(X, Y) = Y^n + a_1(X)Y^{n-1} + \cdots + a_n(X) \in k((X))[Y]$$

be a monic polynomial of degree $n > 0$ *in* Y *with coefficients* $a_1(X), \ldots,$ $a_n(X)$ *in* $k((X))$. *Assume that the characteristic of* k *is zero or that it does not divide* $n!$. *Then there exists a positive integer* m, *which is nondivisible by the characteristic of* k, *such that*

$$F(T^m, Y) = \prod_{i=1}^{n} [Y - \eta_i(T)] \qquad \text{with } \eta_i(T) \in k((T)).$$

We have the following two supplements.

Supplement 1. If $a_i(X) \in k[[X]]$ *for* $i = 1, 2, \ldots, n$ *then* $\eta_i(T) \in k[[T]]$ *for* $i = 1, 2, \ldots, n$.

Supplement 2. If $F(X, Y)$ *is irreducible in* $k((X))[Y]$ *then* $m = n$ *is okay. In fact, in this case,* n *is the least permissible* m. *Moreover, the roots* $\eta_1(T), \ldots, \eta_n(T)$ *can be labeled so that* $\eta_i(T) = \eta_1(\omega^i T)$ *where* ω *is a primitive n-th root of* 1. *(An* n^{th} *root* ω *of* 1 *is said to be* primitive *if all other* n^{th} *roots of* 1 *are powers of it. For example, in the case of* $k = \mathbb{C}$ *we can take* $\omega = e^{2\pi i/n}$.)

Example. Consider $Y^6 - X^2$. This can be factored as $Y^6 - X^2 = (Y^3 - X)(Y^3 + X) = (Y - X^{1/3}) \cdots$. Thus 3 serves as a denominator for the exponent of the relevant fractional power series in X. That is, 3 is a permissible value of m. But we see that any multiple of 3, say 15 for example, would also do the job since we can write $Y^6 - X^2 = (Y - X^{5/15}) \cdots$

REMARK. The set of all exponents m which do the job are multiples of some positive integer. Moreover, $n!$ would always do the job.

History. Newton proved this theorem around 1660 [**New**]. It was revived by Puiseux in 1850 [**Pui**]. The relevant history could be found in G. Chrystal's *Textbook of algebra*, vol. 2, 396 [**Chy**].

We shall use pre-Newton ideas to give a proof of this theorem, which we may call *Shreedharacharya's Proof of Newton's Theorem.*

However, we shall also use a post-Newton lemma, *Hensel's Lemma* (1900). Briefly, Hensel's Lemma says that, given $F \in k[[X]][Y]$, if F factors after putting X equal to 0, then it factors. It can be stated more precisely as follows.

HENSEL'S LEMMA. *Let* k *be a field, and let*

$$F(X, Y) = Y^n + a_1(X)Y^{n-1} + \cdots + a_n(X) \in k[[X]][Y]$$

be a monic polynomial of degree $n > 0$ *in* Y *with coefficients* $a_1(X), \ldots,$ $a_n(X)$ *in* $k[[X]]$. *Assume that* $F(0, Y) = \overline{G}(Y)\overline{H}(Y)$ *where*

$$\overline{G}(Y) = Y^r + \bar{b}_1 Y^{r-1} + \cdots + \bar{b}_r \in k[Y] \quad and$$

$$\overline{H}(Y) = Y^s + \bar{c}_1 Y^{s-1} + \cdots + \bar{c}_s \in k[Y]$$

are monic polynomials of degrees $r > 0$ *and* $s > 0$ *in* Y *with coefficients* $\bar{b}_1, \ldots, \bar{b}_r$ *and* $\bar{c}_1, \ldots, \bar{c}_s$ *in* k *such that* $GCD\left(\overline{G}(Y), \overline{H}(Y)\right) = 1$. (*That is,* $\overline{G}(Y)$ *and* $\overline{H}(Y)$ *are "coprime" which means they have no nonconstant common factor in* $k[Y]$.) *Then there exist unique monic polynomials*

$$G(X, Y) = Y^r + a_1(X)Y^{r-1} + \cdots + a_r(X) \in k[[X]][Y] \ and$$

$$H(X, Y) = Y^s + b_1(X)Y^{s-1} + \cdots + b_s(X) \in k[[X]][Y]$$

of degrees r *and* s *in* Y *with coefficients* $b_1(X), \ldots, b_r(X)$ *and* $c_1(X), \ldots,$ $c_s(X)$ *in* $k[[X]]$ *such that* $G(0, Y) = \overline{G}(Y)$, $H(0, Y) = \overline{H}(Y)$, *and* $F(X, Y) = G(X, Y)H(X, Y)$.

To explain the evolution of these ideas, let us consider the

BINOMIAL THEOREM. *It says that* $(X + 1)^N = X^N + NX^{N-1} + \cdots + 1$ *for any non-negative integer* N. *More generally, it says that*

$$(1 + X)^N = 1 + NX + \frac{N(N-1)}{2!}X^2 + \cdots + \frac{N(N-1)\cdots(N-i+1)}{i!}X^i + \cdots$$

and that this identity holds when N *is any rational number.* (*Actually,* N *could even be any real or complex number.*)

Thus, in particular, if $N = \frac{m}{n}$, where m and n are integers with $n > 0$, then

$$(1+X)^{m/n} = 1 + \frac{m}{n}X + \frac{\frac{m}{n}\left(\frac{m}{n} - 1\right)}{2!}X^2 + \cdots + \frac{\frac{m}{n}\left(\frac{m}{n} - 1\right)\cdots\left(\frac{m}{n} - i + 1\right)}{i!}X^i + \cdots$$

and the power series on the right side can be thought of as a solution for the equation

$$Y^n - (1 + X)^m = 0.$$

Moreover, all the other solutions can be obtained by multiplying the power series by various powers of a primitive n^{th} root ω of 1.

Note that the factorization we get for $Y^n - (1 + X)^m$ is a special case of Hensel's Lemma. By putting $X = 0$ in the equation we get

$$\left(Y^n - 1\right) = \prod_{i=1}^{n}\left(Y - \omega^i\right),$$

where $\omega^i \neq \omega^j$ for $1 \leq i < j \leq n$. So if we split $Y^n - 1$ into two (nonconstant) factors, then they are coprime. Hence, by Hensel's Lemma, we can factor $Y^n - (1 + X)^m$ into two factors. Continuing in this manner we arrive at the factorization of $Y^n - (1 + X)^m$ into linear factors. (Note that whatever m may be in the present situation, 1 would be a permissible value of the m occurring in the relevant case of Newton's Theorem.)

PROOF OF HENSEL'S LEMMA. We can write $F = F(X, Y)$ as a power series in X with coefficients polynomials in Y. Thus

$$F = F_0(Y) + F_1(Y)X + \cdots + F_q(Y)X^q + \cdots ,$$

where $F_0 = F_0(Y) = F(0, Y) \in k[Y]$ is monic with $\deg F_0 = n$, and $F_q = F_q(Y) \in k[Y]$ with $\deg F_q < n$ for all $q > 0$. We want to find

$$G = G_0(Y) + G_1(Y)X + \cdots + G_i(Y)X^i + \cdots , \quad \text{and}$$

$$H = H_0(Y) + H_1(Y)X + \cdots + H_j(Y)X^j + \cdots ,$$

where

$$\begin{cases} G_0 = G_0(Y) = \overline{G}(Y) \in k[Y] \text{ is monic with } \deg G_0 = r, \\ H_0 = H_0(Y) = \overline{H}(Y) \in k[Y] \text{ is monic with } \deg H_0 = s, \\ G_i = G_i(Y) \in k[Y] \text{ with } \deg G_i(Y) < r \text{ for all } i > 0, \text{ and} \\ H_j = H_j(Y) \in k[Y] \text{ with } \deg H_j(Y) < s \text{ for all } j > 0, \end{cases}$$

such that $F = GH$. Note that $F = GH$ means

$$F_q = \sum_{i+j=q} G_i H_j \quad \text{for all } q \geq 0.$$

Let us make induction on q. The case of $q = 0$ being obvious, let $q > 0$. Suppose we have found G_i and H_j in $k[Y]$, with $\deg G_i < r$ and $\deg H_j < s$ for $1 \leq i < q$ and $1 \leq j < q$, satisfying the above equation for all values of q smaller than the given one. Now we need to find G_q and H_q in $k[Y]$, with $\deg G_q < r$ and $\deg H_q < s$, satisfying the equation

$$F_q = \sum_{i+j=q} G_i H_j .$$

The equation can also be written as

$$G_0 H_q + H_0 G_q = U_q ,$$

where

$$U_q = F_q - \sum_{\substack{i+j=q \\ i<q,\, j<q}} G_i H_j \in k[Y] \quad \text{and clearly } \deg U_q < n.$$

Since GCD $(G_0, H_0) = 1$, we can write

$$G_0 H^* + H_0 G^* = 1 \quad \text{with } G^* \in k[Y] \quad \text{and} \quad H^* \in k[Y].$$

Multiplying both sides by U_q we get

$$G_0 H_q^* + H_0 G_q^* = U_q \quad \text{with } G_q^* = U_q G^* \in k[Y] \quad \text{and} \quad H_q^* = U_q H^* \in k[Y].$$

Now by division algorithm, we can write $H_q^* = E_q H_0 + H_q$ where E_q, $H_q \in k[Y]$ with $\deg H_q < s$. It follows that $U_q = G_0 H_q + H_0 G_q$ where $G_q = G_q^* - E_q G_0 \in k[Y]$. Moreover, since $\deg U_q < n$, we see that the G_q thus obtained would automatically satisfy the condition $\deg G_q < r$.

Thus, by induction we have found G_i and H_j, with $\deg G_i < r$ and $\deg H_j < s$ for all $i > 0$ and $j > 0$, such that

$$F_q = \sum_{i+j=q} G_i H_j \quad \text{for all } q \geq 0.$$

So we have proved the existence of G and H having the desired properties.

To prove the uniqueness, it is enough to note that

$$G_0 H_q + G_q H_0 = G_0 h_q + g_q H_0 \quad \text{for any } g_q, h_q \in k[Y]$$
$$\Rightarrow G_0(H_q - h_q) = -(G_q - g_q)H_0$$
$$\Rightarrow G_0 | (G_q - g_q) \quad \text{(because } G_0 \text{ and } H_0 \text{ are coprime)}$$
$$\Rightarrow (G_q - g_q) = 0 \quad \text{(because } \deg(G_q - g_q) < \deg G_0)$$
$$\Rightarrow G_q = g_q \quad \text{and} \quad H_q = h_q.$$

This proves Hensel's Lemma.

Exercise. In case $k = \mathbb{R}$ or $\mathbb{C}$ (or any "valued field"), show that if F is convergent then so are G and H. (Hint: use Weierstrass' M-test or the ratio test.)

History. Hensel proved this lemma while studying p-adic number fields for various prime numbers p [**Hen**]. The p-adic number fields are examples of "valued fields," that is, of "fields with valuations." We can consider the notion of convergence in such fields just as in $\mathbb{R}$ or $\mathbb{C}$.

Now let us complete the proof of Newton's Theorem.

We want to reduce to the case when Hensel's Lemma can be applied. In order to do this we shall use the process of multiplying the roots by a constant δ and the process of increasing the roots by a constant ϵ. If $\alpha_1, \dots, \alpha_n$ are the roots of F, then the multiplication process amounts to finding a polynomial whose roots are $\delta\alpha_1, \dots, \delta\alpha_n$, whereas the addition process amounts to constructing a polynomial whose roots are $\alpha_1 + \epsilon, \dots, \alpha_n + \epsilon$. In this connection (although we shall not use it here), there is yet a third process, namely that of finding a polynomial whose roots are reciprocals of the roots of F. These three processes can be combined in changing roots by a fractional linear transformation $\alpha_i \mapsto \frac{\delta\alpha_i + \epsilon}{\delta'\alpha_i + \epsilon'}$. An excellent reference for such things is Burnside-Panton's *Theory of equations* [**BPa**].

To use the multiplication process on the equation

$$F(X, Y) = Y^n + a_1(X)Y^{n-1} + \dots + a_i(X)Y^{n-i} + \dots + a_n(X),$$

we note that for any nonzero element $\delta = X^d$ we have

$$F^*(X, Y)$$
$$= X^{-dn} F(X, X^d Y)$$
$$= Y^n + a_1(X) X^{-d} Y^{n-1} + \cdots + a_i(X) X^{-di} Y^{n-i} + \cdots + a_n(X) X^{-dn}.$$

We want to arrange matters so that all the coefficients $a_i(X) X^{-di}$ are "power series" in X and at least one of them has a nonzero value at $X = 0$. In other words, we want $\operatorname{ord} a_i(X) X^{-di} \geq 0$ for all i with equality for at least one value of i. This forces us to take

$$d = \min_{1 \leq i \leq n} \left(\frac{\operatorname{ord} a_i(X)}{i} \right).$$

If $F(X, Y) = Y^n$ then we have nothing to show. So assume that $F(X, Y) \neq Y^n$. Now $\operatorname{ord} a_i(X) \neq \infty$ for some i. Hence, for the d defined by the above equation we have $d \neq \infty$. However, the d so found may not be an integer but it is certainly a rational number and can be written as $d = \frac{\lambda}{\mu}$, where λ and μ are integers with $\mu > 0$. This really amounts to taking $X'^{\mu} = X$ and

$$F'(X', Y) = X'^{-\lambda n} F(X'^{\mu}, X'^{\lambda} Y)$$
$$= Y^n + a_1'(X') Y^{n-1} + \cdots + a_i'(X') Y^{n-i} + \cdots + a_n'(X')$$

with $a_i'(X') = a_i(X'^{\mu}) X'^{-\lambda i} \in k[[X']]$ for all i, and $a_i'(0) \neq 0$ for some i.

Actually, we precede the above multiplication process by Shreedhara-charya's trick of completing the nth power. This amounts to making the addition process with $\epsilon = \frac{a_1(X)}{n}$ to get

$$\tilde{F}(X, Y) = F\left(X, Y + \frac{a_1(X)}{n}\right)$$
$$= Y^n + \tilde{a}_1(X) Y^{n-1} + \cdots + \tilde{a}_i(X) Y^{n-i} + \cdots + \tilde{a}_n(X)$$

with $\tilde{a}_i(X) \in k((X))$ for all i, and $\tilde{a}_1(X) = 0$. Now if the coefficient of Y^{n-1} is zero to begin with, then it remains zero after the multiplication process. Thus, by first applying Shreedharacharya's trick and then the multiplication process, we transform F into a monic polynomial, say $\widehat{F}(X', Y)$, of degree n in Y with coefficients in $k[[X']]$ such that the coefficient of Y^{n-1} is (identically) zero but the value of the coefficient of Y^{n-i} at $X' = 0$ is nonzero for some i. Now, since n is nondivisible by the characteristic of k, for any $\gamma \in k$ we have

$$(Y + \gamma)^n = \begin{cases} Y^n & \text{if } \gamma = 0 \\ Y^n + n\gamma Y^{n-1} + \cdots + \gamma^n & \text{with } n\gamma \neq 0 \quad \text{if } \gamma \neq 0. \end{cases}$$

Therefore, $\widehat{F}(0, Y)$ can be split into two nonconstant coprime factors in $k[Y]$. Now Hensel's Lemma can be applied to factor $\widehat{F}(X', Y)$ into polynomials of smaller degrees in Y. Since the characteristic of k does not divide $n!$, we can continue this process (or use induction on the Y-degree of F) to factor $\widehat{F}$ into linear factors. The factorization of $F(X, Y)$ into linear factors can then be obtained by a substitution of the form $X = T^m$. It is also clear that if $a_i(X) \in k[[X]]$ for all i, then $d \geq 0$. It follows that if $a_i(X) \in k[[X]]$ for all i, then $\eta_i(T) \in k[[T]]$ for all i.

This proves Newton's Theorem, except for Supplement 2 which we shall discuss later.

NOTE. Shreedharacharya's method of completing the nth power, at least in its incarnation of completing the square for solving a quadratic equation, dates back to 500 A.D. Its versified version was given in Bhaskaracharya's *Beejaganit* of 1150 [**Bha**]. This verse is reproduced in the preface of my Springer Lecture Notes on "Canonical Desingularization" [**A28**]. You may also see the numerous Shreedharacharya expansions in my paper on "Desingularization of Plane Curves," [**A29**] which is meant to serve as an introduction to "Canonical Desingularization."

More about Newton's Theorem

Let us first review some of the results proved in the last lecture. Let

$$F(X, Y) = Y^n + a_1(X)Y^{n-1} + \cdots + a_n(X) \in R[Y]$$

be a monic polynomial of degree $n > 0$ in Y whose coefficients $a_1(X), \ldots, a_n(X)$ are in a ring R. Then we have (1) Hensel and (2) Newton.

(1) *Hensel.* If $R = k[[X]]$ where k is a field, and if $F(0, Y) = \overline{G}(Y)\overline{H}(Y)$ where

$$\overline{G}(Y) = Y^r + \cdots \in k[Y] \quad \text{and} \quad \overline{H}(Y) = Y^s + \cdots \in k[Y]$$

are nonconstant monic polynomials in Y with coefficients in k such that $\text{GCD}(\overline{G}(Y), \overline{H}(Y)) = 1$, then there exist unique monic polynomials

$$G(X, Y) = Y^r + \cdots \in R[Y] \quad \text{and} \quad H(X, Y) = Y^s + \cdots \in R[Y]$$

in Y with coefficients in R such that: $G(0, Y) = \overline{G}(Y)$, $H(0, Y) = \overline{H}(Y)$, and $F(X, Y) = G(X, Y)H(X, Y)$.

Exercise. Show that Hensel's Lemma is true when $R = $ the ring of power series $k[[X_1, \ldots, X_d]]$ in any number of variables $X_1, \ldots, X_d$ over any field k. Also show that Hensel's Lemma is true when $R = $ the ring of convergent power series ring in $X_1, \cdots, X_d$ with coefficients in k, where $k = \mathbb{R}$ or $\mathbb{C}$ or any field in which convergence of power series makes sense. (*Hint.* Use Weierstrass' M-test. For reference, see my book, *Local analytic geometry* [**A07**].)

(2) *Newton.* If $R = k((X))$ and k is an algebraically closed field whose characteristic does not divide $n!$, then there exists a positive integer m, which is not divisible by the characteristic of k, such that

$$f\left(T^m, Y\right) = \prod_{i=1}^{n} [Y - \eta_i(T)] \quad \text{with} \quad \eta_i(T) \in k((T)).$$

Supplement 1. If $R = k[[X]]$ then $\eta_i(T) \in k[[T]]$ for all $i = 1, \ldots, n$.

Exercise. By examples show that, unlike Hensel, Newton is not true for the power series ring $R = k[[X_1, \cdots, X_d]]$, even when we allow fractional power series (or fractional meromorphic series) in $X_1, \ldots, X_d$. Find conditions when this is true.

REMARK. If ω is any m^{th} root of 1 and we change T to ωT then $F\left(T^m, Y\right)$ remains unchanged. Hence, $\eta_1(\omega T), \ldots, \eta_n(\omega T)$ is simply a permutation of $\eta_1(T), \ldots, \eta_n(T)$ with multiplicities!

To elaborate on this remark, let us divide the roots $\eta_1(T), \ldots, \eta_n(T)$ into "conjugacy classes" or "boxes." In other words, for $1 \le j \le n$, let L'_j be the box of integers defined by putting

$$L'_j = \left\{ \ell : 1 \le \ell \le n \text{ and } \eta_\ell(T) = \eta_j(\omega T) \text{ for some } \omega \text{ with } \omega^m = 1 \right\}.$$

Now for any j and j', the two boxes L'_j and $L'_{j'}$ either coincide or are disjoint. (*Disjoint* means have no element in common.) Therefore, we get a sequence $1 = j_1 < j_2 < \cdots < j_h \le n$ such that, by letting $L_i = L'_{j_i}$, we have a *disjoint partition*

$$\{1, 2, \ldots, n\} = L_1 \amalg L_2 \amalg \cdots \amalg L_h,$$

where $\amalg$ denotes disjoint union. That is, we have

$$\{1, 2, \ldots, n\} = L_1 \cup L_2 \cup \cdots \cup L_h \quad \text{and} \quad L_i \cap L_{i'} = \varnothing \text{ for all } i \ne i'.$$

(As usual, $\varnothing$ is the symbol for the *empty set*, and $\{1, 2, \ldots, n\}$ is the set consisting of $1, 2, \ldots, n$.)

To account for repetitions, for every i with $1 \le i \le h$, we can further divide the box L_i into sub-boxes $L_{i1}, \ldots, L_{ie_i}$ such that

$$L_i = L_{i1} \amalg L_{i2} \amalg \cdots \amalg L_{ie_i}$$

and such that for $1 \le j \le e_i$: (i) for all $\ell \ne \ell'$ in L_{ij} we have $\eta_\ell(T) \ne \eta_{\ell'}(T)$, and (ii) for every ℓ in L_{ij} and every ω with $\omega^m = 1$ we have $\eta_\ell(\omega T) = \eta_{\ell'}(T)$ for some ℓ' in L_{ij}. It is easily seen that for all $1 \le j < j' \le e_i$, the number of elements in L_{ij} equals the number of elements in $L_{ij'}$. Let us call this common number f_i. Let

$$\phi^*_{ij}(T, Y) = \prod_{\ell \in L_{ij}} [Y - \eta_\ell(T)].$$

Then ϕ^*_{ij} is a monic polynomial of degree f_i in Y with coefficients in $k((T))$. Moreover, these coefficients remain unchanged when we replace T by ωT for any ω with $\omega^m = 1$. Therefore, m must be divisible by f_i and we get a monic polynomial $\phi_{ij}(X, Y)$ of degree f_i in Y with coefficients in $k((X))$ such that

$$\phi_{ij}(T^m, Y) = \phi^*_{ij}(T, Y).$$

It is clear that $\phi_{ij}(X, Y)$ is irreducible in $k((X))[Y]$, and we have

$$\phi_{ij}\left(T^{f_i}, Y\right) = \prod_{\ell \in L_{ij}} [Y - y_\ell(T)],$$

where $y_\ell(T)$ in $k((T))$ is such that

$$y_\ell(T^{m/f_i}) = \eta_\ell(T).$$

Now clearly,

$$\phi_{i1} = \phi_{i2} = \cdots = \phi_{ie_i}.$$

So let us write

$$\phi_i = \text{ the common value } = \phi_{i1} = \cdots = \phi_{ie_i}.$$

Thus, we get

$$\sum_{i=1}^{h} e_i f_i = n$$

and

$$F(X, Y) = \prod_{i=1}^{h} \phi_i(X, Y)^{e_i},$$

where $\phi_1(X, Y), \ldots, \phi_h(X, Y)$ are pairwise coprime monic irreducible polynomials of degrees $f_1, \ldots, f_h$ in Y with coefficients in $k((X))$. For all i, j, ℓ with $1 \le i \le h$, $1 \le j \le e_i$, $\ell \in L_{ij}$, we have

$$\phi_{ij}\left(T^{f_i}, Y\right) = \prod_{\kappa=1}^{f_i} \left[Y - y_\ell(\omega_i^\kappa T)\right],$$

where ω_i is any primitive f_i^{th} root of 1.

To prove *Supplement 2*, it suffices to note that in case F is irreducible we must have $h = 1 = e_1$ and $n = f_1$ and $F = \phi_1$.

In the general case, for any fixed i with $1 \le i \le h$, as ℓ varies over L_i, we have the parametrizations

$$\begin{cases} X = T^m \\ Y = \eta_\ell(T), \end{cases}$$

which gives rise to the irredundant parametrizations

$$\begin{cases} X = T^{f_i} \\ Y = y_\ell(T). \end{cases}$$

All these are equivalent to each other because they can be obtained from each other by changing T to ωT, for some ω with $\omega^{f_i} = 1$. (The adjective *irredundant* refers to the fact that if $(T^{f_i}, y_\ell(T)) = (\lambda(\sigma(T)), \mu(\sigma(T))$ for any $\lambda(T), \mu(T), \sigma(T)$ in $k((T))$ with $\text{ord}\,\sigma(T) \ge 1$ then we must have $\text{ord}\,\sigma(T) = 1$). Let B_i be the corresponding "branch." That is, let B_i be the set of all power series $(\lambda(T), \mu(T))$ in T with coefficients in k such that for some $\ell \in L_i$ and some $\sigma(T) \in k[[T]]$ with $\text{ord}\,\sigma(T) = 1$ we have $(\lambda(T), \mu(T)) = (\sigma(T)^{f_i}, y_\ell(\sigma(T)))$.

It can be easily seen that $B_1, B_2, \ldots, B_h$ are distinct branches. That is, if $1 \le i \le j \le h$ and $\ell \in L_i$ and $\kappa \in L_j$ are such that $(T^{f_j}, y_\kappa(T)) = (\sigma(T)^{f_i}, y_\ell(\sigma(T)))$ for some $\sigma(T) \in k[[T]]$ with $\text{ord}\,\sigma(T) = 1$ then we must have $i = j$.

Now suppose that $F(X, Y) \in k[X, Y]$ and consider the plane curve C: $F(X, Y) = 0$. Then $y_1(T), \ldots, y_n(T)$ belong to $k[[T]]$, and for any fixed

i with $1 \leq i \leq h$, we clearly have $y_\ell(0) = y_{\ell'}(0)$ for all ℓ and ℓ' in L_i. Let $\beta_i^* = $ the common value of $y_\ell(0)$ for all $\ell \in L_i$. Let

$$\{1, 2, \ldots, h\} = L_1^* \amalg L_2^* \amalg \cdots \amalg L_s^*$$

be the disjoint partition of $\{1, 2, \ldots, h\}$ into the nonempty subsets $L_1^*, L_2^*, \ldots, L_s^*$, such that for all $1 \leq i \leq i' \leq h$ we have that $\beta_i^* = \beta_{i'}^* \Leftrightarrow i$ and i' belong to the same subset L_j^*. For $1 \leq j \leq s$, let $\beta_j = $ the common value of β_i^* for all $i \in L_j^*$. Let q_j be the number of elements in L_j^*, and note that then $q_1 + q_2 + \cdots + q_s = h$.

We have arranged matters so that $(0, \beta_1), (0, \beta_2), \ldots, (0, \beta_s)$ are exactly all the distinct points of C whose X-coordinate is 0. For each j with $1 \leq j \leq s$, as i varies over the set L_j^*, we get the q_j distinct branches B_i of C at the point $(0, \beta_j)$.

In particular, if C passes through the origin $(0, 0)$, then $\beta_j = 0$ for some j, say $\beta_1 = 0$. Now as i varies over L_1^*, the q_1 branches B_i are exactly all the distinct branches of C at $(0, 0)$. See Figure 13.1.

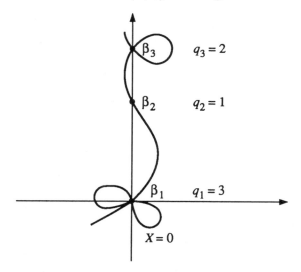

FIGURE 13.1

Later on we shall discuss how, using the Weierstrass Preparation Theorem, the branches of C at the origin can be found directly, without going through all the branches of C at the various points of C whose X-coordinate is zero.

NOTE. For ample illustrations of the power of the Weierstrass Preparation Theorem, see my book, *Local analytic geometry* [A07].

Branches and Valuations

Let k be a field and let $F(X, Y) \in k[X, Y]$ be a (nonconstant) polynomial so that $C : F(X, Y) = 0$ is a plane curve.

Recall that a parametrization of C at a point (α, β) of C is given by

$$\begin{cases} X = \lambda(T) \\ Y = \mu(T), \end{cases}$$

where $\lambda(T)$ and $\mu(T)$ are power series in T with coefficients in k such that $(\lambda(0), \mu(0)) = (\alpha, \beta)$ and $F(\lambda(T), \mu(T)) = 0$. A parametrization $(\lambda^*(T), \mu^*(T))$ of C at (α, β) is said to be *redundant* if $(\lambda^*(T), \mu^*(T)) = (\lambda(\sigma^*(T)), \mu(\sigma^*(T)))$ for some parametrization $(\lambda(T), \mu(T))$ of C at (α, β), and some $\sigma^*(T) \in k[[T]]$ with $\operatorname{ord} \sigma^*(T) > 1$. A parametrization of C at (α, β) is said to be *irredundant* if it is not redundant. Note that if a parametrization $(\lambda(T), \mu(T))$ of C at (α, β) is irredundant then it must be nonconstant. We must have either $\lambda(T) \notin k$ or $\mu(T) \notin k$. A *branch* of C at (α, β) is an equivalence class of irredundant parametrizations of C at (α, β), where two parametrizations $(\lambda(T), \mu(T))$ and $(\lambda'(T), \mu'(T))$ of C at (α, β) are said to be equivalent if $(\lambda'(T), \mu'(T)) = (\lambda(\sigma(T)), \mu(\sigma(T)))$, for some $\sigma(T) \in k[[T]]$ with $\operatorname{ord} \sigma(T) = 1$.

As a consequence of Newton's Theorem we see that if k is an algebraically closed field of characteristic zero and if $F(X, Y)$ is monic in Y, then given any point (α, β) of C (any elements α and β in k with $F(\alpha, \beta) = 0$), there exists at least one and at most finitely many branches of C at (α, β). The condition of F being monic in Y can be relaxed because by a linear transformation we can arrange it to be satisfied. In other words, if k is algebraically closed, then we can always find elements a, b, a', b' in k with $ab' - a'b \neq 0$ such that $F(aX + bY, a'X + b'Y)$ is monic in Y.

Question. Do the power series $\lambda(T)$ and $\mu(T)$ have to be convergent?

Answer. Need not be. For example, for the curve $Y^2 - X^3 = 0$, we have the perfectly nice parametrization

$$\begin{cases} X = T^2 \\ Y = T^3 \end{cases}$$

at the origin. However, say in case $k = \mathbb{C}$, if we multiply T by a suitable unit we can get a parametrization where the power series are divergent. We may take the unit $1 + X + 2^2 X^2 + \cdots + i^i X^i + \cdots$, which diverges everywhere except at $X = 0$. Then

$$\begin{cases} X &= \left[T\left(1 + T + 2^2 T^2 + \cdots\right)\right]^2 \\ Y &= \left[T\left(1 + T + 2^2 T^2 + \cdots\right)\right]^3 \end{cases}$$

gives a parametrization where both the power series are divergent.

REMARK. Having said that the power series $\lambda(T)$ and $\mu(T)$ need not be convergent, we may ask whether they can always be chosen to be convergent. In view of what we have said in connection with Hensel's Lemma, it follows that, in case $k = \mathbb{C}$, given any branch of C at any point of C, we can always find a representative $(\lambda(T), \mu(T))$ of that branch so that the power series $\lambda(T)$ and $\mu(T)$ are convergent. Then $X = \lambda(T)$ and $Y = \mu(T)$ would be a valid parametrization of C in some neighborhood of $T = 0$. It follows that, in case of $k = \mathbb{C}$, given any point of C, an "entire neighborhood" of that point can be "covered" by a finite number of parametrizations. A suitable generalization of this continues to hold for other "valued fields."

Now, reverting to any field k, suppose that F is irreducible (as a polynomial). Then we can consider rational functions on C.

So what are rational functions on C? Until now we have said something like, "on C we can consider polynomial functions $G(x, y)$ with polynomial expressions $G(X, Y)$, as well as rational functions $u = \frac{G(x,y)}{H(x,y)}$ with polynomial expressions $G(X, Y)$ and $H(X, Y)$ such that $H(x, y) \neq 0$, where x and y are related by $F(x, y) = 0$." Now having the notions of college algebra in our hands, we can formalize these matters.

Let N be the ideal of C in the polynomial ring $R = k[X, Y]$, i.e., $N = F(X, Y)R = $ the prime ideal in R generated by the irreducible polynomial $F(X, Y)$. For the residue class map $\Phi : R \to R/N$ we have $\ker \Phi = N$ and $N \cap k = (0)$. Hence, we may "identify" $\Phi(k)$ with k. We need not distinguish between elements of k and their images under Φ. Let $x = \Phi(X)$ and $y = \Phi(Y)$; ($\Phi(X)$ and $\Phi(Y)$ are not polynomials in X and Y!) Now

$$\Phi : R \to R/N = k[x, y] = \{G(x, y) : G(X, Y) \in k[X, Y]\}.$$

We call $k[x, y]$ the *affine coordinate ring* of C or the *ring of polynomial functions* on C.

In other words, any $G(X, Y) \in k[X, Y]$ is a polynomial function on the (X, Y)-plane. By restricting it to points of C we get the polynomial function $G(x, y)$ on C. Two polynomial functions $G(X, Y) \in k[X, Y]$ and $G^*(X, Y) \in k[X, Y]$ on the (X, Y)-plane give rise to the same polynomial functions on C if the difference $G(X, Y) - G^*(X, Y)$ is divisible by $F(X, Y)$. The converse is strictly true only when k is algebraically closed. However, even when k is not algebraically closed, we formally regard $G(x, y)$ to be equal to $G^*(x, y)$ if and only if $G(X, Y) - G^*(X, Y)$ is divisible by $F(X, Y)$. In particular, x and y are polynomial functions on

C. They are connected only by the relation $F(x, y) = 0$ and its multiples. At any rate, the "variables" X, Y are allowed to take any values, whereas the "functions" x, y are allowed to take only such values α, β for which $F(\alpha, \beta) = 0$.

Now since N is a prime ideal in R, the ring $R/N = k[x, y]$ is a domain. Hence, we can take its quotient field. We denote the quotient field of $k[x, y]$ by $k(x, y)$ or by $k(C)$, and we call it the *function field* of C or the *field of rational functions* on C.

Quite generally, if $x_1, \ldots, x_n$ are any *finite number of elements* in an overring K' of a ring k', then by $k'[x_1, \ldots, x_n]$ we denote the subring of K' given by

$$k'[x_1, \ldots, x_n] = \{g(x_1, \ldots, x_n) : g(X_1, \ldots, X_n) \in k'[X_1, \ldots, X_n]\}.$$

A ring of the form $k'[x_1, \ldots, x_n]$ is said to be a *finitely generated ring extension* of k', or an *affine ring* over k'. If K' and k' are fields, then by $k'(x_1, \ldots, x_n)$ we denote the subfield of K' given by

$$k'(x_1, \ldots, x_n) =$$
$$\left\{ \frac{g(x_1, \ldots, x_n)}{h(x_1, \ldots, x_n)} : g(X_1, \ldots, X_n), h(X_1, \ldots, X_n) \in k'[X_1, \ldots X_n] \right.$$
$$\left. \text{and } h(x_1, \ldots, x_n) \neq 0 \right\}.$$

A field of the form $k'(x_1, \ldots, x_n)$ is said to be a *finitely generated field extension* of k' or an *algebraic function field* over k'. Moreover, upon letting $m = \mathrm{trdeg}_{k'} K'$, we may also say that $k'(x_1, \ldots, x_n)/k'$ is an *m-dimensional algebraic function field* or that $k'(x_1, \ldots, x_n)/k'$ is an *m-variable algebraic function field*.

Now clearly $k(C)/k$ is a one-variable algebraic function field. To obtain $k(C)$ in a different manner, consider the localization R_N of R at N where we note that R_N is the set of all rational functions $U(X, Y) = \frac{G(X, Y)}{H(X, Y)}$ with $G(X, Y) \in R$ and $H(X, Y) \in R \setminus N$. It is easily seen that R_N is a Noetherian domain in which NR_N is the only maximal ideal and also the only nonzero prime ideal. Thus R_N is a one-dimensional local domain with $M(R_N) = NR_N$. For the residue class map $\Psi : R_N \to R_N/NR_N$ we have $\ker \Psi = NR_N$ and $NR_N \cap R = N$, and $\Psi(R_N)$ is the quotient field of $\Psi(R)$. Therefore, we may identify $\Psi(R_N)$ with $k(x, y)$.

If you want, you may really disregard the previous gibberish. In concrete terms, all I am trying to say is that you may substitute x, y for X, Y in $U(X, Y) \in k(X, Y)$ to get a rational function $u = U(x, y)$ on C *provided* you can express $U(X, Y)$ as $U(X, Y) = \frac{G(X, Y)}{H(X, Y)}$ with $G(X, Y), H(X, Y) \in k[X, Y]$ such that $H(X, Y)$ is nondivisible by $F(X, Y)$. *Caution*: If $U(X, Y)$ cannot be so expressed, then $U(x, y)$ will not make sense! Now if B is any branch of C at any point of C, then for any representative parametrization $(\lambda(T), \mu(T))$ we have

$$\operatorname{ord}_B u = \operatorname{ord} G(\lambda(T), \mu(T)) - \operatorname{ord} H(\lambda(T), \mu(T)).$$

We also have

$$H(\lambda(T), \mu(T)) \neq 0 \quad \text{and} \quad \operatorname{ord} U(\lambda(T), \mu(T))$$
$$= \operatorname{ord} G(\lambda(T), \mu(T)) - \operatorname{ord} H(\lambda(T), \mu(T)).$$

Therefore, we can directly write down the equation

$$\operatorname{ord}_B u = \operatorname{ord} U(\lambda(T), \mu(T)).$$

So far we have only discussed branches at points at finite distance. Let us now proceed to enlarge our discussion to include branches at points at ∞. For this purpose, we first homogenize F to get

$$f(\mathscr{X}, \mathscr{Y}, \mathscr{Z}) = \mathscr{Z}^d F\left(\frac{\mathscr{X}}{\mathscr{Z}}, \frac{\mathscr{Y}}{\mathscr{Z}}\right) \quad \text{where } d = \deg F.$$

Then we dehomogenize f to get

$$F'(Y', Z') = f(1, Y', Z') \quad \text{with } Y' = \frac{\mathscr{Y}}{\mathscr{X}} \quad \text{and} \quad Z' = \frac{\mathscr{Z}}{\mathscr{X}},$$

and

$$F^*(X^*, Z^*) = f(X^*, 1, Z^*) \quad \text{with } X^* = \frac{\mathscr{X}}{\mathscr{Y}} \quad \text{and} \quad Z^* = \frac{\mathscr{Z}}{\mathscr{Y}}.$$

Now the points at ∞ of C correspond to points of f with $\mathscr{Z} = 0$. Among these we have the points with $\mathscr{X} \neq 0$; they correspond to points of $F'(Y', Z') = 0$ with $Z' = 0$. Finally, the point $\mathscr{Z} = 0 = \mathscr{X}$ is a point at ∞ of C if and only if $F^*(0, 0) = 0$. Then it corresponds to the point $(0, 0)$ of $F^*(X^*, Y^*) = 0$.

For making our discussion applicable to branches at points at ∞, we allow $\lambda(T)$ and $\mu(T)$ to be meromorphic series instead of requiring them to be power series. In greater detail, by a *local parametrization* of C we mean a pair of meromorphic series $(\lambda(T), \mu(T))$ in T with coefficients in k such that $F(\lambda(T), \mu(T)) = 0$. A local parametrization $(\lambda^*(T), \mu^*(T))$ of C is said to be *redundant* if $(\lambda^*(T), \mu^*(T)) = (\lambda(\sigma^*(T)), \mu(\sigma^*(T)))$ for some local parametrization $(\lambda(T), \mu(T))$ of C and some $\sigma^*(T) \in k[[T]]$ with $\operatorname{ord} \sigma^*(T) > 1$. A local parametrization of C is said to be *irredundant* if it is not redundant. Note that if a local parametrization $(\lambda(T), \mu(T))$ of C is irredundant then it must be nonconstant. We must have either $\lambda(T) \notin k$ or $\mu(T) \notin k$. A *branch* of C is an equivalence class of irredundant local parametrizations of C, where two local parametrizations $(\lambda(T), \mu(T))$ and $(\lambda'(T), \mu'(T))$ of C are said to be equivalent if $(\lambda'(T), \mu'(T)) = (\lambda(\sigma(T)), \mu(\sigma(T)))$ for some $\sigma(T) \in k[[T]]$ with $\operatorname{ord} \sigma(T) = 1$. Given any branch B of C and given any rational function $u = U(x, y)$ on C where $U(X, Y) = \frac{G(X, Y)}{H(X, Y)}$ with $G(X, Y), H(X, Y) \in k[X, Y]$ such that

$H(X, Y)$ is nondivisible by $F(X, Y)$, we define the order of u at B by putting

$$\operatorname{ord}_B u = \operatorname{ord} U(\lambda(T), \mu(T))$$

where $(\lambda(T), \mu(T))$ is any representative of B. Note that then $\operatorname{ord}_B u$ depends only on B and u and not on the particular representative $(\lambda(T), \mu(T))$ of B. Also note that $U(\lambda(T), \mu(T))$ makes sense because it can be seen that $H(\lambda(T), \mu(T)) \neq 0$. Finally note that if $u = U'(x, y)$ where $U'(X, Y) = \frac{G'(X, Y)}{H'(X, Y)}$ with $G'(X, Y), H'(X, Y) \in k[X, Y]$ such that $H'(X, Y)$ is nondivisible by $F'(X, Y)$, then $U(\lambda(T), \mu(T)) = U'(\lambda(T), \mu(T))$. Therefore, $U(\lambda(T), \mu(T))$ depends only on $(\lambda(T), \mu(T))$ and u, and so $(\lambda(T), \mu(T))$ *induces* the k-homomorphism $k(C) \to k((T))$, which sends every $u \in k(C)$ to the corresponding $U(\lambda(T), \mu(T))$, and which is easily seen to be injective. (For any overrings K' and K^* of a ring k', by a k'-*homomorphism* $K' \to K^*$ we mean a ring homomorphism $K' \to K^*$, which sends every element of k' to itself.)

Let us now introduce the concept of the center of a branch of C. Firstly, if B is any branch of C such that $\operatorname{ord}_B x \geq 0$ and $\operatorname{ord}_B y \geq 0$, then clearly there exist unique elements α and β in k such that $\operatorname{ord}_B(x - \alpha) > 0$ and $\operatorname{ord}_B(y - \beta) > 0$. Now $F(\alpha, \beta) = 0$. That is, (α, β) is a point of C (at finite distance), and obviously B is a branch of C at (α, β). In this case, we say that (α, β) is the *center* of B, or that B is *centered* at (α, β). Note that for every representative $(\lambda(T), \mu(T))$ of B we now have that $(\lambda(T), \mu(T))$ is an irredundant parametrization of $F(X, Y) = 0$ at (α, β). Conversely, if (α, β) is a point of C (at finite distance) and B is a branch of C at (α, β), then clearly B is a branch of C such that $\operatorname{ord}_B x \geq 0$ and $\operatorname{ord}_B y \geq 0$, and such that $\operatorname{ord}_B(x - \alpha) > 0$ and $\operatorname{ord}_B(y - \beta) > 0$. Secondly, if B is any branch of C such that $\operatorname{ord}_B y \geq \operatorname{ord}_B x < 0$, then $x \neq 0$. Upon letting $y' = y/x$ and $z' = 1/x$ we have $\operatorname{ord}_B y' \geq 0$, $\operatorname{ord}_B z' > 0$, and $\operatorname{ord}_B(y' - \beta') > 0$ for a unique element β' in k. Now $F'(\beta', 0) = 0$. The point with homogeneous coordinates $(1, \beta', 0)$ is a point of C at ∞. In this case, we say that $(1, \beta', 0)$ is the *center* of B, or that B is *centered* at $(1, \beta', 0)$. Note that for every representative $(\lambda(T), \mu(T))$ of B we now have that $(\mu(T)/\lambda(T), 1/\lambda(T))$ is an irredundant parametrization of $F'(Y', Z') = 0$ at $(\beta', 0)$. Finally, if B is any branch of C such that $\operatorname{ord}_B x > \operatorname{ord}_B y < 0$, then $y \neq 0$. Upon letting $x^* = x/y$ and $z^* = 1/y$ we have $\operatorname{ord}_B x^* > 0$ and $\operatorname{ord}_B z^* > 0$. Now $F^*(0, 0) = 0$. That is, the point with homogeneous coordinates $(0, 1, 0)$ is a point of C at ∞. In this case, we say that $(0, 1, 0)$ is the *center* of B, or that B is *centered* at $(0, 1, 0)$. Note that for every representative $(\lambda(T), \mu(T))$ of B we now have that $(\lambda(T)/\mu(T), 1/\mu(T))$ is an irredundant parametrization of $F^*(X^*, Z^*) = 0$ at $(0, 0)$.

Given any point of C at ∞, by homogenizing F and then suitably dehomogenizing the resulting homogeneous polynomial, we can "bring" that point to finite distance. Therefore, as a consequence of Newton's Theorem we see that, if k is an algebraically closed field of characteristic zero then, given any point of C, at finite distance or at ∞, there exists at least one and at most a finite number of branches of C centered at that point.

Again reverting to any field k, given any branch B of C, we get the map $\operatorname{ord}_B : k(C) \to \mathbb{Z} \cup \{\infty\}$ such that for all u and u' in $k(C)$ we have

$$\operatorname{ord}_B(uu') = \operatorname{ord}_B u + \operatorname{ord}_B u' \quad \text{and} \quad \operatorname{ord}_B(u + u') \geq \min(\operatorname{ord}_B u, \operatorname{ord}_B u').$$

For any u in $k(C)$ we have

$$\operatorname{ord}_B u = \infty \Leftrightarrow u = 0.$$

Moreover, we have

$$\operatorname{ord}_B c = 0 \quad \text{for all } 0 \neq c \in k,$$

and we have

$$u \in k(C) \quad \text{with } \operatorname{ord}_B u \geq 0 \Rightarrow \operatorname{ord}_B(u - \gamma) > 0 \quad \text{for some } \gamma \in k.$$

(Note that by convention $\infty +$ any integer $= \infty = \infty + \infty$ and $\infty >$ every integer.)

Exercise. Given any branch B of C, show that there exists $t \in k(C)$ such that $\operatorname{ord}_B t = 1$. Also show that, given any such $t \in k(C)$, there exists a unique representative $(\lambda(T), \mu(T))$ of B such that the k-homomorphism $k(C) \to k((T))$ induced by $(\lambda(T), \mu(T))$ sends t to T.

Thus, for any branch B of C, the map $\operatorname{ord}_B : k(C) \to \mathbb{Z} \cup \{\infty\}$ is a residually rational discrete valuation of $k(C)/k$ in the following sense.

A *discrete valuation* of a field K is a surjective map $V : K \to \mathbb{Z} \cup \{\infty\}$, such that for all u and u' in K we have

$$V(uu') = V(u) + V(u') \quad \text{and} \quad V(u + u') \geq \min(V(u), V(u')),$$

and for any u in K we have

$$V(u) = \infty \Leftrightarrow u = 0.$$

By a *uniformizing parameter* of V we mean an element t in K such that $V(t) = 1$. Given a discrete valuation V of an overfield K of k, we say that V is *trivial* on k, or that V is a discrete valuation of K/k (of K *over* k), to mean that

$$V(c) = 0 \text{ for all } 0 \neq c \in k.$$

We say that V is *residually rational* over k to mean that

$$u \in K \quad \text{with } V(u) \geq 0 \Rightarrow V(u - \gamma) > 0 \quad \text{for some } \gamma \in k.$$

If V is trivial on k and V is residually rational over k, we may indicate this by saying that V is a residually rational discrete valuation of K/k, and we note that in that case

$$u \in K \quad \text{with } V(u) \geq 0 \Rightarrow V(u - \gamma) > 0 \quad \text{for a unique } \gamma \in k.$$

Thus, any branch B of C gives rise to the residually rational discrete valuation ord_B of $k(C)/k$. Conversely, every residually rational discrete valuation of $k(C)/k$ comes from a branch of C. That is, it is equal to ord_B for some branch B of C. Now we used Newton's Theorem to prove the existence of branches centered at any point of C (at least when k is an algebraically closed field of characteristic zero). We shall now use the thoughts of Newton's follower, Taylor, to sketch a proof of the said converse.

If you studied calculus, you must have heard about the Taylor expansion of a function which, by successive approximations, expands the given function in a power series. To mimic this procedure, let K be an overfield of k, and let t be a uniformizing parameter of a residually rational discrete valuation V of K/k. Given any $0 \neq u \in K$, let $e = V(u)$. Then $V(ut^{-e}) = 0$. Hence, there exists a unique $0 \neq c_e \in k$ such that $V(ut^{-e} - c_e) > 0$. Now $V(ut^{-e-1} - c_e t^{-1}) \geq 0$, and hence there exists a unique $c_{e+1} \in k$ such that $V(ut^{-e-1} - c_e t^{-1} - c_{e+1}) > 0$, and so on. This gives us a sequence of elements $c_e, c_{e+1}, \ldots$ in k such that $V(u - c_e t^e - c_{e+1} t^{e+1} - \cdots - c_{e+i} t^{e+i}) > e + i$ for all $i \geq 0$. Now $c_e T^e + c_{e+1} T^{e+1} + \cdots$ may be called the *Taylor expansion* of u at V relative to t. By sending every $u \in K$ to its Taylor expansion at V relative to t we get an injective k-homomorphism $K \to k((T))$. (We take 0 to be the Taylor expansion of 0.) In case $K = k(C)$, let $\lambda(T)$ and $\mu(T)$ be the respective Taylor expansions of x and y at V relative to t, and let B be the totality of pairs $(\lambda(\sigma(T)), \mu(\sigma(T)))$ as $\sigma(T)$ varies over all elements of $k((T))$ whose order is 1. Now B is a branch of C such that $V = \mathrm{ord}_B$. Note that the k-homomorphism $K \to k((T))$ of the last sentence now coincides with the k-homomorphism of the above example.

It now follows that $B \mapsto \mathrm{ord}_B$ gives a bijection of the set of all branches of C onto the set of all residually rational discrete valuations of $k(C)/k$. Therefore, given any residually rational discrete valuation V of $k(C)/k$, we can define the *center* of V on C to be the center of the corresponding branch of C. In case k is an algebraically closed field (of any characteristic), it can be shown that every discrete valuation of $k(C)$ is trivial on k and residually rational over k. Moreover, in this case, the above Taylor expansion argument can be used to circumvent Newton's Theorem and prove that every point of C, at finite distance or at ∞, is the center of at least one and at most a finite number of discrete valuations of $k(C)/k$.

Algebraic varieties. The idea of affine coordinate rings and function fields can be carried over to algebraic varieties. We have already considered plane curves and hypersurfaces in n-space. Generalizing these, we get the idea of an *algebraic variety* $W : f_1(X_1, \ldots, X_n) = f_2(X_1, \ldots, X_n) = \cdots = 0$ in n-space as the common solutions of a bunch of polynomial equations over a field k. Again, generalizing the idea of an irreducible plane curve or an irreducible hypersurface, we say that W is *irreducible* if the ideal N^* generated by $f_1, f_2, \ldots$ in $R_n = k[X_1, \ldots, X_n]$ is a prime ideal. Geometrically speaking, this means that the (algebraic) variety W cannot be expressed as a finite union of proper subvarieties. By a *subvariety* of W we mean an algebraic variety which is contained in W. By a *proper subvariety* of W we mean a subvariety of W that is different from W. Analogous to the case of plane curves, we have the residue class map $R_n \to R_n/N^* = k[x_1, \ldots, x_n]$ where x_i is the function on W *induced* by X_i. That is, x_i is the image of X_i under the said residue class map. We define

$$k[W] = \text{the } \textit{affine coordinate ring} \text{ of } W$$
$$\text{or the } \textit{ring of polynomial functions} \text{ on } W$$
$$= k[x_1, \ldots, x_n].$$

Assuming W to be irreducible, we define

$$k(W) = \text{the } \textit{function field} \text{ of } W \text{ or the } \textit{field of rational functions} \text{ on } W$$
$$= k(x_1, \ldots, x_n),$$

where $k(x_1, \ldots, x_n)$ is the quotient field of $k[x_1, \ldots, x_n]$. We note that $k(W)$ may also be thought of as the residue field $(R_n)_{N^*}/M((R_n)_{N^*})$ of the *local ring* $(R_n)_{N^*}$ of W on the affine n-space $\mathbb{A}_k^n$. At any rate, the variety W gives rise to the overring $k[x_1, \ldots x_n]$ of k. Conversely, given any elements $x_1, \ldots, x_n$ in an overring of k, we can easily get a variety W by first letting $f_1, f_2, \ldots$ be a set of generators of the kernel of the k-homomorphism of $k[X_1, \ldots, X_n]$ into the said overring, which sends X_i to x_i for $1 \leq i \leq n$ and then taking W to be the variety defined by the equations $f_1 = f_2 = \cdots = 0$.

Dimension and codimension. We start by noting that any given (algebraic) variety W in n-space can be uniquely written as a finite union $W_1 \cup \cdots \cup W_r$ of irreducible subvarieties $W_1, \ldots, W_r$ with $W_i \not\subset W_j$ for all $i \neq j$. The irreducible subvarieties $W_1, \ldots, W_r$ are called the *irreducible components* of W. The *dimension* of W is defined by putting

$$\dim W = \max_{1 \leq i \leq r} \text{trdeg}_k k(W_i).$$

We note that then

$$\dim W = \max_{1 \leq i \leq r} \dim W_i.$$

Now we may define a *curve* to be a variety of dimension 1, a *surface* to be a variety of dimension 2, a *solid* to be a variety of dimension 3, ..., and so on. Moreover we define the *codimension* of W in the ambient space to be $n - m$ where $m = \dim W$. We observe that a hypersurface is a variety of pure codimension 1 in the following sense. If $\dim W_i = \dim W = m$ for $1 \leq i \leq r$, then we may say W is *pure m-dimensional*, or *pure $(n - m)$-codimensional*. It can be seen that the dimension of W coincides with the (Krull) dimension of the affine coordinate ring $k[W]$ of W, i.e.,

$$\dim W = \dim k[W].$$

We note that this gives a characterization of the dimension of W without decomposing it into irreducible components.

Local rings of points and irreducible subvarieties. To every point $P = (\alpha_1, \ldots, \alpha_n)$ of an irreducible variety W in n-space over k, we associate the local ring $\mathscr{L}(P, W)$, which we call the *local ring* of P on W and define by putting

$$\mathscr{L}(P, W) = k[W]_{\mathscr{I}(P, W)}.$$

Here, $\mathscr{I}(P, W)$ is a maximal ideal in $k[W]$, which we call the *prime ideal* of P on W and define by putting

$$\mathscr{I}(P, W) = \{g(x_1, \ldots, x_n) : g(X_1, \ldots, X_n) \in R_n$$
$$\text{such that } g(\alpha_1, \ldots, \alpha_n) = 0\},$$

where $R_n = k[X_1, \ldots, X_n]$. More generally, for any subvariety E of W, we introduce the *ideal* $\mathscr{I}(E, W)$ of E on W by putting

$$\mathscr{I}(E, W) = \{g(x_1, \ldots, x_n) : g(X_1, \ldots, X_n) \in R_n \text{ such that}$$
$$g(\alpha_1, \ldots, \alpha_n) = 0 \text{ for all } (\alpha_1, \ldots, \alpha_n) \in E\}.$$

We note that E is irreducible iff $\mathscr{I}(E, W)$ is a prime ideal in $k[W]$. In that case we introduce the local ring $\mathscr{L}(E, W)$, which we call the *local ring* of E on W and define by putting

$$\mathscr{L}(E, W) = k[W]_{\mathscr{I}(E, W)}.$$

Note that points of W are special cases of irreducible subvarieties of W.

NOTE. It was essentially by using the idea of discrete valuations that Dedekind [**DWe**] and Kronecker [**Kro**] unified algebraic number theory and algebraic curve theory. In effect, every $0 \neq u \in \mathbb{Q}$ can be uniquely written as

$$u = \pm \prod_{\text{primes } p} p^{V_p(u)} \quad \text{with } V_p(u) \in \mathbb{Z},$$

where the product is over all prime numbers p (and where the infinite product makes sense because for all except finitely many p we have $V_p(u) = 0$ and hence $p^{V_p(u)} = 1$). For every prime number p, the map $V_p : \mathbb{Q} \to \mathbb{Z} \cup \{\infty\}$ is a discrete valuation of $\mathbb{Q}$. (We take $V_p(0) = \infty$.) It can be seen that there are no other discrete valuations of $\mathbb{Q}$. For a proof of this and other related matter, see Ostrowski [**Os1**], where valuations were formally introduced. Analogously, considering the X-axis, every $0 \neq u \in k(X)$ can be uniquely written as

$$u = c \prod_p p^{V_p(u)} \quad \text{with } 0 \neq c \in k \quad \text{and} \quad V_p(u) \in \mathbb{Z},$$

where the product is over all nonconstant monic irreducible polynomials $p = p(X) \in k[X]$. For every nonconstant monic irreducible polynomial p, the map $V_p : k(X) \to \mathbb{Z} \cup \{\infty\}$ is a discrete valuation of $k(X)/k$ (again $V_p(0) = \infty$). This time there is exactly one more discrete valuation of $k(X)/k$ that we may denote by V_∞ and which is given by $V_\infty(u) = -\deg u$ for all $u \in k(X)$. Here, by the degree of a rational function we mean the degree of its numerator minus the degree of its denominator. Similarly, the case of an algebraic plane curve is a generalization of this case of the X-axis. Likewise, the discussion of $\mathbb{Q}$ generalizes to an "algebraic number field." Note that what we have been calling a plane curve may also be called an *algebraic plane curve* to stress the fact that its defining equation $F(X, Y) = 0$ is a polynomial equation. Similarly, for any power series $\phi(X, Y)$, the curve given by $\phi(X, Y) = 0$ may be called an *analytic plane curve*.

Divisors of Functions and Differentials

Let $F(X, Y) \in k[X, Y]$ be an irreducible polynomial where k is an algebraically closed field, and consider the plane curve $C : F(X, Y) = 0$.

In discussing the zeros and poles of functions and differentials on C, first we took a naive approach by considering points of C. Then we indicated a refinement by using branches of C instead of points of C. Now, having set up a one-to-one correspondence between branches of C and discrete valuations of the function field $k(C)$ of C, we shall carry out the said refinement in terms of discrete valuations of $k(C)$. Note that $k(C) = k(x, y)$ where x and y are the rational functions on C "induced" by X and Y respectively. That is, x and y are the respective images of X and Y under the residue class map $k[X, Y] \to k[X, Y]/(F(X, Y)k[X, Y])$.

Let us start by recalling that a discrete valuation of $k(C)$ is a map $V : k(C) \to \mathbb{Z} \cup \{\infty\}$ such that for all u and u' in $k(C)$ we have

$$V(uu') = V(u) + V(u') \quad \text{and} \quad V(u + u') \geq \min(V(u), V(u')).$$

For any u in $k(C)$ we have

$$V(u) = \infty \Leftrightarrow u = 0.$$

For some $t \in k(C)$ we have $V(t) = 1$. Any such element t is called a uniformizing parameter of V. We have noted that every discrete valuation V of $k(C)$ is automatically trivial on k and residually rational over k. That is,

$$V(c) = 0 \quad \text{for all } 0 \neq c \in k$$

and

$$u \in k(C) \quad \text{with } V(u) \geq 0 \Rightarrow V(u - \gamma) > 0 \quad \text{for a unique } \gamma \in k.$$

Given any discrete valuation V of $k(C)$, we have defined the center of V on C thus. First, if $V(x) \geq 0$ and $V(y) \geq 0$ then there exist unique elements α and β in k such that $V(x - \alpha) > 0$ and $V(x - \beta) > 0$. Now $F(\alpha, \beta) = 0$. That is, $P = (\alpha, \beta)$ is a point of C (at finite distance). We call it the center of V on C. Note that $(\alpha, \beta, 1)$ are homogeneous coordinates of the point P. We recall that the multiplicity of P on C, is given by

$$\text{mult}_P C = \text{ord}_{(\alpha, \beta)} F(X, Y)$$

$$= \text{ord}\, G(X, Y) \quad \text{where } G(X, Y) = F(X + \alpha, Y + \beta),$$

and we define the *multiplicity* of V on C, or the *multiplicity* of C at V, by putting

$$\text{mult}_V C = \min(V(x - \alpha), V(y - \beta)).$$

Second, if $V(y) \geq V(x) < 0$ then $x \neq 0$. Upon letting $y' = y/x$ and $z' = 1/x$, we have $V(y') \geq 0$, $V(z') > 0$ and $V(y' - \beta') > 0$ for a unique element β' in k. Now upon letting $F'(Y', Z') = Z'^n F(1/Z', Y'/Z')$ (where $n = \deg F$) we have $F'(\beta', 0) = 0$. That is, the point with homogeneous coordinates $(1, \beta', 0)$ is a point of C at ∞. We call it the center of V on C. In case $\beta' \neq 0$, put $P' = (1, \beta', 0)$ and note that the multiplicity of P' on C is given by

$$\text{mult}_{P'} C = \text{ord}_{(\beta', 0)} F(Y', Z')$$
$$= \text{ord}\, G'(Y', Z') \quad \text{where } G'(Y', Z') = F'(Y' + \beta', Z').$$

Let us define the *multiplicity* of V on C, or the *multiplicity* of C at V, by putting

$$\text{mult}_V C = \min(V(y' - \beta'), V(z')).$$

Similarly, in case $\beta' = 0$, let us put $Q' = (1, 0, 0)$ and note that the multiplicity of Q' on C is given by

$$\text{mult}_{Q'} C = \text{ord}\, F'(Y', Z').$$

Again let us define the *multiplicity* of V on C, or the *multiplicity* of C at V, by putting

$$\text{mult}_V C = \min(V(y'), V(z')).$$

Finally, if $V(x) > V(y) < 0$ then $y \neq 0$. Upon letting $x^* = x/y$ and $z^* = 1/y$ we have $V(x^*) > 0$ and $V(z^*) > 0$. Now upon letting $F^*(X^*, Z^*) = Z^{*n} F(X^*/Z^*, 1/Z^*)$ (where $n = \deg F$) we have $F^*(0, 0) = 0$. The point Q^* with homogeneous coordinates $(0, 1, 0)$ is a point of C at ∞. We call it the center of V on C. Recall that the multiplicity of Q^* on C is given by

$$\text{mult}_{Q^*} C = \text{ord}\, F^*(X^*, Z^*).$$

We define the *multiplicity* of V on C, or the *multiplicity* of C at V, by putting

$$\text{mult}_V C = \min(V(x^*), V(z^*)).$$

We may depict all this in Figure 15.1, where the "hypotenuse" is the line at ∞.

At any rate, given any discrete valuation V of $k(C)$, the center of V is a certain point of C, which may be at finite distance or at ∞. Conversely, given any point Q of C, at finite distance or at ∞, it is the center on C of at least one and at most finitely many discrete valuations of $k(C)$. A more precise version of this converse is given by the formula

$$\text{mult}_Q C = \sum_{V \text{ centered at } Q} \text{mult}_V C,$$

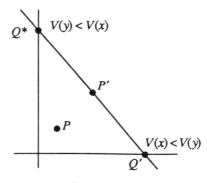

FIGURE 15.1

where the sum is over all discrete valuations V of $k(C)$ whose center on C is Q. It follows that

Q is a simple point of C

$\Leftrightarrow \mathrm{mult}_Q C = 1$

$\Leftrightarrow$ there is exactly one discrete valuation V of $k(C)$ centered at Q

and for that V we have $\mathrm{mult}_V C = 1$.

In greater detail, in view of the Implicit Function Theorem, we have a description where $F'(Y', Z')$, $F^*(Y^*, Z^*)$, y', z', x^*, z^* are as above and where subscripts denote partial derivatives.

Suppose that $Q = P = (\alpha, \beta)$ where α and β are elements in k such that $F(\alpha, \beta) = 0$. Now

P is a simple point of C

$\Leftrightarrow F_X(\alpha, \beta) \neq 0$ or $F_Y(\alpha, \beta) \neq 0$

$\Leftrightarrow$ there is exactly one discrete valuation V of $k(C)$ centered at P,

and for that V we have either $V(x - \alpha) = 1$ or $V(y - \beta) = 1$.

Moreover, if $F_X(\alpha, \beta) \neq 0$, then P is a simple point of C, and for the unique discrete valuation V of $k(C)$ centered at Q we have $V(x - \alpha) = 1$. That is, $x - \alpha$ is a uniformizing parameter at V, and so x is a "uniformizing coordinate" at V and we are in the nonvertical tangent case. Likewise, if $F_X(\alpha, \beta) \neq 0$, then P is a simple point of C, and for the unique discrete valuation V of $k(C)$ centered at P we have $V(y - \beta) = 1$. That is, $y - \beta$ is a uniformizing parameter at V, and so y is a "uniformizing coordinate" at V, and we are in the nonhorizontal tangent case. See Figure 15.2 on page 112.

Next suppose that $Q = P' = (1, \beta', 0)$, where β' is an element in k such that $F'(\beta', 0) = 0$. (Note that in the notation $(1, \beta', 0) = P'$ or Q' according as $\beta' \neq 0$ or $\beta' = 0$. At present we are letting $(1, \beta', 0) = P'$ in

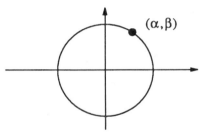

FIGURE 15.2

both the cases.) Now

P' is a simple point of C

⇔ $F'_{Y'}(\beta', 0) \neq 0$ or $F'_{Z'}(\beta', 0) \neq 0$

⇔ there is exactly one discrete valuation V of $k(C)$ centered at P',

and for that V we have either $V(y' - \beta') = 1$ or $V(z') = 1$.

Moreover, if $F'_{Z'}(\beta', 0) \neq 0$, then P' is a simple point of C, and for the unique discrete valuation V of $k(C)$ centered at P' we have $V(y' - \beta') = 1$. That is, $y' - \beta'$ is a uniformizing parameter at V. Likewise, if $F'_{Y'}(\beta', 0) \neq 0$, then P' is a simple point of C. For the unique discrete valuation V of $k(C)$ centered at P' we have $V(z') = 1$. That is, z' is a uniformizing parameter at V.

Finally, suppose that $Q = Q^* = (0, 1, 0)$ and $F^*(0, 0) = 0$. Now

Q^* is a simple point of C

⇔ $F^*_{X^*}(0, 0) \neq 0$ or $F^*_{Z^*}(0, 0) \neq 0$

⇔ there is exactly one discrete valuation V of $k(C)$ centered at Q^*,

and for that V we have either $V(x^*) = 1$ or $V(z^*) = 1$.

Moreover, if $F^*_{Z^*}(0, 0) \neq 0$, then Q^* is a simple point of C, and for the unique discrete valuation V of $k(C)$ centered at Q^* we have $V(x^*) = 1$; x^* is a uniformizing parameter at V. Likewise, if $F^*_{X^*}(0, 0) \neq 0$, then Q^* is a simple point of C, and for the unique discrete valuation V of $k(C)$ centered at Q^* we have $V(z^*) = 1$; z^* is a uniformizing parameter at V.

Since $F(\alpha, \beta) = F_X(\alpha, \beta) = F_Y(\alpha, \beta) = 0$ for every singular point (α, β) of C at finite distance, and since C has only finitely many points at ∞, it follows that C has at most a finite number of singular points at finite distance or at ∞. Since every simple point of C is the center of only one discrete valuation of $k(C)$, we conclude that all except finitely many points of C, at finite distance or at ∞, have the property of being the center of a unique discrete valuation of $k(C)$. An example of a singular point having this property is the cuspidal point that the curve $Y^2 - X^3 = 0$ has at the origin. On the other hand, the nodal singularity that the curve $Y^2 - X^2 - X^3 = 0$ has at the origin does not have this property because it is the center of two discrete valuations arising from the two branches the curve has at the origin. This is, in fact, how we can tell when a double point is a cusp or a node. A cusp is the center of only one discrete valuation; a node is the center of two.

Birational viewpoint. The curve C gives rise to the field $k(C)$. We consider discrete valuations of $k(C)$ and note that $k(C)/k$ is a one-variable algebraic function field, i.e., a finitely generated field extension of transcendence degree 1. Alternatively we could start with a one-variable algebraic function field K/k and study discrete valuations of it. Such a one-variable algebraic function field K/k gives rise to several curves C'. We have $K = k(C')$ for various irreducible plane curves C'. Now a discrete valuation V of K has a respective center on each such C'. Any such C' is called a "model" of K/k. This is the birational viewpoint.

Divisors. We talked about a divisor on C as a finite set of points on C with multiplicities. Then we indicated a refinement of it as a finite set of branches of C with multiplicities. To complete this refinement, we now introduce the notion of a divisor of K where $K = k(C)$, or as said above, K/k could be any one-variable algebraic function field over the algebraically closed "ground field" k. Briefly, a divisor of K is a finite set of discrete valuations of K with multiplicities. More precisely, a *divisor* of K is a map $D : \mathscr{R}(K) \to \mathbb{Z}$ such that $D(V) = 0$ for all except finitely many V in $\mathscr{R}(K)$. By $\mathscr{R}(K)$ we denote the set of all discrete valuations of K. We are using the notation $\mathscr{R}(K)$ to remind ourselves of the Riemann surface associated with the curve C. Indeed, in case $K = k(C)$ with $k = \mathbb{C}$, the said Riemann surface can be obtained by suitably "topologizing" $\mathscr{R}(K)$. For this viewpoint see Chevalley [**Ch4**] or Weyl [**Wey**]. We define the *degree* of D by the equation

$$\deg D = \sum_{V \in \mathscr{R}(K)} D(V).$$

For any other divisor D' of K we define the sum $D + D'$ as the divisor of K given by

$$(D + D')(V) = D(V) + D'(V) \quad \text{for all } V \in \mathscr{R}(K).$$

With the obvious definition of the zero divisor (not zerodivisor!), the set of all divisors of K becomes an additive abelian group. It is the "free" abelian group "generated" by $\mathscr{R}(K)$. We may denote this group by $\mathscr{D}(K)$. For any D and D' in $\mathscr{D}(K)$ we define

$$D \geq D' \Leftrightarrow D(V) \geq D'(V) \quad \text{for all } V \in \mathscr{R}(K),$$

and for any $D \in \mathscr{D}(K)$ we define

$$D \text{ is } positive \Leftrightarrow D \geq \text{the zero divisor.}$$

That is,

$$D \text{ is } positive \Leftrightarrow D(V) \geq 0 \quad \text{for all } V \in \mathscr{R}(K).$$

By $\mathscr{D}_+(K)$ we denote the set of all *positive divisors* of K. We note that every $D \in \mathscr{D}(K)$ can uniquely be expressed as

$$D = D_0 - D_\infty \quad \text{with } D_0 \text{ and } D_\infty \text{ in } \mathscr{D}_+(K).$$

Functions. To complete the refinement of the notion of the divisor of a function, given any $0 \neq u \in K$, we define the *divisor* (u) of u to be the member of $\mathscr{D}(K)$ given by

$$(u)(V) = V(u) \quad \text{for all } V \in \mathscr{R}(K).$$

We let $(u)_0$ and $(u)_\infty$ be the unique members of $\mathscr{D}_+(K)$ such that

$$(u) = (u)_0 - (u)_\infty.$$

We note that $(u)_0$ and $(u)_\infty$ may respectively be called the *divisor of zeros* of u and the *divisor of poles* of u. We say that u has a *zero* (resp. *pole*) of order $V(u)$ (resp. $-V(u)$) at those $V \in \mathscr{R}(K)$ for which $V(u) > 0$ (resp. $V(u) < 0$). Now here is the latest incarnation of the

THEOREM ON ZEROS OF FUNCTIONS. *For any* $0 \neq u \in K$ *we have*

$$\deg(u)_0 = \deg(u)_\infty \quad \text{or equivalently} \quad \deg(u) = 0.$$

If $K = k(C)$ and if

$$F(X, Y) = Y^n + a_1(X)Y^{n-1} + \cdots + a_n(X) \quad \text{with } a_i(X) \in k[X],$$

then the above theorem as applied to the function $u = x$ amounts to saying that

$$\deg(x)_0 = \text{number of zeros of } x = n = \deg(x)_\infty = \text{number of poles of } x.$$

To elucidate the equation

$$\deg(x)_0 = \text{number of zeros of } x = n,$$

let $(0, \beta_1), \ldots, (0, \beta_s)$ be all the distinct points of C whose X-coordinate is 0. Let $V_{i1}, \ldots, V_{iq_i}$ be the discrete valuations of $k(C)$ centered at $(0, \beta_i)$ on C. Let $e_{ij} = V_{ij}(x)$. Then

$$\sum_{i=1}^{s} \sum_{j=1}^{q_i} e_{ij} = \text{number of zeros of } x = n.$$

Moreover, it can be seen that

$$F(X, Y) = \prod_{i=1}^{s} \prod_{j=1}^{q_i} F_{ij}(X, Y),$$

where the $F_{ij}(X, Y)$ are pairwise distinct irreducible monic polynomials of degree e_{ij} in Y with coefficients in $k[[X]]$ such that

$$F_{ij}(0, Y) = (Y - \beta_i)^{e_{ij}}$$

and the discrete valuation V_{ij} of $k(C)$ "corresponds to" F_{ij}. In case k is of characteristic zero, the above factorization of F can be obtained by Newton's Theorem.

By taking $x = t$ in the above discussion we get the following

CRITERION OF RATIONALITY. C *is rational if and only if there exists a rational function* $t \in K = k(C)$ *such that* $\deg(t)_0 = 1$. *Then any such* t *is a "global parameter" of* C, *i.e.,* $k(C) = k(t)$.

In the criterion, we have used the definition of rationality according to which C is *rational* $\Leftrightarrow k(C) = k(t)$. This is *a priori* stronger than the usual definition according to which C is *rationally parametrizable* $\Leftrightarrow x = \phi(t)$ and $y = \psi(t)$ for some $\phi(t)$ and $\psi(t)$ in $k(t)$. That is, $\Leftrightarrow$ there exist two elements $\phi(t)$ and $\psi(t)$ in $k(t)$, at least one of which is not in k, such that $F(\phi(t), \psi(t)) = 0$. Let us say that C is *unirational* to mean that C is rationally parametrizable. Now if C is rational then clearly it is unirational, because the unirationality of C means $k(C) \subset k(t)$ whereas the rationality of C means the stronger condition $k(C) = k(t)$. The converse of this is the following classical Theorem of Lüroth. For a proof of it, see Severi [**Se2**] or van der Waerden [**Wa2**].

LÜROTH'S THEOREM. *If* C *is unirational, then it is rational.* *In other words, if* $k(C) \subset k(t)$ *then* $k(C) = k(t')$ *for some* $t' \in k(t)$.

Differentials. So far we have said that a differential is an expression of the form udv which transforms like an integrand. We can formalize this by saying that a *differential* of K is an equivalence class of pairs (u, v) of elements of K under the equivalence relation: $(u, v) \sim (u', v') \Leftrightarrow u\frac{dv}{dt} = u'\frac{dv'}{dt}$ for some (and hence every) separating transcendence t of K/k. We write udv for the equivalence class containing (u, v). In other words, a differential of K is an expression of the form udv with u and v in K. For any other differential $u'dv'$ of K we have $udv = u'dv' \Leftrightarrow u\frac{dv}{dt} = u'\frac{dv'}{dt}$ for some (and hence every) separating transcendence t of K/k. (If k was not algebraically closed, we should be talking about differentials of K/k. However, as hinted above, assuming k to be algebraically closed, k is determined by K.) Following the customary procedure of these lectures, we shall now define the new terms: equivalence relation, equivalence class, separating transcendence, and derivative ($= \frac{dv}{dt}$).

An *equivalence relation* on a set Ω is a "binary" relation $a \sim b$ which holds between some element a, b of Ω and which is such that for all a, b, c in Ω we have: (reflexive) $a \sim a$; (symmetric) $a \sim b \Rightarrow b \sim a$; and (transitive) $a \sim b$ and $b \sim c \Rightarrow a \sim c$. By putting in one "box" all elements equivalent to each other, we get a disjoint partition of Ω into nonempty subsets called *equivalence classes*. Conversely, given any disjoint partition of Ω into nonempty subsets, we get an equivalence relation by defining two elements to be equivalent to each other iff they are in the same subset; "iff" is a standard abbreviation for "if and only if." A typical example of equivalence classes (and hence of an equivalence relation) is provided by the residue classes of a ring modulo an ideal.

In view of the "usual equation" $F_X(x, y)dx + F_Y(x, y)dy = 0$, we see that if $F_Y(x, y) \neq 0$ then $\frac{dy}{dx} = \frac{-F_X(x,y)}{F_Y(x,y)}$, and now disregarding the question of the meaning of the "usual equation," we take this to be the definition of $\frac{dy}{dx}$, which in turn gives an obvious meaning to $\frac{dz}{dx}$ for every $z \in k(C)$ and then we have $\frac{dz}{dx} \in k(C)$. If $F_Y(x, y) \neq 0$, then we say that x is a separating transcendence of $k(C)/k$. Likewise, if $F_X(x, y) \neq 0$, then we say that y is a separating transcendence of $k(C)/k$. In that case we can find $\frac{dz}{dy} \in k(C)$ for every $z \in k(C)$. It can be shown that either x or y is a separating transcendence of $k(C)/k$. (If k is of zero characteristic, then x as well as y is a separating transcendence of $k(C)/k$, provided C is not a line parallel to one of the coordinate axis.) More generally, by a *separating transcendence* of K/k we mean $t \in K \setminus k$ such that for every $z \in K$ we can find $H(T, Z) \in k[T, Z]$ with $H(t, z) = 0 \neq H_Z(t, z)$. Then we define $\frac{dz}{dt} = \frac{-H_T(t, z)}{H_Z(t, z)} \in K$.

Given any $V \in \mathscr{R}(K)$ and given any uniformizing parameter t of V, it can be shown that t is a separating transcendence of K/k. It can also be shown that for every other uniformizing parameter t' of V we have $V(\frac{dt'}{dt}) = 0$. Therefore, for any differential $u dv$ of K we can unambiguously define $V(u dv)$ by putting

$$V(u dv) = V(u \frac{dv}{dt}).$$

It can be shown that if $\delta(T) \in k((T))$ and $\theta(T) \in k((T))$ are the respective Taylor expansions of u and v at V relative to t, then $\delta(T)\frac{d\theta(T)}{dT}$ is the Taylor expansion of $u\frac{dv}{dt}$ at V relative to t. Hence, in particular,

$$V(u \frac{dv}{dt}) = \operatorname{ord} \delta(T) \frac{d\theta(T)}{dT}.$$

It follows that if $K = k(C)$ and B is the branch of C with $\operatorname{ord}_B = V$, then for any representative $(\lambda(T), \mu(T))$ of B we have

$$V(u \frac{dv}{dt}) = \operatorname{ord} \delta'(T) \frac{d\theta'(T)}{dT},$$

where $\delta'(T) \in k((T))$ and $\theta'(T) \in k((T))$ are respective images of u and v under the k-homomorphism $k(C) \to k((T))$ induced by $(\lambda(T), \mu(T))$.

Let us denote the set of all differentials of K by $\mathscr{D}^*(K)$. Now if t is any separating transcendence of K/k, then every member of $\mathscr{D}^*(K)$ can uniquely be expressed as $z dt$ with $z \in K$. Hence, in a natural manner, $\mathscr{D}^*(K)$ becomes a vector space over K with $[\mathscr{D}^*(K) : K] = 1$. Given any $w = u dv \in \mathscr{D}^*(K)$ and $0 \neq w' = u'dv' \in \mathscr{D}^*(K)$, we have $w = z dt$ with $z \in K$ and $w' = z'dt$ with $0 \neq z' \in K$. So we may define $w/w' \in K$ by putting $w/w' = z/z'$. Since $\mathscr{D}^*(K)$ is a vector space over K, we can also regard it as a vector space over k, but now we would have $[\mathscr{D}^*(K) : k] = \infty$.

To recast the concept of the divisor of a differential in the language of valuations, given any $0 \neq w \in \mathscr{D}^*(K)$, we define the *divisor* (w) of w to be the member of $\mathscr{D}(K)$ such that

$$(w)(V) = V(w) \text{ for all } V \in \mathscr{R}(K).$$

We let $(w)_0$ and $(w)_\infty$ be the unique members of $\mathscr{D}_+(K)$ such that

$$(w) = (w)_0 - (w)_\infty.$$

We note that $(w)_0$ and $(w)_\infty$ may respectively be called the *divisor of zeros* of w and the *divisor of poles* of w. We say that w has a *zero* (resp. *pole*) of order $V(w)$ (resp. $-V(w)$) at those $V \in \mathscr{R}(K)$ for which $V(w) > 0$ (resp. $V(w) < 0$). For every $0 \neq w' \in \mathscr{D}^*(K)$ we obviously have

$$(w) - (w') = (w/w').$$

Therefore, because of the theorem on zeros of functions, the integer $\deg(w)$ depends only on K and not on the particular differential w. It can also be seen that $1 + (1/2)\deg(w)$ is a non-negative integer. So we may define

$$g(K) = \text{ the } genus \text{ of } K = 1 + (1/2)\deg(w).$$

Upon letting $g = g(K)$, we get the latest incarnation of

THEOREM ON ZEROS OF DIFFERENTIALS. *For every* $0 \neq w \in \mathscr{D}^*(K)$ *we have* $\deg(w) = 2g - 2$. *If* $K = k(C)$ *then we have* $g(C) = g$.

Here is an

ALTERNATIVE CRITERION OF RATIONALITY. *We have* $K = k(t)$ *for some* $t \in K \setminus k \Leftrightarrow g = 0 \Leftrightarrow \deg(dt) = -2$ *for some* $t \in K \setminus k \Leftrightarrow \deg(dt) = -2$ *for all* $t \in K \setminus k$. *In particular, by taking* $K = k(C)$ *we see that* C *is rational iff* $g(C) = 0$.

Let us close by stating the famous

RIEMANN-ROCH THEOREM. *Given any* $D \in \mathscr{D}(K)$, *let* $L(D)$ *be the set of all functions having poles at most in* D *and let* $I(D)$ *be the set of all differentials having zeros at least in* D. *That is,* $L(D) = \{z \in K : (z) + D \geq 0\}$ *and* $I(D) = \{w \in \mathscr{D}^*(K) : (w) - D \geq 0\}$. *Then* $L(D)$ *and* $I(D)$ *are finite dimensional vector spaces over* k *and upon letting* $l(D) = [L(D) : k]$ *and* $i(D) = [I(D) : k]$ *we have* $l(D) = \deg D - g + 1 + i(D)$. *Moreover, if* $\deg D > 2g - 2$ *then* $i(D) = 0$.

NOTE. So what are differentials? Nobody really knows. Perhaps that is why they are so useful!

Weierstrass Preparation Theorem

Again let k be a field and consider

$$F'(X, Y) = Y^{n'} + a_1'(X)Y^{n'-1} + \cdots + a_{n'}'(X) \in k[[X]][Y].$$

In case k is algebraically closed and the characteristic of k does not divide $n'!$, by Newton's Theorem we can factor $F'(X, Y)$ into linear factors

$$F'\left(T^m, Y\right) = \prod_{i=1}^{n'} [Y - \eta_i(T)] \quad \text{with } 0 < m \in \mathbb{Z} \quad \text{and} \quad \eta_i(T) \in k[[T]].$$

Then, collating these factors into conjugacy classes etc., we get

$$F'(X, Y) = \prod_{j=1}^{s} \prod_{i=1}^{q_j} \psi_{ji}(X, Y)^{e_{ji}}$$

where $\psi_{ji}(X, Y)$ are monic irreducible polynomials in Y with coefficients in $k[[X]]$ such that

$$\psi_{ji}(0, Y) = (Y - \beta_j)^{e_{ji}}$$

where $\beta_1, \ldots, \beta_s$ are pairwise distinct elements in k with $\beta_1 = 0$. Thus,

$$\prod_{i=1}^{q_1} \psi_{1i}(X, Y)^{e_{1i}} \quad \text{and} \quad \prod_{j=2}^{s} \prod_{i=1}^{q_j} \psi_{ji}(X, Y)^{e_{ji}}$$

give the separation of the branches of F' into those at the origin and those at the points on the Y-axis other than the origin. To get such a separation directly, we can use either Hensel's Lemma, or alternatively the theorem of Weierstrass, which does the same job in the more general case when $F'(X, Y)$ is replaced by a power series $F(X, Y) \in k[[X, Y]]$. Once again k is a field of any characteristic which need not be algebraically closed.

WEIERSTRASS PREPARATION THEOREM (WPT). *Given any*

$$F(X, Y) = \sum a_{ij} X^i Y^j \in k[[X, Y]] \quad \text{with } a_{ij} \in k \quad \text{and} \quad F(0, Y) \neq 0,$$

let $d = \operatorname{ord} F(0, Y)$. *There exist unique elements* $G(X, Y)$ *and* $H(X, Y)$ *in* $k[[X, Y]]$ *such that* $F(X, Y) = G(X, Y)H(X, Y)$ *and* $G(0, 0) \neq 0$ *and*

$$H(X, Y) = Y^d + c_1(X)Y^{d-1} + \cdots + c_d(X) \text{ with } c_i(X) \in k[[X]] \text{ and } c_i(0) = 0$$
for $1 \leq i \leq d$.

The statement as well as the proof of WPT is very parallel to Hensel's Lemma and its proof. To bring out the parallelism between the statements of Weierstrass and Hensel, we draw Table 16.1, which compares them.

TABLE 16.1.

Weierstrass	Hensel
$F(X, Y) = \sum a_{ij}X^iY^j$	$F'(X,Y) = Y^{n'} + a'_1(X)Y^{n'-1} + \cdots + a'_{n'}(X)$
ord $_Y F(0, Y) = d \neq \infty$	$F'(X, Y)$ is monic in Y
$F(0, Y) = \underbrace{(a + bY + \ldots)}_{\overline{G}(Y)} \underbrace{Y^d}_{\overline{H}(Y)}$	$F'(0, Y) = \underbrace{(Y^{e'} + \ldots)}_{\overline{G}'(Y)} \underbrace{(Y^{d'} + \cdots)}_{\overline{H}'(Y)}$
$0 \neq a \in k$ and $b \in k$.	$GCD\left(\overline{G}', \overline{H}'\right) = 1$.
Then there is a unique factorization	*Then there is a unique factorization*
$F(X, Y) = G(X, Y)H(X, Y)$	$F'(X, Y) = G'(X, Y)H'(X, Y)$
$G(X, Y) = a + a'X + bY + \cdots$	$G'(X, Y) = Y^{e'} + b'_1(X)Y^{e'-1} + \cdots$
$G(0, 0) = a \neq 0$ and $a' \in k$	$G'(0, Y) = \overline{G}'(Y)$
$H(X, Y) = Y^d + c_1(X)Y^{d-1} + \cdots$	$H'(X, Y) = Y^{d'} + c'_1(X)Y^{d'-1} + \cdots$
$c_1(0) = \cdots = c_d(0) = 0$.	$H'(0, Y) = \overline{H}'(Y)$.

REMARKS

(1) Note that the "unique factorization" in Weierstrass is not in the sense of unique factorization in a UFD. What it is doing is similar to the following:

$$-6 = (-1)6 \quad \text{or} \quad 4X^2 + 2X + 1 = 4\left(X^2 + \frac{1}{2}X + \frac{1}{4}\right).$$

In other words we are simply taking the "natural" unit out.

(2) A polynomial of the type

$$H(X, Y) = Y^d + c_1(X)Y^{d-1} + \cdots + c_d(X)$$
$$\text{where } c_i(X) \in k[[X]] \text{ with } c_i(0) = 0 \text{ for } 1 \leq i \leq d$$

is called a *distinguished polynomial*.

(3) As in Hensel, if $F(X, Y)$ is convergent then so are $G(X, Y)$ and $H(X, Y)$ obtained in Weierstrass. In general, if a formal problem has a *unique* solution then, say by using the Weierstrass M-test, it can usually be shown that: input convergent $\Rightarrow$ output convergent.

(4) From the Weierstrass Preparation Theorem, one can easily deduce that the power series ring $k[[X, Y]]$ is a UFD.

(5) Weierstrass Preparation Theorem continues to hold if the coefficients $a_j(X) = \sum a_{ij} X^i \in k[[X]]$ are replaced by $a_j(X_1, \ldots, X_e) \in k[[X_1, \ldots, X_e]]$ for any positive integer e. Then items (3) and (4) remain valid. In particu-

lar, the formal power series ring in any finite number of variables is a UFD, and so is the convergent power series ring.

Reference. The general reference for the Weierstrass Preparation Theorem and related matters is Abhyankar, *Local analytic geometry*, (1964) [A07]. The power of WPT is amply exhibited in this book. It also gives a quick introduction to commutative algebra.

PROOF OF WEIERSTRASS PREPARATION THEOREM. The proof is exactly analogous to that of Hensel's Lemma. First we write

$$F(X, Y) = F_0(Y) + F_1(Y)X + F_2(Y)X^2 + \cdots + F_q(Y)X^q + \cdots$$
$$\text{with } F_q(Y) \in k[[Y]].$$

(So here $F_q(Y)$ need not be polynomials as in Hensel. They are allowed to be power series instead.) Now we want to find power series

$$G(X, Y) = G_0(Y) + G_1(Y)X + \cdots + G_i(Y)X^i + \cdots, \text{ and}$$
$$H(X, Y) = \underbrace{H_0(Y)}_{Y^d} + H_1(Y)X + \cdots + H_j(Y)X^j + \cdots$$

where

$$\begin{cases} G_0(Y) \in k[[Y]] \text{ with } G_0(0) \neq 0 \text{ and } G_i(Y) \in k[[Y]] \text{ for } i > 0, \text{ and} \\ H_0(Y) = Y^d \text{ and } H_j(Y) \in k[Y] \text{ with } \deg H_j(Y) < d \text{ for } j > 0 \end{cases}$$

so that $F(X, Y) = G(X, Y)H(X, Y)$. Now $F(X, Y) = G(X, Y)H(X, Y)$ means

$$F_q(Y) = \sum_{i+j=q} G_i(Y)H_j(Y) \quad \text{for all } q \geq 0.$$

Let us make induction on q. Now $\operatorname{ord} F(0, Y) = d$. Hence, upon letting $H_0(Y) = Y^d$ there exists a unique $G_0(Y) \in k[[Y]]$ with $G_0(0) \neq 0$ such that $F_0(Y) = F(0, Y) = G_0(Y)H_0(Y)$. Thus having settled the case of $q = 0$, let $q > 0$ and suppose we have found $G_i(Y) \in k[[Y]]$ for $1 \leq i \leq q$ and $H_j(Y) \in k[Y]$ with $\deg H_j(Y) < d$ for $1 \leq j \leq q$, satisfying the above equation for all values of q smaller than the given one. Now we need to find $G_q(Y) \in k[[Y]]$ and $H_q(Y) \in k[Y]$ with $\deg H_q(Y) < d$, satisfying the equation

$$F_q(Y) = \sum_{i+j=q} G_i(Y)H_j(Y).$$

The equation can be written as

$$G_0(Y)H_q(Y) + H_0(Y)G_q(Y) = U_q(Y) = F_q(Y) - \sum_{\substack{i+j=q \\ i<q, j<q}} G_i(Y)H_j(Y) \in k[[Y]].$$

Because $G_0(Y)$ is a unit in $k[[Y]]$, we can "uniquely" write

$$U_q(Y) = G_0(Y)H_q(Y) + H_0(Y)G_q(Y),$$

where

$$H_q(Y) = \left(\text{sum of the terms of degree} < d \text{ in } \frac{U_q(Y)}{G_0(Y)} \right) \in k[Y]$$

$$\text{with } \deg H_q(Y) < d$$

and

$$G_q(Y) = \left[\frac{U_q(Y)}{G_0(Y)} - H_q(Y) \right] Y^{-d} G_0(Y) \in k[[Y]].$$

Thus by induction we have found $G_i(Y) \in k[[Y]]$ for all $i > 0$ and $H_j(Y) \in k[Y]$ with $\deg H_j(Y) < d$ for all $j > 0$ such that

$$F_q(Y) = \sum_{i+j=q} G_i(Y) H_j(Y) \quad \text{for all } q \geq 0.$$

We have proved the existence of $G(X, Y)$ and $H(X, Y)$ having the desired properties.

By another obvious induction, the above "uniquely" yields the uniqueness of $G(X, Y)$ and $H(X, Y)$.

Exercise. Prove Hensel's Lemma and Weierstrass Preparation Theorem for several variables.

REMARK. Weierstrass [**Wei**] originally proved this theorem by using contour integration. That proof could be found in the book cited previously, *Local analytic geometry* [**A07**].

Intersection multiplicity. We defined the intersection multiplicity of a curve $\widehat{G}(X, Y) = 0$ at a branch B of $\widehat{F}(X, Y) = 0$ centered at a point P of $\widehat{F}(X, Y) = 0$ by putting

$$I(B, \widehat{G}; P) = \text{ord } \widehat{G}(\lambda(T), \mu(T)),$$

where $(\lambda(T), \mu(T))$ is a representative of the branch B. Using this we get the intersection multiplicity of the curves $\widehat{F}(X, Y) = 0$ and $\widehat{G}(X, Y) = 0$ at a point P of $\widehat{F}(X, Y) = 0$ as

$$I(\widehat{F}, \widehat{G}; P) = \sum_B I(B, \widehat{G}; P),$$

where the sum is taken over all the branches B of $\widehat{F}(X, Y) = 0$ centered at P. However, this definition is not symmetric. To give a symmetric definition we bring P to the origin "0" by translation. Now by the Weierstrass Preparation Theorem (WPT) we write

$$\widehat{F}(X, Y) = \delta(X, Y) F^*(X, Y)$$
$$\widehat{G}(X, Y) = \epsilon(X, Y) G^*(X, Y)$$

where δ and ϵ are units in $k[[X, Y]]$ and F^* and G^* are distinguished (monic) polynomials in Y with coefficients in $k[[X]]$. Then we define

$$I(\widehat{F}, \widehat{G}; 0) = \text{ord }_X \text{Res }_Y(F^*, G^*),$$

where we write ord_X instead of ord to indicate that $\operatorname{Res}_Y(F^*, G^*)$ is a power series in X.

Note that the *Y-resultant* of two polynomials

$$F^* = a_0 Y^n + a_1 Y^{n-1} + \cdots + a_n,$$
$$G^* = b_0 Y^m + b_1 Y^{m-1} + \cdots + b_m$$

in Y is defined as the determinant of an $(m+n)$ by $(m+n)$ matrix given by:

$$\operatorname{Res}_Y(F^*, G^*) = \det \begin{pmatrix} a_0 & a_1 & \cdot & \cdot & \cdot & \cdot & a_n & 0 & \cdot & \cdot & \cdot & \cdot & 0 \\ 0 & a_0 & a_1 & \cdot & \cdot & \cdot & \cdot & a_n & 0 & \cdot & \cdot & \cdot & 0 \\ \cdot & \cdot & & & & & & & & & & & \cdot \\ \cdot & \cdot & & & & & & & & & & & \cdot \\ 0 & 0 & \cdot & \cdot & \cdot & a_0 & a_1 & \cdot & \cdot & \cdot & \cdot & \cdot & a_n \\ b_0 & b_1 & \cdot & \cdot & \cdot & \cdot & b_m & 0 & \cdot & \cdot & \cdot & \cdot & 0 \\ 0 & b_0 & b_1 & \cdot & \cdot & \cdot & \cdot & b_m & 0 & \cdot & \cdot & \cdot & 0 \\ \cdot & \cdot & & & & & & & & & & & \cdot \\ \cdot & \cdot & & & & & & & & & & & \cdot \\ 0 & 0 & \cdot & \cdot & \cdot & b_0 & b_1 & \cdot & \cdot & \cdot & \cdot & \cdot & b_m \end{pmatrix}.$$

The basic property of the resultant says that $\operatorname{Res}_Y(F^*, G^*) = 0 \Leftrightarrow$ either F^* and G^* have a common solution or $a_0 = 0 = b_0$.

Geometrically, the aphorism is: *Resultant is the projection of intersection.* See Figure 16.1.

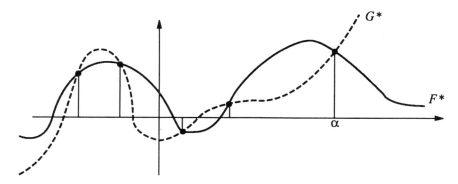

FIGURE 16.1

If a_i's and b_j's are polynomials in X, the Y-resultant $\operatorname{Res}_Y(F^*, G^*)$ is a polynomial in X and the values α of X for which $\operatorname{Res}(F^*, G^*)(\alpha) = 0$ correspond to the common points (α, β) of the curves F^* and G^* whose X-coordinate is α. Moreover, the order of $\operatorname{Res}_Y(F^*, G^*)$ at α gives the number of common points of F^* and G^* lying above $X = \alpha$, counted properly. It equals the sum of the intersection multiplicities of F^* and G^* at those points.

More generally, the coefficients a_i and b_j could even be functions of two or more variables; F^* and G^* could be surfaces or hypersurfaces. The resultant would again give the "intersection locus" of F^* and G^* measured properly.

To take care of common points at ∞, we can homogenize and then take the resultant.

Incidentally, a variation of WPT is the

WEIERSTRASS DIVISION THEOREM (WDT). *Let there be given any positive integer e and any $f(X_1, \ldots, X_e, Y)$ and $h(X_1, \ldots, X_e, Y)$ in $k[[X_1, \ldots, X_e, Y]]$. Assume $f(0, \ldots, 0, Y) \neq 0$, and $d = \operatorname{ord}_Y f(0, \ldots, 0, Y)$ (d is sometimes called the Weierstrass degree of $f(X_1, \ldots, X_e, Y)$ and denoted by* wideg f). *Then there exist unique $g \in k[[X_1, \ldots, X_e, Y]]$ and $r \in k[[X_1, \ldots, X_e]][Y]$ such that*

$$h = gf + r \quad and \quad \deg_Y r < d.$$

Now WPT and WDT can be deduced from each other. For example, let us show the

PROOF OF WDT $\Rightarrow$ WPT. Given

$$f(X_1, \ldots, X_e, Y) \in k[[X_1, \ldots, X_e, Y]]$$

such that $\operatorname{ord}_Y f(0, \ldots, 0, Y) = d \neq \infty$, by taking $h = Y^d$ in WDT we find $g \in k[[X_1, \ldots, X_e, Y]]$ and $r \in k[[X_1, \ldots, X_e]][Y]$ such that

$$Y^d - r = gf \quad and \quad \deg_Y r < d.$$

After substituting $X_1 = \cdots = X_e = 0$ in both sides of the equation we get

$$\operatorname{ord}_Y \text{LHS} = \operatorname{ord}_Y \left[Y^d - r(0, \ldots, 0, Y) \right] \leq d = \operatorname{ord}_Y f(0, \ldots, 0, Y),$$

but

$$\operatorname{ord}_Y \text{RHS} = \operatorname{ord}_Y g(0, \ldots, 0, Y) + \operatorname{ord}_Y f(0, \ldots, 0, Y).$$

Hence,

$$\operatorname{ord}_Y g(0, \ldots, 0, Y) = 0 \quad and \quad \operatorname{ord}_Y \left[Y^d - r(0, \ldots, 0, Y) \right] = d.$$

Therefore,

$$g(0, \ldots, 0, 0) \neq 0 \quad and \quad r(0, \ldots, 0, Y) = 0.$$

Dividing the equation $Y^d - r = gf$ by the unit g we obtain $f = g'h'$, where $g' = g^{-1} = a$ unit in $k[[X_1, \ldots, X_e, Y]]$, and $h' = Y^d - r = a$ distinguished polynomial in Y with coefficients in $k[[X_1, \ldots, X_e]]$. This proves the existence part of WPT. Uniqueness follows readily.

Exercise. Show that WPT $\Rightarrow$ WDT.

NOTE. Karl Weierstrass, the prince of analysis, was a great algebraist. Indeed, his preparation theorem prepares us so well!

Intersection Multiplicity

Having proved the Weierstrass Preparation Theorem (WPT), we defined the intersection multiplicity of plane curves $C : F(X, Y) = 0$ and $D : G(X, Y) = 0$ over a field k as follows: By WPT write

$$F(X, Y) = \delta(X, Y)F^*(X, Y) \quad \text{with } \delta \text{ a unit and } F^* \text{ distinguished in } Y,$$

$$G(X, Y) = \epsilon(X, Y)G^*(X, Y) \quad \text{with } \epsilon \text{ a unit and } G^* \text{ distinguished in } Y,$$

and then define the intersection multiplicity of C and D at the origin "0" by the equation

$$I(C, D; 0) = I(F, G; 0) = \text{ord}_X \text{Res}_Y(F^*, G^*).$$

In case k is algebraically closed, our previous definition was

$$I(F, G; 0) = \sum_{\substack{\text{branches } B \text{ of } C \text{ at } 0}} I(B, G; 0) = \sum_{\substack{\text{discrete valuations} \\ V \text{ of } k(C) \\ \text{centered at } 0 \text{ on } C}} V(G(x, y))$$

where, in the second equation, C is assumed irreducible, and x and y are the functions induced by X and Y on C. The said second equation follows from the bijection $B \mapsto \text{ord}_B$. The first part of the definition can be written as a double sum as well:

$$I(F, G; 0) = \sum_{\substack{\text{branches } B \text{ of } C \text{ at } 0}} I(B, G; 0) = \sum_{\substack{\text{branches } B \text{ of } C \text{ at } 0 \\ \text{branches } \Gamma \text{ of } D \text{ at } 0}} I(B, \Gamma; 0).$$

However we haven't yet defined $I(B, \Gamma; 0)$; we will do that in a moment.

Assuming C to be irreducible and k to be algebraically closed, given any point P of C, we have the bijections: branches of C at $P \leftrightarrow$ valuations of $k(C)/k$ centered at $P \leftrightarrow$ places of C at P. As indicated, the first bijection is given by $B \mapsto \text{ord}_B$. The second bijection has been hinted at previously. Here, a "place" of C at P means an irreducible power series factor of F at P.

More precisely, in case P is the origin "0", but without any assumptions on F or k, we write

$$F^*(X, Y) = \prod_{i=1}^{h} \phi_i(X, Y)^{e_i},$$

where $\phi_1, \ldots, \phi_h$ are "distinct" irreducible factors of F^* in $k[[X, Y]]$. By "distinct" we mean that the ideals in $k[[X, Y]]$ generated by $\phi_1, \ldots, \phi_h$ are pairwise distinct principal prime ideals. The irreducible factors $\phi_1, \ldots, \phi_h$ (if you prefer, the ideals $\phi_1 k[[X, Y]], \ldots, \phi_h k[[X, Y]]$) are called the *irreducible analytic components* of C at P. To each of them we associate a *place* of C at P. In the general case, when P is any point of C, at finite distance or at ∞, we "bring" P to the origin and then use the description to define *places* of C at P.

Analogous to the case of a polynomial, given any nonzero nonunit $\phi(X, Y) \in k[[X, Y]]$, by a *parametrization* of the "analytic curve" Λ : $\phi(X, Y) = 0$ *at the origin* we mean a pair $(\lambda(T), \mu(T))$ of elements in $k[[T]]$ with $\lambda(0) = \mu(0) = 0$ such that $\phi(\lambda(T), \mu(T)) = 0$. By a *branch* of Λ *at the origin* we mean an equivalence class of irredundant parametrizations of Λ, where irredundancy and equivalence have the same meaning as in the case of algebraic curves. It can be shown that if ϕ is irreducible in $k[[X, Y]]$, then it has exactly one branch at the origin. Now the branches of the irreducible factors $\phi_1, \ldots, \phi_h$ of F in $k[[X, Y]]$ are precisely the branches of C centered at the origin. (It can easily be seen that if F is irreducible in $k[X, Y]$ then $e_1 = e_2 = \cdots = e_h = 1$.)

We considered branches of ϕ only at the origin because in a given power series, for the variables we can substitute only such power series which are devoid of constant terms. Thus, given any power series

$$f(X_1, \ldots, X_m) = \sum a_{i_1 \cdots i_m} X_1^{i_1} \cdots X_m^{i_m} \in k[[X_1, \ldots, X_m]]$$

$$\text{with } a_{i_1 \cdots i_m} \in k,$$

we can make a substitution

$$X_i = \theta_i(Y_1, \ldots, Y_q) \in k[[Y_1, \ldots, Y_q]]$$

$$\text{with } \theta_i(0, \ldots, 0) = 0 \quad \text{for} \quad 1 \leq i \leq m$$

to get

$$f(\theta_1(Y_1, \ldots, Y_q), \ldots, \theta_m(Y_1, \ldots, Y_q))$$
$$= \sum a_{i_1 \cdots i_m} \theta_1(Y_1, \ldots, Y_q)^{i_1} \cdots \theta_m(Y_1, \ldots, Y_q)^{i_m} \in k[[Y_1, \ldots, Y_q]].$$

This makes sense because no infinite summations of constants are involved.

REMARK. The analysis ability to substitute constants in a power series is lost in algebra. Instead, one has gained the ability to substitute other power series, provided they do not have constant terms.

Given any other nonzero nonunit $\psi(X, Y) \in k[[X, Y]]$, by WPT we write

$$\phi(X, Y) = \delta^*(X, Y)\phi^*(X, Y) \quad \text{with } \delta^* \text{ a unit and } \phi^* \text{ distinguished in } Y,$$

$$\psi(X, Y) = \epsilon^*(X, Y)\psi^*(X, Y) \quad \text{with } \epsilon^* \text{ a unit and } \psi^* \text{ distinguished in } Y.$$

We define the *intersection multiplicity* of ϕ and ψ at the origin "0" by the equation

$$I(\phi, \psi; 0) = \text{ord}_X \text{Res}_Y(\phi^*, \psi^*).$$

(Here we have implicitly assumed that a suitable rotation of coordinates has been performed to make sure that $\phi(0, Y) \neq 0 \neq \psi(0, Y)$; the same applies to the case of algebraic curves F and G.) This definition of intersection multiplicity applies in particular to the pairs ϕ_i and ψ_j where

$$G(X, Y) = \prod_{j=1}^{h'} \psi_j(X, Y)^{e'_j}$$

is a factorization of G into "distinct" irreducible factors in $k[[X, Y]]$. Thus we have fulfilled the promise of defining $I(B, \Gamma; 0)$. Namely, because B and Γ are the branches of C and D at the origin, they correspond to ϕ_i and ψ_j for some i and j.

In these definitions of intersection multiplicity, we have not given equal treatment to X and Y. So we may ask whether

$$\text{ord}_X \text{Res}_Y(\phi^*, \psi^*) = \text{ord}_Y \text{Res}_X(\phi', \psi')$$

where by WPT we write

$\phi(X, Y) = \delta'(X, Y)\phi'(X, Y)$ with δ' a unit and ϕ' distinguished in X,

$\psi(X, Y) = \epsilon'(X, Y)\psi'(X, Y)$ with ϵ' a unit and ψ' distinguished in X.

More generally, we can ask whether the above definition of intersection multiplicity remains unchanged if we make an analytic coordinate change. The answer is yes; it does remain unchanged because of the invariantive characterization according to which

$$I(F, G; 0) = [k((X, Y)) : k((F, G))].$$

The equation was indeed the starting point in Chevalley's pioneering paper, "Intersections of algebraic and algebroid varieties," Trans. Amer. Math. Soc., (1945), [Ch3]. To lay the groundwork for this paper, Chevalley wrote two preparatory papers, "Theory of local rings," Annals of Math., (1943), [Ch1] and "Ideals in rings of power series," Trans. Amer. Math. Soc., (1944), [Ch2]. Out of this, the local rings paper turned out to be chronologically the second of the trilogy of three fundamental papers on local rings. The other two are Krull, "Dimension theory of local rings," *Crelle* Journal, (1938), [Kr1] and Cohen, "Structure and ideal theory of complete local rings," Trans. Amer. Math. Soc., (1946), [Coh]. Unlike other authors, Cohen exhaustively discusses the unequicharacteristic case in which a local ring of zero characteristic has residue field of nonzero characteristic.

Yet another definition of intersection multiplicity. For nonzero nonunits ϕ and ψ in $R = k[[X, Y]]$, it can be shown that

$$I(\phi, \psi; 0) = [k((X, Y)) : k((\phi, \psi))] = \text{length}_R(\phi, \psi)R,$$

where the *R-length* of the ideal $(\phi, \psi)R$ is the maximum of the lengths of chains of ideals between $(\phi, \psi)R$ and R, i.e., the largest integer e such that there exist distinct ideals $J_1, J_2, \ldots, J_e$ in R with

$$(\phi, \psi)R = J_0 \subset J_1 \subset J_2 \subset \cdots \subset J_e = R.$$

Here, as well as previously, we are implicitly assuming that ϕ and ψ have no nonunit common factor. Equivalently, $(\phi, \psi)R$ is primary for the maximal ideal $M(R)$.

Hypersurfaces in m-space. More generally, for any positive integer m, let $R = k[[X_1, \ldots, X_m]]$ and let $f_1, \ldots, f_m$ be nonzero nonunits in R such that the origin "0" is an isolated point of intersection of the analytic hypersurfaces $f_1 = 0, \ldots, f_m = 0$, such that the ideal $(f_1, \ldots, f_m)R$ is primary for the maximal ideal $M(R)$. Now the *intersection multiplicity* of the hypersurfaces $f_1, \ldots, f_m$ at the origin is defined by

$$I(f_1, \ldots, f_m; 0) = \mathrm{mult}_R(f_1, \ldots, f_m)R = \mathrm{length}_R(f_1, \ldots, f_m)R,$$

where the length of any ideal in any ring is defined as before. This definition of the multiplicity mult_R, of the primary ideal $(f_1, \ldots, f_m)R$, as its length, is a "correct" definition for primary ideals generated by the right number of elements, that is, by as many elements as the dimension of the local ring R. For any ideal J in R that is primary for $M(R)$, without any condition on the number of generators of J, the *multiplicity* of J in R is defined by the equation $\mathrm{mult}_R J = e$, where the positive integer e is obtained by noting that, for the *Hilbert function* $\mathscr{H}_{(R,J)}(d)$ of (R, J) defined by

$$\mathscr{H}_{(R,J)}(d) = \mathrm{length}_R J^d,$$

there exists a unique polynomial $\mathscr{H}^*_{(R,J)}(d)$, called the *Hilbert polynomial* of (R, J), such that

$$\mathscr{H}_{(R,J)}(d) = \mathscr{H}^*_{(R,J)}(d) \quad \text{for large } d,$$

and for this polynomial we have

$$\mathscr{H}^*_{(R,J)}(d) = \frac{ed^m}{m!} + \bullet d^{m-1} + \cdots + \bullet d + \bullet.$$

Note that for the power series ring $R = k[[X_1, \ldots, X_m]]$ we have $\dim R = m$ and $M(R) = (X_1, \ldots, X_m)R$. Hence R is *regular* in the following sense.

Regular local rings. For any local ring R we define

emdim R = the *embedding dimension* of R

= the smallest number of elements that generate $M(R)$.

Note that we always have

$$\mathrm{emdim}\, R \geq \dim R,$$

and by definition,

$$R \text{ is } regular \Leftrightarrow \mathrm{emdim}\, R = \dim R.$$

Now the geometric significance of regular local rings is due to the fact that a point of an irreducible algebraic variety is (say, by definition) a *simple point* iff its local ring is regular. (A point of a variety, which need not be irreducible, is (again, by definition) a *simple point* iff it lies only on one irreducible component of the variety and is a simple point of that irreducible component. A point is *singular* iff it is not simple.) At any rate, whether the point is simple or not, the dimension of the said local ring coincides with the dimension of the variety. Moreover, the embedding dimension of the said local ring is the smallest dimension of a "local space" in which a "neighborhood" of the point on the variety can be embedded. More generally, let us consider the local ring of an irreducible subvariety of a given irreducible variety. Now the dimension of this local ring equals the codimension of the subvariety, i.e., the excess of the dimension of the ambient variety over the dimension of the subvariety. Finally, the said local ring is regular iff the subvariety is not contained in the singular locus of the ambient variety. That is, it doesn't consist entirely of singular points of the ambient variety. To say that a variety is *nonsingular* means that it has no singular points.

Cohen-Macaulay rings. As we said, the local ring of a simple point on an irreducible algebraic variety is a regular local ring. Generalizing this fact, if the variety is a *complete intersection*, (if it is defined by as many equations as its codimension in the ambient space), then the local ring of any point on the variety is Cohen-Macaulay. A local ring R of dimension m is said to be *Cohen-Macaulay* if there exists a *regular sequence* $g_1, \ldots, g_m$ in R, that is, a sequence of elements $g_1, \ldots, g_m$ in $M(R)$ such that g_i is a nonzerodivisor in $R/(g_1, \ldots, g_{i-1})R$ for $1 \leq i \leq m$. The above definition of the multiplicity $\text{mult}_R J$ remains valid for a primary ideal J in an arbitrary local ring R of dimension m. Moreover, if R is Cohen-Macaulay and J is generated by a regular sequence in R, then $\text{mult}_R J = \text{length}_R J$.

NOTE. Like so much of modern ideal theory, the ideas of regular sequences, complete intersections, multivariate resultants, and so on, originated in the small (112 pages!) but highly charged "Cambridge Tract" of F. S. Macaulay, published in 1916 [**Mac**]. The reading of this tract, although very rewarding, is no joke. Once, as a public service, I gave an expository lecture [**A23**] in which it took me sixteen pages to explain what Macaulay says in less than one page. Namely, there is no higher limit to the number of generators required to generate the prime ideal of an irreducible space curve. See page 36 of [**Mac**]. In Wolfgang Krull's encyclopedia article [**Kr2**] you can see the very high praise accorded to the importance of Macaulay's ideas by another master ideal-theorist. The encyclopedia article of Krull was recommended to me by Irving Cohen, another master craftsman of modern commutative algebra. In oral conversations, my own teacher, Oscar Zariski, who did so much to harness the power of algebra to the service of geometry, always spoke very highly of Macaulay's work. Indeed, Zariski studied Macaulay's tract so thoroughly that I saw Zariski's personal copy separated into worn-out individual pages. In his own time, Macaulay's work was well recognized in Europe. It remained relatively unknown in England. Emmy Noether had difficulty locating him on her visit to England. Fortunately, in our own time, even a symbolic algebra package has been named after Macaulay.

LECTURE 18

Resolution of Singularities of Plane Curves

Let $C : F(X, Y) = 0$ be an algebraic plane curve over a field k. Then around any point $P = (\alpha, \beta)$ of C we can write

$$F(X, Y) = \sum b_{ij}(X - \alpha)^i (Y - \beta)^j \quad \text{with } b_{ij} \in k.$$

We recall that then

$$\text{mult}_P C = \text{ord}_P F = \min \ i + j \quad \text{such that } b_{ij} \neq 0.$$

Also recall that if $\text{mult}_P C = d$ then P is called a d-fold point or a point of multiplicity d. Moreover, P is a simple point or a singular point according as $\text{mult}_P C = 1$ or $\text{mult}_P C > 1$.

Equivalently, we can characterize the multiplicity as

$$\text{mult}_P C = \min \ i + j \quad \text{such that } \left. \frac{\partial^{i+j} F}{\partial X^i \partial Y^j} \right|_{(\alpha, \beta)} \neq 0.$$

Note that, in case of char $k \neq 0$, this characterization is true only if $\text{mult}_P C < \text{char } k$. However, the singular locus is always given by $F = F_X = F_Y = 0$. It consists of points at which the Implicit Function Theorem cannot be applied.

Suppose curve C has a certain number of singular points $P, P', \ldots, P''$. See Figure 18.1.

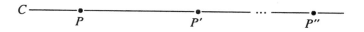

FIGURE 18.1

Then by *blowing-up* we can obtain *singularity trees* that look like that shown in Figure 18.2 on page 132.

By blowing-up we mean applying *quadratic transformations* with centers at $P, P', \ldots, P''$, and so on.

The quadratic transformation (QDT) with *origin as center* is the transformation of the (X, Y)-plane onto the (X', Y')-plane defined by $X' = X$ and $Y' = Y/X$ or inversely defined by $X = X'$ and $Y = X'Y'$. See Figure

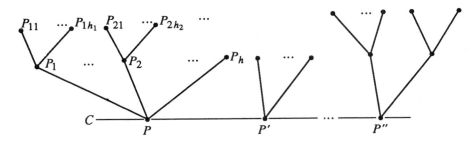

FIGURE 18.2

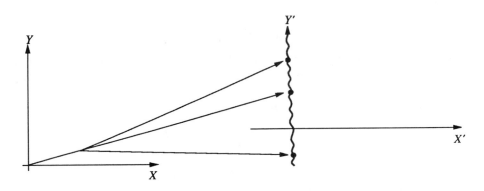

FIGURE 18.3

18.3. Under this transformation, the origin is mapped onto the entire Y'-axis, i.e., the line $X' = 0$, which we call the *exceptional line*. To see this, we can put $X' = 0$ in the reverse equations $X = X'$ and $Y = X'Y'$ to get $X = 0$ and $Y = 0$. Alternatively, at the origin in the (X, Y)-plane we have the indeterminate form $Y' = \frac{0}{0}$. By approaching it along the line of slope m, (along the line $Y = mX$), we get

$$Y' = \frac{Y}{X} = \frac{mX}{X} = m \quad \text{for all } X \neq 0, \quad \text{and hence} \quad \lim_{\substack{X \to 0 \\ \text{along } Y = mX}} Y' = m.$$

Thus the points on the Y'-axis correspond to lines through the origin in the (X, Y)-plane. In this correspondence, the point whose Y'-coordinate is m corresponds to the line whose slope is m. So we may actually think of the Y'-axis as the "m-axis" or the "slope-axis." At any rate, the line $Y = mX$ is transformed into $Y - mX = X'Y' - mX' = X'(Y' - m) = 0$, i.e., into the exceptional line $X' = 0$ together with the line $Y' = m$ that is parallel to the X'-axis and that meets the exceptional line in the point whose Y'-coordinate equals the slope of the original line. The line $Y' = m$ is called the *proper transform* of the line $Y = mX$.

More generally, to see the effect of a quadratic transformation on the curve C we proceed thus. Given a point P of C, by a translation we can bring it to the origin. Suppose we have already done this, i.e., assume that $P = (0, 0)$ and $F(0, 0) = 0$. So let

$$d = \text{mult}_P C \quad \text{with } P = (0, 0)$$

and write

$$F(X, Y) = F_d(X, Y) + F_{d+1}(X, Y) + \cdots + F_j(X, Y) + \cdots$$

where $F_j(X, Y)$ is a homogeneous polynomial of degree j. The QDT transforms the curve $F(X, Y)$ into the curve

$$
\begin{aligned}
F\left(X', X'Y'\right) \\
&= F_d\left(X', X'Y'\right) + F_{d+1}\left(X', X'Y'\right) + \cdots + F_j(X', X'Y') + \cdots \\
&= X'^d F_d\left(1, Y'\right) + X'^{d+1} F_{d+1}\left(1, Y'\right) + \cdots + X'^j F_j(1, Y') + \cdots \\
&= X'^d \left[F_d\left(1, Y'\right) + X' F_{d+1}\left(1, Y'\right) + \cdots + X'^{j-d} F_j(1, Y') + \cdots \right] \\
&= X'^d F'(X', Y')
\end{aligned}
$$

where

$$F'\left(X', Y'\right) = F_d\left(1, Y'\right) + X' F_{d+1}\left(1, Y'\right) + \cdots + X'^{j-d} F_j(1, Y') + \cdots .$$

The curve $F(X', X'Y') = 0$ is called the *total transform* of C. It factors into the curve $C' : F'(X', Y') = 0$, which is called the *proper transform* of C, and the exceptional line $E : X' = 0$. The exceptional line meets the proper transform in a certain number of points which are the common solutions of $X' = 0$ and $F_d(1, Y') = 0$. To find these, first by factoring $F_d(X, Y)$ we get

$$F_d(X, Y) = m^* \prod_{i=1}^{h} (Y - m_i X)^{e_i}$$

$$\text{with } e_i > 0 \quad \text{and} \quad m^* \neq 0 \quad \text{and} \quad m_i \neq m_j \quad \text{for } i \neq j.$$

Now the lines $Y = m_1 X, \ \ldots, \ Y = m_h X$ are the *tangent lines* to C at the origin, with "tangential multiplicities" $e_1, \ldots, e_h$. The above factorization of $F_d(X, Y)$ gives the factorization

$$F_d\left(1, Y'\right) = m^* \prod_{i=1}^{h} \left(Y' - m_i\right)^{e_i},$$

and so E meets C' at the points $P_i = (0, m_i)$ for $1 \leq i \leq h$. For the intersection multiplicity of E with C' at P_i we have $I(E, C'; P_i) = e_i \geq d_i$ where $d_i = \text{mult}_{P_i} C'$. See Figure 18.4 on page 134.

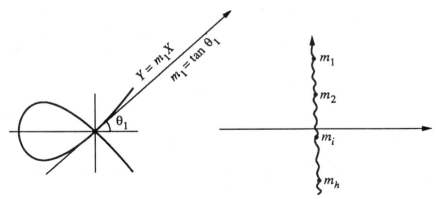

FIGURE 18.4

To explain the relation between e_i and d_i more clearly, we note that $X' = X$. We translate the point P_i to the origin by putting $Y^{(i)} = Y' - m_i$ and $F^{(i)}(X, Y^{(i)}) = F'(X, Y^{(i)} + m_i)$, so that in the $(X, Y^{(i)})$-plane, the point P_i is at the origin $(0, 0)$, and the curve C' is given by $F^{(i)}(X, Y^{(i)}) = 0$. Now

$$F^{(i)}(X, Y^{(i)}) = F_d(1, Y^{(i)} + m_i)$$
$$+ XF_{d+1}(1, Y^{(i)} + m_i) + \cdots + X^{j-d}F_j(1, Y^{(i)} + m_i) + \cdots.$$

Hence,

$$F^{(i)}(0, Y^{(i)}) = F_d(1, Y^{(i)} + m_i).$$

Clearly,

$$\operatorname{ord} F_d(1, Y^{(i)} + m_i) = e_i.$$

Therefore,

$$I(E, C'; P_i) = \operatorname{ord} F^{(i)}(0, Y^{(i)}) = e_i.$$

Obviously,

$$d_i = \operatorname{mult}_{P_i} C' = \operatorname{ord} F^{(i)}(X, Y^{(i)}) \le \operatorname{ord} F^{(i)}(0, Y^{(i)}) = I(E, C'; P_i).$$

Hence,

$$\operatorname{mult}_P C' = d_i \le e_i = I(E, C'; P_i).$$

We have tacitly assumed that $X = 0$ is not a tangent line of C at P. This can of course be arranged by a rotation. Moreover, we have also assumed that $F_d(X, Y)$ factors into linear factors, which would certainly be so in case k is algebraically closed. In the general case, we can factor $F_d(X, Y)$ into powers of distinct irreducible homogeneous factors whose degrees need not be 1, and so on. For a detailed working out in such a general case, see the Desingularization Paper [A29].

Our task is to show that after a finite number of quadratic transformations, the multiplicity is everywhere 1. Now since $d_i \le e_i$ for $1 \le i \le h$, and clearly

$e_1 + \cdots + e_h = d$, we see that

$$d_1 + d_2 + \cdots + d_h \leq e_1 + \cdots + e_h = d.$$

It follows that

$$d_i \leq d \quad \text{for } 1 \leq i \leq h$$

and

$$d_i = d \quad \text{for some } i \Leftrightarrow h = 1 \quad \text{and} \quad d_1 = e_1 = d$$

$\Leftrightarrow$ the exceptional line E meets proper transform C' in exactly one point P_1, and E is not tangent to C' at P_1.

Thus, if after one QDT there is no drop in multiplicity, then the exceptional line $X = 0$ meets the proper transform C' only in one point $P_1 = (0, m_1)$. At that point $\text{mult}_{P_1} C' = d = \text{mult}_P C$ and the line $X = 0$ is not tangent to C' at P_1.

Assuming there is no drop in multiplicity after one QDT, let us write C_1, Q_1, n_1 for C', P_1, m_1 respectively, and let us put $Y_1 = \frac{Y}{X} - n_1$ to bring Q_1 to the origin in the (X, Y_1)-plane. Since the line $X = 0$ is not tangent to C_1 at Q_1, we can apply a QDT to the (X, Y_1)-plane with center at the origin and obtain the proper transform C_2 of C_1 in the $(X, \frac{Y_1}{X})$-plane. If again there is no drop in multiplicity as we go from C_1 to C_2, then by repeating the previous argument we see that the "new" exceptional line $X = 0$ meets C_2 only in one point $Q_2 = (0, n_2)$. At that point $\text{mult}_{Q_2} C_2 = d$ and the line $X = 0$ is not tangent to C_2 at Q_2. Assuming there is no drop in multiplicity even after the second QDT, let us put $Y_2 = \frac{Y_1}{X} - n_2$ to bring Q_2 to the origin in the (X, Y_2)-plane. Since the line $X = 0$ is not tangent to C_2 at Q_2, we can apply a QDT to the (X, Y_2)-plane and obtain the proper transform C_3 of C_2 in the $(X, \frac{Y_2}{X})$-plane. Again, if there is no drop in multiplicity as we go from C_2 to C_3, then by the previous argument we see that the "newer" exceptional line $X = 0$ meets C_3 only in one point $Q_3 = (0, n_3)$. At that point $\text{mult}_{Q_3} C_3 = d$ and the line $X = 0$ is not tangent to C_3 at Q_3, and so on.

Now this process either stops after a finite number of steps, say after s steps, or it repeats infinitely often, in which case we take $s = \infty$. Thus we have found a sequence $(n_i, Y_i, C_i, Q_i)_{0 < i < s}$, where s is either a positive integer or ∞. Upon letting $C_0 = C$ and $Y_0 = Y$, for $0 < i < s$ we have $n_i \in k$, $Y_i = \frac{Y_{i-1}}{X} - n_i$. Here, C_i is a curve in the $(X, \frac{Y_{i-1}}{X})$-plane, which is the proper transform of C_{i-1} obtained by applying a QDT to the (X, Y_{i-1})-plane with center at the origin. The exceptional line $X = 0$ meets C_i only at the point $Q_i = (0, n_i)$, $\text{mult}_{Q_i} C_i = d$, and the line $X = 0$ is not tangent to C_i at Q_i. In case $s \neq \infty$, upon letting $R_1, \ldots, R_l$ to be the points where the "last" exceptional line meets the proper transform C_s of C_{s-1}

in the $(X, \frac{Y_{s-1}}{X})$-plane obtained by applying a QDT to the (X, Y_{s-1})-plane with center at the origin, we have $\text{mult}_{R_j} C_s < d$ for $1 \leq j \leq l$.

By making the "analytic coordinate change"

$$Y^* = Y - n_1 X - n_2 X^2 - \cdots - n_i X^i - \cdots ,$$

(where the expansion is infinite if $s = \infty$ and stops at $n_{s-1} X^{s-1}$ if $s < \infty$), we get a "good" coordinate system (X, Y^*) for the (X, Y)-plane. In this new coordinate system, C is given by $F^*(X, Y^*) = 0$ where

$$F^*(X, Y^*) = F(X, Y) = F(X, Y^* + n_1 X + n_2 X^2 + \cdots + n_i X^i + \cdots).$$

Upon writing

$$F^*(X, Y^*) = b_0^*(X, Y^*)Y^{*d} + b_1^*(X)Y^{*d-1} + \cdots + b_j^*(X)Y^{*d-j} + \cdots + b_d^*(X)$$

with

$$b_0^*(X, Y^*) \in k[[X, Y]] \quad \text{and} \quad b_j^*(X) \in k[[X]] \quad \text{for } 1 \leq j \leq d$$

we can easily see that

$$b_0^*(0, 0) \neq 0.$$

Upon letting

$$e = \min_{1 \leq j \leq d} \frac{\text{ord } b_j^*(X)}{j}$$

we can easily see that

$$s = \text{ the integral part of } e.$$

(If $e < \infty$ then s is the largest integer which is $\leq e$, and if $e = \infty$ then $s = \infty$). Note that for

$$\delta^*(X, Y) = b_0^*(X, Y - n_1 X - n_2 X^2 - \cdots)$$

we have $\delta^*(X, Y) \in k[[X, Y]]$ with $\delta^*(0, 0) \neq 0$.

The conclusions reached so far can be stated as a

PROPOSITION. *If $s \neq \infty$ then after s QDTs we have a reduction in multiplicity. On the other hand, if $s = \infty$, i.e., if there is an infinite QDT sequence of d-fold points emanating from a d-fold point P of C (which we may take to be the origin), then $F(X, Y) = \delta^*(X, Y)Y^{*d}$, where $\delta^*(X, Y) \in k[[X, Y]]$ with $\delta^*(0, 0) \neq 0$, and $Y^* \in k[[X, Y]]$ with $\text{ord } Y^* = 1$, i.e., the curve C has a d-fold component through P, which means that C has the analytic decomposition $C = \Delta \cup \Gamma$ where $\Delta: \delta^*(X, Y) = 0$ is an analytic curve not passing through P, and $\Gamma: Y^* = 0$ is an analytic curve having a simple point at P.*

Let us take note of the following

Fact about multiple components. If the algebraic plane curve C has no multiple components as an algebraic curve then it has no multiple components as an analytic curve. In other words, if the polynomial $F(X, Y)$ is not divisible by the square of any nonconstant polynomial, then it is not divisible by the square of any nonunit power series.

Now an algebraic plane curve C which is free of multiple components has only finitely many singularities. Therefore, by augmenting the discussion by an obvious induction, we obtain a proof of the

THEOREM OF RESOLUTION OF SINGULARITIES OF PLANE CURVES. *The singularities of any algebraic plane curve devoid of multiple components can be resolved by a finite number of quadratic transformations emanating from the various singular points of the curve.*

The above discussion applies to the case of an analytic plane curve, i.e., to the case when $F(X, Y) \in k[[X, Y]]$. Hence, we get the

THEOREM OF RESOLUTION OF SINGULARITIES OF ANALYTIC PLANE CURVES. *The singularities of any analytic plane curve devoid of (analytic) multiple components can be resolved by a finite number of quadratic transformations emanating from the origin.*

In the resolution theorems, we only require that the eventual proper transform of the curve be nonsingular. A careful analysis of the proof yields the following

TOTAL RESOLUTION FOR PLANE CURVES. *By applying a finite number of quadratic transformations to any given (algebraic or analytic) plane curve, it can be arranged that its total transform has only normal crossings, i.e., at every point $\widehat{P}$, the equation of the curve has the form $\hat{\delta}(\widehat{X}, \widehat{Y})\widehat{X}^a\widehat{Y}^b$ where $(\widehat{X}, \widehat{Y})$ is a "local" coordinate system with respect to which $\widehat{P}$ is at the origin, a and b are non-negative integers, and $\hat{\delta}(\widehat{X}, \widehat{Y}) = 0$ is an analytic curve not passing through $\widehat{P}$.*

Taylor expansion by resolution. If $d = 1$, i.e., if P is a simple point of C, then the above proposition gives a way of "solving" the equation $F(X, Y) = 0$. The solution is

$$Y = n_1 X + n_2 X^2 + \cdots + n_i X^i + \cdots .$$

Equivalently, we get the local parametrization

$$\begin{cases} X = T \\ Y = n_1 T + n_2 T^2 + \cdots \end{cases}$$

giving the (unique) branch of C at $P = (0, 0)$. In the general case, i.e., when $d > 1$, we can first resolve the singularity at P by a finite number of quadratic transformations and then follow it up by the above procedure to get all the branches of C at P. Contrary to Newton's Theorem, this procedure of finding the branches of C works in any characteristic.

Can't do better. If $s = \infty$ and $d > 1$, then by the proposition, $F(X, Y) = \delta^*(X, Y)Y^{*d}$. We can keep making transformations of the type $(X, Y) \to$

$(X, \frac{Y^*}{X}) \to (X, \frac{Y^*}{X^2}) \to (X, \frac{Y^*}{X^3}) \to \cdots$, and the multiplicity of the resulting points (which are all the time at the origin) never drops. Clearly in this case we should regard C as "resolved" at P since there isn't much that we can do here.

Alternative way of finding the number of steps. The number of steps s is sometimes called the *integral magnitude* of C at P, and the (possibly infinite) rational number e is then called the *magnitude* of C at P. As an alternate method of finding e (and hence s), by Weierstrass Preparation Theorem we can write $F(X, Y) = \delta(X, Y)\tilde{F}(X, Y)$ where $\delta(X, Y)$ is a unit in $k[[X, Y]]$ and $\tilde{F}(X, Y) = Y^d + b_1(X)Y^{d-1} + \cdots + b_j(X)Y^{d-j} + \cdots + b_d(X)$ is a distinguished polynomial of degree d in Y, i.e., $b_j(X) \in k[[X]]$ with $b_j(0) = 0$. If the characteristic of k does not divide d, by Shreedharacharya's method of completing the dth power we can write

$$\tilde{F}(X, Y) = \tilde{Y}^d + \tilde{b}_1(X)\tilde{Y}^{d-1} + \cdots + \tilde{b}_j(X)\tilde{Y}^{d-j} + \cdots + \tilde{b}_d(X),$$

where

$$\tilde{Y} = Y + \frac{b_1(X)}{d} \quad \text{and} \quad \tilde{b}_j(X) \in k[[X]] \text{ for } 1 \leq j \leq d \quad \text{and} \quad \tilde{b}_1(X) = 0.$$

It turns out that

$$e = \min_{1 \leq j \leq d} \frac{\operatorname{ord} \tilde{b}_j(X)}{j}.$$

For further details see the desingularization paper [A29], where no assumption is made on the characteristic of k. Also note the analogy of e with the rational number d occurring in the proof of Newton's Theorem.

Analytic irreducibility. In the previous notation, without assuming d to be nondivisible by the characteristic of k, we clearly have

$$b_j(X) = X^j b_j'(X) \quad \text{with } b_j'(X) \in k[[X]] \quad \text{for } 1 \leq j \leq d$$

and

$$\prod_{i=1}^{h} (Y' - m_i)^{e_i} = (1/m^*)F_d(1, Y') = Y'^d + \sum_{j=1}^{d} b_j'(0)Y'^{d-j}.$$

Hence,

$$F'(X', Y') = \delta'(X', Y')\tilde{F}'(X', Y')$$

where

$$\delta'(X', Y') = \delta(X', X'Y') \in k[[X', Y']] \quad \text{with } \delta(0, 0) \neq 0$$

and

$$\tilde{F}'(X', Y') = \tilde{Y}'^d + \sum_{j=1}^{d} b_j'(X)\tilde{Y}'^{d-j}.$$

By Hensel's Lemma it follows that if $h > 1$ then

$$\tilde{F}'(X', Y') = \hat{G}'(X', Y')\hat{H}'(X', Y'),$$

where

$$\widehat{G}'(X',\,Y') = Y'^{e_1} + \sum_{j=1}^{e_1} \hat{b}_j(X')Y'^{e_1-j} \quad \text{with } \hat{b}_j(X') \in k[[X']]$$

and

$$\widehat{H}'(X',\,Y') = Y'^{e_2+\cdots+e_h} + \sum_{j=1}^{e_2+\cdots+e_h} \hat{c}_j(X')Y'^{e_2+\cdots+e_h-j} \quad \text{with } \hat{c}_j(X') \in k[[X']].$$

Hence,

$$\widetilde{F}(X,\,Y) = \widehat{G}(X,\,Y)\widehat{H}(X,\,Y),$$

where

$$\widehat{G}(X,\,Y) = Y^{e_1} + \sum_{j=1}^{e_1} \hat{b}_j(X)X^j Y^{e_1-j}$$

and

$$\widehat{H}(X,\,Y) = Y^{e_2+\cdots+e_h} + \sum_{j=1}^{e_2+\cdots+e_h} \hat{c}_j(X)X^j Y^{e_2+\cdots+e_h-j}.$$

Thus, if C is analytically irreducible at P, (if $F(X,\,Y)$ is irreducible in $k[[X,\,Y]]$), then C has only one tangent line at P, i.e., we have $h = 1$. Assuming that $h = 1$, by making the linear transformation $Y \to Y - m_1 X$, we can arrange matters so that $m_1 = 0$. Now if

$$\widetilde{F}'(X',\,Y') = \widehat{G}^*(X',\,Y')\widehat{H}^*(X',\,Y'),$$

where

$$\widehat{G}^*(X',\,Y') = Y'^{\rho} + \sum_{j=1}^{\rho} \hat{b}_j^*(X')Y'^{\rho-j} \quad \text{with } \hat{b}_j^*(X') \in k[[X']]$$

and

$$\widehat{H}^*(X',\,Y') = Y'^{\sigma} + \sum_{j=1}^{\sigma} \hat{c}_j^*(X')Y'^{\sigma-j} \quad \text{with } \hat{c}_j^*(X') \in k[[X']]$$

and

$$\rho > 0 \quad \text{and} \quad \sigma > 0 \quad \text{with } \rho + \sigma = d,$$

then

$$\widetilde{F}(X,\,Y) = \widehat{G}^{**}(X,\,Y)\widehat{H}^{**}(X,\,Y),$$

where

$$\widehat{G}^{**}(X,\,Y) = Y^{\rho} + \sum_{j=1}^{\rho} \hat{b}_j^*(X)X^j Y^{\rho-j}$$

and

$$\widehat{H}^{**}(X,\,Y) = Y^{\sigma} + \sum_{j=1}^{\sigma} \hat{c}_j^*(X)X^j Y^{\sigma-j}.$$

It follows that C is analytically irreducible at $P = (0, 0) \Rightarrow C'$ is analytically irreducible at $P_1 = (0, m_1)$. Instead of using Hensel's Lemma to prove that if C is analytically irreducible at $P = (0, 0)$ then C has only one tangent line at P, let us, by mimicking the proof of Hensel's Lemma, give a

Direct proof of unitangency. Assume that $F_d(X, Y) = \overline{G}(X, Y)\overline{H}(X, Y)$ where $\overline{G}(X, Y)$ and $\overline{H}(X, Y)$ are homogeneous polynomials of positive degrees ρ and σ respectively, such that GCD $(\overline{G}(X, Y), \overline{H}(X, Y)) = 1$; that is, GCD $(\overline{G}(1, Y), \overline{H}(1, Y)) = 1$. Note that the Y-degrees of $\overline{G}(X, Y)$ and $\overline{H}(X, Y)$ must be equal to their degrees, that is, equal to ρ and σ respectively. Recall that

$$F(X, Y) = F_d(X, Y) + F_{d+1}(X, Y) + \cdots + F_{d+q}(X, Y) + \cdots.$$

We shall find homogeneous polynomials $G_{\rho+i}(X, Y)$ and $H_{\sigma+j}(X, Y)$ of degrees $\rho+i$ and $\sigma+j$ for $0 \le i < \infty$ and $0 \le j < \infty$ such that $G_\rho(X, Y) = \overline{G}(X, Y)$ and $H_\sigma(X, Y) = \overline{H}(X, Y)$, and such that upon letting

$$G(X, Y) = G_\rho(X, Y) + G_{\rho+1}(X, Y) + \cdots + G_{\rho+i}(X, Y) + \cdots$$
$$H(X, Y) = H_\sigma(X, Y) + H_{\sigma+1}(X, Y) + \cdots + H_{\sigma+j}(X, Y) + \cdots$$

we have $F(X, Y) = G(X, Y)H(X, Y)$. That is,

$$F_{d+q}(X, Y) = \sum_{i+j=q} G_{\rho+i}(X, Y)H_{\sigma+j}(X, Y) \quad \text{for all } q \ge 0.$$

By induction on q we shall find $G_{\rho+i}(X, Y)$ and $H_{\sigma+j}(X, Y)$ for $1 \le i \le q$ and $1 \le j \le q$ satisfying the conditions. The case of $q = 0$ being obvious, let $q > 0$ and suppose we have found $G_{\rho+i}(X, Y)$ and $H_{\sigma+j}(X, Y)$ for $1 \le i < q$ and $1 \le j < q$. Now we want to find $G_{\rho+q}(X, Y)$ and $H_{\sigma+q}(X, Y)$ such that

$$F_{d+q}(X, Y) = \sum_{i+j=q} G_{\rho+i}(X, Y)H_{\sigma+j}(X, Y),$$

that is, such that

$$G_\rho(X, Y)H_{\sigma+q}(X, Y) + H_\sigma(X, Y)G_{\rho+q}(X, Y) = U_{d+q}(X, Y),$$

where

$$U_{d+q}(X, Y) = F_{d+q}(X, Y) - \sum_{\substack{i+j=q \\ i<q, j<q}} G_{\rho+i}(X, Y)H_{\sigma+j}(X, Y).$$

Since GCD $(G_\rho(1, Y), H_\sigma(1, Y)) = 1$, we can find polynomials $\widetilde{G}^*(Y)$ and $\widetilde{H}^*(Y)$ such that

$$G_\rho(1, Y)\widetilde{H}^*(Y) + H_\sigma(1, Y)\widetilde{G}^*(Y) = 1.$$

Now multiplying throughout by $U_{d+q}(1, Y)$ we get

$$G_\rho(1, Y)\widetilde{H}^*_{\sigma+q}(Y) + H_\sigma(1, Y)\widetilde{G}^*_{\rho+q}(Y) = U_{d+q}(1, Y),$$

where

$$\widetilde{G}^*_{\rho+q}(Y) = \widetilde{G}^*(Y)U_{d+q}(1,Y) \quad \text{and} \quad \widetilde{H}^*_{\sigma+q}(Y) = \widetilde{H}^*(Y)U_{d+q}(1,Y).$$

By the division algorithm we can write $\widetilde{H}^*_{\sigma+q}(Y) = E_q(Y)H_\sigma(1,Y) + \widetilde{H}_{\sigma+q}(Y)$
where $E_q(Y)$ and $\widetilde{H}_{\sigma+q}(Y)$ are polynomials with $\deg \widetilde{H}_{\sigma+q}(Y) \le \sigma + q$. It follows that

$$G_\rho(1,Y)\widetilde{H}_{\sigma+q}(Y) + H_\sigma(1,Y)\widetilde{G}_{\rho+q}(Y) = U_{d+q}(1,Y),$$

where $\widetilde{G}_{\rho+q}(Y) = \widetilde{G}^*_{\rho+q}(Y) + E_q(Y)G_\rho(1,Y)$. Moreover, since $\deg U_{d+q}(Y)$
$\le d + q$, we see that the $\widetilde{G}_{\rho+q}(Y)$ thus obtained would automatically satisfy
$\deg \widetilde{G}_{\rho+q}(Y) \le \rho + q$. So we have

$$\deg \widetilde{G}_{\rho+q}(Y) \le \rho + q \quad \text{and} \quad \deg \widetilde{H}_{\sigma+q}(Y) \le \sigma + q.$$

Now upon letting

$$G_{\rho+q}(X,Y) = X^{\rho+q}\widetilde{G}_{\rho+q}\left(\frac{Y}{X}\right) \quad \text{and} \quad H_{\sigma+q}(X,Y) = X^{\sigma+q}\widetilde{H}_{\sigma+q}\left(\frac{Y}{X}\right)$$

we get homogeneous polynomials $G_{\rho+q}(X,Y)$ and $H_{\sigma+q}(X,Y)$ of degrees
$\rho + q$ and $\sigma + q$ such that

$$F_{d+q}(X,Y) = \sum_{i+j=q} G_{\rho+i}(X,Y)H_{\sigma+j}(X,Y).$$

This completes the induction on q. Thus we have given two proofs of the

TANGENT LEMMA. *If the plane curve C, which may be algebraic or analytic, is analytically irreducible at the point P, then C has only one tangent line at P, and the proper transform C' of C is analytically irreducible at the corresponding point P_1.*

Curves on nonsingular surfaces. The proof of resolution of singularities of plane curves works for curves on any nonsingular surface.

Nontangential parameters. In the previous discussion we noted that if the line $X = 0$ is nontangential to C at the point P, and if after a quadratic transformation the multiplicity doesn't drop, then the exceptional line is nontangential to the proper transform of C at the point corresponding to P. This phenomenon also occurs in the more general case. For a surface $\phi(X,Y,Z) = 0$, if originally the line $X = Y = 0$ is nontangential to the surface, and if after a quadratic transformation the multiplicity doesn't drop, then the corresponding line is nontangential to the transformed surface. The same works for hypersurfaces in m-space for any m. For more details see the resolution book [A12].

Discriminant. We shall now give a proof of the fact about multiple components of a plane curve. Indeed, we shall more generally prove it for a hypersurface in m-space where m is any positive integer. We start by noting that by the *Y-discriminant* of a polynomial $f(X_1, \ldots, X_r, Y)$ in Y

whose coefficients are functions of $X_1, \ldots, X_r$, we mean the Y-resultant of f and its Y-derivative f_Y. We denote this by $\mathrm{Disc}_Y f$. Now the equation $f = 0$ defines Y as an n-valued function of $X_1, \ldots, X_r$ where n is the Y-degree of f. The Y-discriminant $\mathrm{Disc}_Y f$ is a function $D(X_1, \ldots, X_r)$ of $X_1, \ldots, X_r$. More precisely, if the coefficients of f belong to a field K, (if $f \in K[Y]$), then $D(X_1, \ldots, X_r) = \mathrm{Disc}_Y f \in K$. The values $(\alpha_1, \ldots, \alpha_r)$ of $(X_1, \ldots, X_r)$ for which Y has less than n values are exactly the points of the *discriminant locus*. They are precisely those values of $X_1 \ldots, X_r$ for which $D(\alpha_1, \ldots, \alpha_r) = 0$. We have already met the discriminant locus while discussing Riemann surfaces. Moreover, $\mathrm{Disc}_Y f = 0$ (i.e., $\mathrm{Disc}_Y f$ is "identically zero," i.e., $\mathrm{Disc}_Y f$ is the zero element of K) iff f has multiple factors, i.e., iff $f \in g^2 K[Y]$ for some nonconstant polynomial $g = g(Y) \in K[Y]$. Let us also take note of the product formula for discriminants, which says that for any $f_1 \in K[Y]$ and $f_2 \in K[Y]$,

$$\mathrm{Disc}_Y(f_1 f_2) = (\mathrm{Disc}_Y f_1)(\mathrm{Disc}_Y f_2)(\mathrm{Res}_Y(f_1, f_2))^2.$$

Now given an algebraic hypersurface $S : \tilde{f}(X_1, \ldots, X_r, Y) = 0$ in m-space over the field k where $m = r + 1$, S is said to have "algebraic" multiple components if $\tilde{f}(X_1, \ldots, X_r, Y)$ is divisible by the square of a nonconstant polynomial $\tilde{g}(X_1, \ldots, X_r, Y)$; assuming that a suitable rotation has been made, we may suppose that $\tilde{f}$ is monic in Y. Then we see that S is devoid of algebraic multiple components $\Leftrightarrow \mathrm{Disc}_Y \tilde{f} \neq 0$. Likewise, given an analytic hypersurface $\hat{S} : \hat{f}(X_1, \ldots, X_r, Y) = 0$ in "local" m-space over the field k, $\hat{S}$ is said to have "analytic" multiple components if the power series $\hat{f}(X_1, \ldots, X_r, Y)$ is divisible by the square of a nonunit power series $\hat{g}(X_1, \ldots, X_r, Y)$. Again, assuming that a suitable rotation has been made, we may suppose that $\hat{f}(0, \ldots, 0, Y) \neq 0$. Now by WPT, $\hat{f}(X_1, \ldots, X_r, Y) = f'(X_1, \ldots, X_r, Y) f^*(X_1, \ldots, X_r, Y)$ where $f'(X_1, \ldots, X_r, Y)$ is a unit in $k[[X_1, \ldots, X_r, Y]]$, and $f^*(X_1, \ldots, X_r, Y)$ is a distinguished polynomial in Y with coefficients in $k[[X_1, \ldots, X_r]]$. We see that $\hat{S}$ is devoid of analytic multiple components $\Leftrightarrow \mathrm{Disc}_Y f^* \neq 0$. Now in case of $\hat{f} = \tilde{f}$ we have $f' \in k[[X_1, \ldots, X_r]][Y]$. By the product formula we see that $\mathrm{Disc}_Y \tilde{f} \neq 0 \Rightarrow \mathrm{Disc}_Y f^* \neq 0$. *Therefore, if the algebraic hypersurface S is devoid of algebraic multiple components, then it is devoid of analytic multiple components.*

Chevalley and Zariski. By letting a suitable element play the role similar to the role played by the discriminant in the previous discussion, Chevalley obtained a theorem to the effect that any algebraic local domain is analytically unramified. Here, by an *algebraic local domain* we mean a local domain R that is the localization of a finitely generated ring extension of a field at some prime ideal. In other words, R is the local ring of a point, or an irreducible subvariety of an irreducible algebraic variety. Moreover, R is said to be *analytically unramified* if its "completion" has no nonzero *nilpotent*, i.e., a nonzero element, some power of which is zero. As a companion to Chevalley's Theorem, Zariski proved that if an algebraic local domain is "normal" then its completion is a normal domain. For a proof of these theorems of Chevalley and Zariski, see Volume II, page 320, of [ZSa].

Completion. Having used the notion of the completion of a local ring R, we had better explain it. In brief, to go from R to its *completion* $\widehat{R}$ is like going from $\mathbb{Q}$ to $\mathbb{R}$. Pedantically, $\widehat{R}$ may be defined as the set of all equivalence classes of cauchy sequences of elements from R, where a sequence of elements $x_1, x_2, \ldots$ from R is said to be *cauchy* if $x_i - x_{i+1} \in M(R)^{a(i)}$ where $a(1), a(2), \ldots$ is a sequence of positive integers which tends to ∞ as i tends to ∞. Two cauchy sequences $x_1, x_2, \ldots$ and $y_1, y_2, \ldots$ of elements from R are said to be *equivalent* if $(x_i - y_i) \in M(R)^{b(i)}$ where $b(1), b(2), \ldots$ is a sequence of positive integers which tends to ∞ as i tends to ∞. It can be seen that $\widehat{R}$ is a local ring with $\dim \widehat{R} = \dim R$ and $\operatorname{emdim} \widehat{R} = \operatorname{emdim} R$. Hence, in particular, $\widehat{R}$ is regular $\Leftrightarrow$ R is regular. By identifying every $x \in R$ with the equivalence class of the cauchy sequence $x, x, \ldots$ we may regard R as a subring of $\widehat{R}$. Then $M(R)\widehat{R} = M(\widehat{R})$. It can be shown that if $\operatorname{char} R = \operatorname{char} R/M(R)$ then R is regular $\Leftrightarrow$ $\widehat{R}$ is isomorphic to $k[[X_1, \ldots, X_m]]$ where $k = R/M(R)$ and $m = \dim R$. In particular, if $R = k[X_1, \ldots X_m]_N$ where k is a field and $N = (X_1, \ldots X_m)k[X_1, \ldots X_m]$ then $\widehat{R} = k[[X_1, \ldots, X_m]]$.

Normality. Again, having used the concept of normality, let us define it. By the *integral closure* A^* of a domain A in an overdomain B, we mean the set of all elements in B which are integral over A. Clearly A^* is a ring between A and B. A is said to be *integrally closed* in B if $A^* = A$. A is said to be *normal* if it is integrally closed in its quotient field. By the *conductor* of A in B we mean the largest ideal in A which remains an ideal in B. Equivalently, it may be defined as the set of all $c \in A$ such that $cB \subset A$, i.e., $cb \in A$ for all $b \in B$. By the *conductor* of A we mean the conductor of A in A', where A' is the integral closure of A in its quotient field. Clearly, A is normal $\Leftrightarrow$ the conductor of A is the unit ideal A. In case A is the affine coordinate ring $k[C]$ of an irreducible algebraic plane curve C over a field k, the conductor $\mathfrak{C}$ of A is what we talked about in the note following the lecture on functions and differentials on a curve.

Exercise. Show that a regular local ring is a normal domain. Also show that a one-dimensional local domain is normal iff it is regular.

NOTE. The cusp $Y^2 - X^3 = 0$ can be generalized to a higher cusp $Y^d - X^v = 0$. This is an example of a singularity with only "one characteristic pair" (d, v). We extend this notion to the curve $C : F(X, Y) = 0$ to call the pair (d, de) to be the *first characteristic pair* of C at $P = (0, 0)$, where d is the multiplicity of C at P and e is the magnitude of C at P. For further discussion, see the desingularization paper [A29].

Infinitely Near Singularities

Again let $C : F(X, Y) = 0$ be an algebraic plane curve over a field k. Given any point P of C we bring it to the origin $(0, 0)$ by translation. By rotation we arrange that $X = 0$ is not tangent to C at P. Now by a quadratic transformation (QDT) with origin as center we map the (X, Y)-plane onto the (X', Y')-plane defined by $X' = X$ and $Y' = Y/X$ or inversely defined by $X = X'$ and $Y = X'Y'$. See Figure 19.1 .

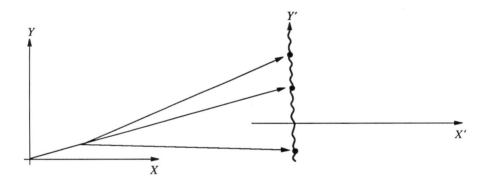

FIGURE 19.1

To see the effect of this QDT on C, we write

$$F(X, Y) = F_d(X, Y) + F_{d+1}(X, Y) + \cdots + F_j(X, Y) + \cdots \text{ with } d = \text{mult }_P C,$$

where $F_j(X, Y)$ is a homogeneous polynomial of degree j. We call $F_d(X, Y)$ the *initial form* of $F(X, Y)$. Now substituting X' for X and $X'Y'$ for Y we get

$$F\left(X', X'Y'\right) = X'^d F'(X', Y'),$$

where

$$F'\left(X', Y'\right) = F_d\left(1, Y'\right) + X'F_{d+1}\left(1, Y'\right) + \cdots + X'^{j-d}F_j(1, Y') + \cdots .$$

The total transform $F(X', X'Y') = 0$ of C factors into the exceptional line $E : X' = 0$ and the proper transform $C' : F'(X', Y') = 0$ of C. Factoring

$F_d(X, Y)$ we get

$$F_d(X, Y) = m^* \prod_{i=1}^{h} (Y - m_i X)^{e_i}$$

with $e_i > 0$ and $m^* \neq 0$ and $m_i \neq m_j$ for $i \neq j$

where the lines $Y = m_1 X, \ldots, Y = m_h X$ are the tangent lines to C at the origin. By putting $X = 1$ and $Y = Y'$ in the above factorization we get

$$F_d\left(1, Y'\right) = m^* \prod_{i=1}^{h} \left(Y' - m_i\right)^{e_i}.$$

We see that E meets C' at the points $P_i = (0, m_i)$ for $1 \leq i \leq h$. Let $d_i =$ mult$_{P_i} C'$, and note that we have $d_1 + \cdots + d_h \leq d$. Following Max Noether, points of the exceptional line E are called *points in the first neighborhood* of P. In turn, the points $P_1, \ldots, P_h$ are called the *points of C in the first neighborhood* of P. We also put mult$_P C = d_i$, and we call P_i a d_i-fold point of C. Again, P_i is a *simple* or *singular* point of C according as $d_i = 1$ or $d_i > 1$. See Figure 19.2.

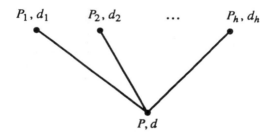

FIGURE 19.2

Clearly the points $P_1, \ldots, P_h$ of C in the first neighborhood of P are in a one-to-one correspondence with the tangents of C at P.

Now we can iterate this procedure. A point that is in the first neighborhood of some point in the first neighborhood of P is called a *point in the second neighborhood* of P. It should now be clear which are the *points of C in the second neighborhood* of P, what are their multiplicities, and so on. See Figure 19.3.

Inductively, a point in the first neighborhood of some point in the $(N-1)$-th neighborhood of P is called a *point in the Nth neighborhood* of P. The totality of the points in the various neighborhoods of P are called *points infinitely near* to P. Again, it should now be clear which are the *points of C infinitely near* to P, what are their multiplicities, and so on.

Finally, letting P vary over the (X, Y)-plane, we get all the *infinitely near points* of the (X, Y)-plane. To distinguish them from the infinitely

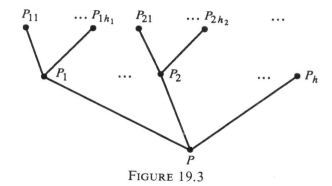

FIGURE 19.3

near points, the usual points of the (X, Y)-plane may be called *ordinary* or *distinct*.

The theorem of resolution of singularities can be paraphrased into a new

THEOREM. *There exists a positive integer N such that all the points in the N-th or higher neighborhood of any point of C, are simple points of C. Alternatively, singularities of C, ordinary as well as infinitely near, are finite in number.*

Another version of the theorem is as follows:

There does not exist an infinite quadratic sequence of d-fold points, with $d > 1$. In other words, there does not exist an infinite sequence $P \to Q_1 \to Q_2 \to \cdots$ such that Q_i is a point in the i-th neighborhood of P and $\mathrm{mult}_{Q_i} C = \mathrm{mult}_P C = d > 1$ for all $i > 0$.

In both of these versions, C is assumed to be devoid of multiple components.

The above description of infinitely near points clearly holds for any analytic plane curve, i.e., when $F(X, Y) \in k[[X, Y]]$.

Using the idea of infinitely near points, several basic results in the theory of plane curves may now be sharpened thus:

Bezout's Theorem refined. Let $C_n : F(X, Y) = 0$ and $D_m : G(X, Y) = 0$ be plane curves of degree n and m respectively. Let $(R_j)_{j=1,2,\ldots}$ be all the common points of C and D, distinct as well as infinitely near, and let s_j and r_j be the respective multiplicities of C and D at R_j. Then, assuming C and D to be devoid of common components, we have $\sum_j r_j s_j = mn$.

More precisely, for any ordinary point P we have $\sum'^P r_j s_j = I(C, D; P)$ where $\sum'^P$ is the sum over all j for which R_j is either equal to P or infinitely near to P. This continues to hold if C and D are analytic plane curves and $P = (0, 0)$. (The reference to the degrees m and n is to be disregarded.)

Noether's Fundamental Theorem refined. With notation as in Refined Bezout, a plane curve $H = 0$ can be expressed in the form $H = AF + BG$, with plane curves $A = 0$ and $B = 0$ having multiplicities at least $r_j - 1$ and

$s_j - 1$ at R_j for all j, if and only if H has multiplicity at least $r_j + s_j - 1$ at R_j for all j. This time, having stated the refined version first, we may note that the unrefined version simply says that if H passes through all the points of intersection of F and G then $H = AF + BG$ for suitable A and B. At any rate, this theorem is fundamental in the sense that Max Noether deduced the Riemann-Roch Theorem and numerous other results about plane curves from it. For details see [A33], [Cay], [No2], [Sco].

Genus Formula refined. Assuming C to be an irreducible algebraic plane curve of degree n, let $(T_i)_{i=1,2,\ldots}$ be all the singular points of C, distinct as well as infinitely near, at finite distance as well as at ∞, and let $\nu_i = \text{mult}_{T_i} C$. Now as a precise genus formula we have

$$g(C) = \frac{(n-1)(n-2)}{2} - \delta(C),$$

where by *definition*

$$\delta(C) = \text{number of double points of } C \text{ counted properly} = \sum_i \frac{\nu_i(\nu_i - 1)}{2}.$$

Clearly

$$\delta(C) = \sum_P \delta(P),$$

where the sum is over all ordinary points P of C, at finite distance or at ∞. Also, by *definition*,

$\delta(P) = $ number of double points of C counted properly and *clustered at* P

$$= \sum^P \frac{\nu_i(\nu_i - 1)}{2},$$

where $\sum^P$ is the sum over all i for which T_i is either equal to P or infinitely near to P; this definition remains valid if C is an analytic curve and P is the origin. Since $g(C) \geq 0$, we get the precise singularity bound (for irreducible curves)

$$\delta(C) = \sum_i \frac{\nu_i(\nu_i - 1)}{2} \leq \frac{(n-1)(n-2)}{2}.$$

As usual, so also in this lecture, whenever convenient we tacitly assume the ground field k to be algebraically closed. If k were not algebraically closed, then for any point of C, distinct or infinitely near, we must take into account the "degree" of that point over the field k. At any rate, assuming k to be algebraically closed, to obtain yet another genus formula, let DOF (C) be the set of all *differentials* of $k(C)$ *of first kind*, that is, the set of all differentials of $k(C)$ having no poles. Now obviously DOF (C) is a vector space over k. By applying the Riemann-Roch Theorem to the "zero" divisor we immediately see that

$$[\text{DOF}(C) : k] = g(C).$$

Historically speaking, this seems to have been the earliest definition of genus. It is due to Jacobi (1832, [**Jac**]). If instead of assuming C to be irreducible, we only assume C to be free from multiple components, then we get the precise singularity bound (for curves devoid of multiple components)

$$\delta(C) = \sum_i \frac{\nu_i(\nu_i - 1)}{2} \leq \frac{n(n-1)}{2}.$$

Conductor. In the notation of refined genus, let $\mathfrak{C}(P)$ be the conductor of the local ring $R(P)$ of any ordinary point P of C and let $R'(P)$ be the integral closure of $R(P)$ in its quotient field. As a consequence of Noether's Fundamental Theorem, it can be shown that

$$\text{length}_{R(P)}\mathfrak{C}(P) = \frac{1}{2}\text{length}_{R'(P)}\mathfrak{C}(P) = \delta(P).$$

Again, this continues to hold if C is an irreducible analytic curve, P is the origin, and for the local ring $R(P)$ we take the residue class ring

$$k[[X, Y]]/F(X, Y)k[[X, Y]].$$

The equation immediately yields the following equation where $\mathfrak{C}$ is the conductor of the affine coordinate ring A of C, A' is the integral closure of A in its quotient field, and $\sum_{\text{finite}}$ is the sum over all i for which T_i is either at finite distance or infinitely near to a point at finite distance:

$$\text{length}_A\mathfrak{C} = \frac{1}{2}\text{length}_{A'}\mathfrak{C} = \delta(C)_{\text{finite}} \quad \text{with} \quad \delta(C)_{\text{finite}} = \sum_{\text{finite}} \frac{\nu_i(\nu_i - 1)}{2}.$$

By the *divisor of singularities* of C we mean the divisor $\text{DOS}(C)$ of $k(C)$ such that for every discrete valuation V of $k(C)$, upon letting P to be the center of V on C, we have

$$\text{DOS}(C)(V) = \min\{V(z) : 0 \neq z \in \mathfrak{C}(P)\}.$$

Now $k(C) = k(x, y)$ where x and y are the functions on C induced by X and Y respectively. The *different* of $k(C)/k(x)$ is the divisor $\mathfrak{D}$ of $k(C)$ such that for every discrete valuation V of $k(C)$ we have

$$\mathfrak{D}(V) = V(\frac{d\tau}{dt}),$$

where t is a uniformizing parameter of V and τ is uniformizing parameter of the restriction of V to $k(x)$, i.e., $\tau \in k(x)$ is such that

$$V(\tau) = \min\{V(z) : 0 \neq z \in k(x) \text{ with } V(z) > 0\}.$$

The different $\mathfrak{D}$ gives the ideal $\mathfrak{D}'$ in A' such that for every discrete valuation V of $k(C)$, for which $V(x) \geq 0 \leq V(y)$, we have

$$\mathfrak{D}'R_V = M(R_V)^{\mathfrak{D}(V)}.$$

In the note following the lecture on functions and differentials on a curve, we described an equation proved by Dedekind (see page 376 of [**Ded**] and Chapter III of [**ASa**]), which can now be precisely stated as

Dedekind's Theorem. $F_Y(x, y)A' = \mathfrak{C}\mathfrak{D}'$.

Adjoint. With notation as in refined genus and conductor, an algebraic plane curve $D : G(X, Y) = 0$ is said to be *adjoint* to C if $\text{mult}_{T_i} D \geq \nu_i - 1$ for all i. Moreover, D is said to be *adjoint* to C at P, where P is an ordinary point of C, if $\text{mult}_{T_i} D \geq \nu_i - 1$ for all those i for which T_i is either equal to P or infinitely near to P. Again, this definition remains valid if C and D are analytic curves and P is the origin. Clearly, D is adjoint to C iff it is adjoint to C at every ordinary point of C, at finite distance or at ∞. As another consequence of Noether's Fundamental Theorem, it can be seen that if P is any ordinary point of C at finite distance, then

$$D \text{ is adjoint to } C \text{ at } P \Leftrightarrow G(x, y) \in \mathfrak{C}(P).$$

Note that this remains valid for an ordinary point P of C at ∞ provided we "homogenize" and "dehomogenize" $G(x, y)$ appropriately. Moreover, it also remains valid if C and D are analytic curves and P is the origin and the functions x, y are interpreted as the images of X, Y under the residue class map $k[[X, Y]] \to k[[X, Y]]/F(X, Y)k[[X, Y]]$. It follows that D is adjoint to C at every ordinary point of C at finite distance $\Leftrightarrow G(x, y) \in \mathfrak{C}$. To take better care of points at ∞, we can use homogeneous coordinates. So let R_m^* be the set of all homogeneous polynomials $\Gamma(\mathcal{X}, \mathcal{Y}, \mathcal{Z})$ of degree m in $\mathcal{X}, \mathcal{Y}, \mathcal{Z}$ with coefficients in k. Note that R_m^* is a k-vector space with $[R_m^* : k] = \frac{(m+1)(m+2)}{2}$. For any $\Gamma \in R_m^*$ with $\Gamma(x, y, 1) \neq 0$, by (Γ) we denote the divisor such that for every discrete valuation V of $k(C)$ we have

$$(\Gamma)(V) = \begin{cases} V(\Gamma(x, y, 1)) & \text{if } V(x) \geq 0 \leq V(y) \\ V(\Gamma(1, y/x, 1/x)) & \text{if } V(y) \geq V(x) < 0 \\ V(\Gamma(x/y, 1, 1/y)) & \text{if } V(x) > V(y) < 0. \end{cases}$$

Let $\text{ADJ}_m(C)$ be the set of all homogeneous adjoint polynomials of degree m, i.e., the set of all $\Gamma \in R_m^*$ such that either $\Gamma(x, y, 1) = 0$ or the curve $\Gamma = 0$ is adjoint to C. Note that then $\text{ADJ}_m(C)$ is a subspace of the k-vector space R_m^*. Now the relation between adjoints and conductor can be paraphrased by saying that for any $\Gamma \in R_m^*$ with $\Gamma(x, y, 1) \neq 0$ we have

$$\Gamma \in \text{ADJ}_m(C) \Leftrightarrow (\Gamma) \geq \text{DOS}(C).$$

As a consequence of Noether's Fundamental Theorem, it can be shown that

$$[\text{ADJ}_{n-3}(C) : k] = g(C).$$

Max Noether deduced Jacobi's genus formula from the equation by exhibiting a specific isomorphism between the two vector spaces $\text{DOF}(C)$ and $\text{ADJ}_{n-3}(C)$. He also showed that if $0 \neq w \in \text{DOF}(C)$ and $0 \neq \Gamma \in \text{ADJ}_{n-3}(C)$ correspond under the said isomorphism then

$$(w) = (\Gamma) - \text{DOS}(C).$$

Singularity bounds. In the above notation, we sketch a proof of the singularity bounds by considering adjoints of the next two higher degrees, i.e., adjoints of degree $n-2$ and adjoints of degree $n-1$. The case of $n \leq 2$ being obvious, we may suppose that $n > 2$. Given any ordinary point P of C with mult $_P C = \nu > 1$, by bringing P to the origin we see that mult $_P D \geq \nu - 1$ for $D : G(X, Y) = \sum c_{ij} X^i Y^j = 0$ iff $c_{ij} = 0$ for all $i + j < \nu - 1$, which amounts to imposing $1 + 2 + \cdots + (\nu - 1) = \frac{\nu(\nu-1)}{2}$ linear conditions. For a double point P, we have $\nu = 2$ and so we get $\frac{\nu(\nu-1)}{2} = 1$. A double point P imposes 1 "adjoint condition." Hence, $\frac{\nu(\nu-1)}{2}$ double points $P_1, \ldots, P_{\frac{\nu(\nu-1)}{2}}$ impose $\frac{\nu(\nu-1)}{2}$ "adjoint conditions." Thus, the adjoint conditions imposed by a ν-fold point equals the adjoint conditions imposed by $\frac{\nu(\nu-1)}{2}$ double points. This *explains* why, in the genus formula or in the singularity bound formula, a ν-fold point is counted as $\frac{\nu(\nu-1)}{2}$ double points. It can be argued that an infinitely near ν-fold point also imposes $\frac{\nu(\nu-1)}{2}$ adjoint conditions. Thus $\delta(C)$ equals the total number of adjoint conditions imposed by C, and hence we get the *estimate*

$$[\text{ADJ}_m(C) : k] \geq \frac{(m + 1)(m + 2)}{2} - \delta(C).$$

Assuming C to be free from multiple components, instead of assuming C to be irreducible, let us sketch a proof of the corresponding singularity bound. Note that the relevant portion of the above considerations remains valid in this more general case; now either $F_X(X, Y) = 0$ or $F_Y(X, Y) = 0$ is a curve of degree $n - 1$ having no common component with C and it is easily seen to be adjoint to C, and hence by Refined Bezout we have

$$\sum_i \nu_i(\nu_i - 1) \leq n(n - 1)$$

and upon dividing by 2 we get the singularity bound (for curves devoid of multiple components)

$$\delta(C) = \sum_i \frac{\nu_i(\nu_i - 1)}{2} \leq \frac{n(n - 1)}{2}.$$

Henceforth let us reinstate the assumption of C being irreducible. By taking $m = n - 1$ in the above estimate we have

$$[\text{ADJ}_{n-1}(C) : k] \geq \frac{n(n + 1)}{2} - \delta(C) > \frac{n(n - 1)}{2} - \delta(C).$$

Hence, if $\delta(C) = \frac{n(n-1)}{2}$, then we we can find $\Gamma^* \in R^*_{n-1}$ such that Γ^* is adjoint to C and passes through e^* simple points of C where

$$e^* = \frac{n(n + 1)}{2} - 1 - \delta(C) \geq 0.$$

By Refined Bezout we get

$$n(n - 1) \geq e^* + \sum_i \nu_i(\nu_i - 1),$$

but

$$e^* + \sum_i \nu_i(\nu_i - 1) = \frac{n(n+1)}{2} - 1 + \delta(C) = \frac{n(n+1)}{2} + \frac{n(n-1)}{2} - 1$$

$$= n(n-1) + (n-1)$$

$$> n(n-1),$$

which is a contradiction. Therefore, $\delta(C) < \frac{n(n-1)}{2}$. By taking $m = n-2$ in the above estimate we have

$$[\mathrm{ADJ}_{n-2}(C) : k] \geq \frac{n(n-1)}{2} - \delta(C).$$

Hence, if $\delta(C) > \frac{(n-1)(n-2)}{2}$, then we can find $\Gamma \in R_{n-2}^*$ such that Γ is adjoint to C and passes through e simple points of C where

$$e = \frac{n(n-1)}{2} - 1 - \delta(C) \geq 0.$$

Then by Refined Bezout we get

$$n(n-2) \geq e + \sum_i \nu_i(\nu_i - 1),$$

but

$$e + \sum_i \nu_i(\nu_i - 1) = \frac{n(n-1)}{2} - 1 + \delta(C) \geq \frac{n(n-1)}{2} + \frac{(n-1)(n-2)}{2}$$

$$= n(n-2) + 1$$

$$> n(n-2),$$

which is a contradiction. This establishes the singularity bound (for irreducible curves)

$$\delta(C) = \sum_i \frac{\nu_i(\nu_i - 1)}{2} \leq \frac{(n-1)(n-2)}{2}.$$

Rational parametrization. In the above discussion, if $\delta(C) = \frac{(n-1)(n-2)}{2}$, then we can fix e' simple points $T_1', \ldots, T_{e'}'$ of C where

$$e' = \frac{n(n-1)}{2} - 2 - \delta(C) \geq 0.$$

We can find two linearly independent members Γ' and Γ'' of $\mathrm{ADJ}_{n-2}(C)$ passing through $T_1', \ldots, T_{e'}'$, and then, upon letting $\widehat{\Gamma} = \Gamma' + t\Gamma''$, we have that $\widehat{\Gamma}$ is an adjoint of C of degree $n-2$ which passes through $T_1', \ldots, T_{e'}'$ and

$$e' + \sum_i \nu_i(\nu_i - 1) = \frac{n(n-1)}{2} - 2 + \delta(C) = \frac{n(n-1)}{2} - 2 + \frac{(n-1)(n-2)}{2}$$

$$= n(n-2) - 1.$$

Therefore, by Refined Bezout, $\widehat{\Gamma}$ meets C in exactly one free point U_t whose coordinates must be rational functions of t. Thus we have verified the fact that C is rational iff $g(C) = 0$. At the same time we have given a constructive method of rationally parametrizing a curve of genus zero. (The case of $n \leq 2$ being trivial, we tacitly assume $n > 2$.)

Taylor Resultant. The singularities of C are "set theoretically" given by $F = F_X = F_Y = 0$, and "counting properly" they are given by the conductor $\mathfrak{C}$. This is fine for an implicitly defined curve C. But if C is explicitly given by a polynomial parametrization $X = p(t)$ and $Y = q(t)$, then it should be possible to find its singularities "set theoretically" as well as "counting properly" in an explicit manner. With this conviction, around 1972 I defined the *Taylor t-Resultant* of any two polynomials $p(t)$ and $q(t)$ by putting

$$\mathrm{Tres}_t(p(t), q(t))$$
$$= \mathrm{Res}_\tau \left(p'(t) + \frac{p''(t)}{2!}\tau + \frac{p'''(t)}{3!}\tau^2 + \cdots , \; q'(t) + \frac{q''(t)}{2!}\tau + \frac{q'''(t)}{3!}\tau^2 + \cdots \right)$$

where primes denote derivatives. Using Dedekind's Theorem, I proved the following

THEOREM (Abhyankar). *For any two nonconstant polynomials $p(t)$ and $q(t)$ in t with coefficients in a field k of characteristic zero, upon letting $\Delta(t) = \mathrm{Tres}_t(p(t), q(t))$ and upon letting $\mathfrak{C}$ to be the conductor of $A = k[p(t), q(t)]$ in $A' = k[t]$, we have $\mathfrak{C} = \Delta(t)A'$. Hence in particular, assuming $(p(t), q(t))$ to be a parametrization of the irreducible algebraic plane curve $C : F(X, Y) = 0$, i.e., assuming $F(p(t), q(t)) = 0$, we have the following:*

(1) *$(p(t), q(t))$ is a faithful parametrization of $C \Leftrightarrow k(t) = k(p(t), q(t)) \Leftrightarrow \Delta(t) \neq 0$.*

(2) *If $\Delta(t) \neq 0$ then: C is nonsingular at finite distance $\Leftrightarrow k[t] = k[p(t), q(t)] \Leftrightarrow \Delta(t) \in k$.*

(3) *If $\Delta(t) = \gamma^* \prod_{j=1}^l (t - \gamma_j)^{\epsilon_j}$ where $0 \neq \gamma^* \in k$ and $\gamma_1, \ldots, \gamma_l$ are distinct elements in k and $\epsilon_1, \ldots, \epsilon_l$ are positive integers, and if α and β are elements in k such that $P = (\alpha, \beta)$ is a point of C, i.e., such that $F(\alpha, \beta) = 0$, then P is a singular point of $C \Leftrightarrow (\alpha, \beta) = (p(\gamma_j), q(\gamma_j))$ for some j. Moreover, if P is a singular point of C then*

$$\sum^{(\alpha, \beta)} \epsilon_j = \sum^P \nu_i(\nu_i - 1) = \mathrm{length}_{R'(P)} \mathfrak{C}(P) = 2\,\mathrm{length}_{R(P)} \mathfrak{C}(P),$$

where $\sum^{(\alpha, \beta)}$ is the sum over those j for which $(\alpha, \beta) = (p(\gamma_j), q(\gamma_j))$, and $\sum^P$ is the sum over those i for which the point T_i either equals P or is infinitely near to P.

(4) *If $\Delta(t) = \gamma^* \prod_{j=1}^l (t - \gamma_j)^{\epsilon_j}$ where $0 \neq \gamma^* \in k$ and $\gamma_1, \ldots, \gamma_l$ are distinct elements in k and $\epsilon_1, \ldots, \epsilon_l$ are positive integers, then*

$$\deg \Delta(t) = \sum_{\text{finite}} \nu_i(\nu_i - 1) = 2\,\delta(C)_{\text{finite}} = \mathrm{length}_{A'} \mathfrak{C} = 2\,\mathrm{length}_A \mathfrak{C},$$

where $\sum_{\text{finite}}$ is the sum over all i for which T_i is either at finite distance or infinitely near to a point at finite distance.

Examples. Consider $C : Y^2 - a^2 X^2 - X^3 = 0$. If $a \neq 0$ then the initial form factors into two distinct factors as $Y^2 - a^2 X^2 = (Y - aX)(Y + aX)$. The curve is the nodal cubic depicted in Figure 19.4(a). By a QDT with center at $P = (0, 0)$ this becomes $X'^2 Y'^2 - a^2 X'^2 - X'^3 = X'^2 [Y'^2 - a^2 - X']$. Hence, for the proper transform we get Figure 19.4(b), which is a parabola intersecting the exceptional line in the two points $P_1 = (0, a)$ and $P_2 = (0, -a)$.

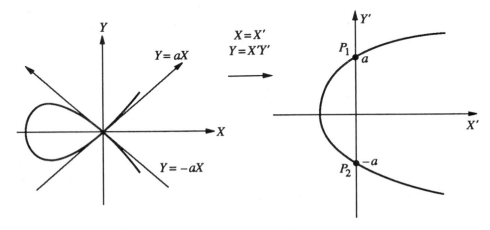

FIGURE 19.4

In this case the singularity tree looks like Figure 19.5.

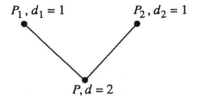

FIGURE 19.5

If $a = 0$ then the curve is the cuspidal cubic $Y^2 - X^3 = 0$ depicted in Figure 19.6(a), having only one tangent at the origin $P = (0, 0)$. By the QDT this becomes $X'^2 Y'^2 - X'^3 = X'^2 [Y'^2 - X']$. For the proper transform we get the figure on the right, a parabola intersecting the exceptional line only in one point. This is the "new" origin, and there the exceptional line is actually tangent to the parabola. See Figure 19.6(b).

The singularity tree in this case looks like Figure 19.7.

FIGURE 19.6

$P_1, 1$

$P, 2$

FIGURE 19.7

Now we can consider $Y^2 - X^5 = 0$, which is a cusp with a sharper beak. This first transforms to $X'^2[Y'^2 - X'^3]$ and then to $X''^4[Y''^2 - X'']$. See Figure 19.8.

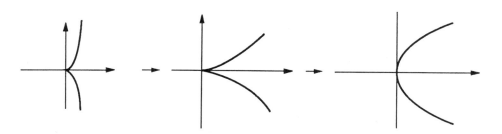

FIGURE 19.8

The singularity tree looks like Figure 19.9 on page 156.

Similarly, we can consider $Y^2 - X^7 = 0$, or more generally $C : Y^2 - X^{2s+1} = 0$. After each transformation the exponent drops by 2. In this case

$P_{11},1$

$P_1,2$

$P,2$

<figure>FIGURE 19.9</figure>

the picture is shown below.

$\bullet \underbrace{P_{11\ldots1}}_{s\text{ times}}, 1$

$\bullet P_{11\ldots1}, 2$

$\vdots$

$\bullet P_{11}, 2$

$\bullet P_1, 2$

$\bullet P, 2$

There are s double points at the origin, including those which are infinitely near. Let us check the genus formula for this curve $C : F(X, Y) = 0$ with $F(X, Y) = Y^2 - X^{2s+1}$. It is obvious that this curve is rational. In fact it has the polynomial parametrization

$$\begin{cases} X = p(t) = t^2 \\ Y = q(t) = t^{2s+1}. \end{cases}$$

Now the degree of C is n where $n = 2s + 1$. Hence we should have $\delta(C) = \frac{(n-1)(n-2)}{2} = s(2s - 1)$. Homogenizing F we get $f(\mathscr{X}, \mathscr{Y}, \mathscr{Z}) = \mathscr{Y}^2 \mathscr{Z}^{2s-1} - \mathscr{X}^{2s+1}$. Assuming that neither 2 nor $2s + 1$ is divisible by the characteristic of k, we have $f_{\mathscr{X}} = -(2s + 1)\mathscr{X}^{2s}$, $f_{\mathscr{Y}} = 2\mathscr{Y}\mathscr{Z}^{2s-1}$, $f_{\mathscr{Z}} = (2s - 1)\mathscr{Z}^{2s}\mathscr{Y}^2$. Hence, in homogeneous coordinates, $P = (0, 0, 1)$ and $P^* = (0, 1, 0)$ are the only distinct singularities of C. At the origin $P = (0, 0, 1)$ there are s double points. Hence we must have $\delta(P^*) = s(2s - 1) - s = 2s(s - 1)$. To take P^* to the origin, we put $X^* = \frac{\mathscr{X}}{\mathscr{Y}}$ and $Z^* = \frac{1}{\mathscr{Y}}$, and $F^*(X^*, Z^*) = f(X^*, 1, Z^*) = Z^{*2s-1} - X^{*2s+1}$. Thus $\text{mult}_{P^*} C = 2s - 1$, and after one QDT with center at P^* we get

$$F^*(X^{*'}, X^{*'}Z^{*'}) = X^{*'2s-1}[Z^{*'2s-1} - X^{*'2}].$$

So, applying the previous considerations at the origin in the $(X^{*'}, Z^{*'})$-plane, as the singularity tree at P^* we get

- $\underbrace{P^*_{11...1}}_{s-1 \text{ times}}$, 1

- $P^*_{11...1}$, 2

$\vdots$

- P^*_{11} , 2
- P^*_1 , 2
- P^* , $2s - 1$.

Hence $\delta(P^*) = \frac{(2s-1)(2s-2)}{2} + (s - 1) = 2s(s - 1)$ as expected.

Trying out the Taylor Resultant on this example we get

$$\text{Tres}_t(t^2, t^{2s+1})$$

$$= \text{Res}_\tau \left(2t + \tau, \ (2s + 1)t^{2s} + \frac{(2s + 1)(2s)}{2!}t^{2s-1}\tau + \cdots + \tau^{2s} \right) = t^{2s}$$

Hence,

$$\delta(P) = \delta(C)_{\text{finite}} = s,$$

which agrees with the above calculation in terms of QDTs. Likewise we can use the Taylor Resultant to calculate $\delta(P^*)$ by considering an appropriate polynomial parametrization of $Z^{*2s-1} - X^{*2s+1}$ in the (X^*, Z^*)-space.

To bring the singularities which C has at P and P^* under one umbrella, we can consider the curve $C^* : Y^u - X^v = 0$, where u and v are positive integers. This curve has the polynomial parametrization

$$\begin{cases} X = p^*(t) = t^u \\ Y = q^*(t) = t^v. \end{cases}$$

Exercise. By calculating the Taylor t-Resultant of $p^*(t)$ and $q^*(t)$, show that the parametrization of C^* is faithful iff u and v are coprime. If they are coprime then $\delta(C^*)_{\text{finite}} = \delta((0, 0)) = \frac{(u-1)(v-1)}{2}$. Also verify this by finding the singularity tree.

Problem. Generalize the idea of Taylor Resultant for explicitly finding the singularities of a surface $\phi(X, Y, Z) = 0$ having a polynomial parametrization

$$\begin{cases} X = p(t, \tau) \\ Y = q(t, \tau) \\ Z = r(t, \tau). \end{cases}$$

NOTE. This lecture has run a bit longer than the previous lectures, but then you may assume that some of the pages are infinitely near to the others! Max Noether, the inventor of the suggestive concept of infinite nearness, has indeed been called the father of algebraic geometry. For instance, see the foreword to van der Waerden's book on algebraic geometry [**Wa1**]. In the preface to his book, van der Waerden says that he thought of writing that book as an introduction to Zariski's precious Ergebnisse monograph on algebraic surfaces [Za3]. In a conversation in 1954, I. S. Cohen recommended this book of van der Waerden to me as a geometric antidote to the overly algebraic treatment which algebraic geometry was beginning to acquire. Returning to Max Noether's greatness, let us quote from the introduction to Lefschetz's autobiographical page [Le2]:

> In its early phase (Abel [Ab1] [Ab2], Riemann [Rie], Weierstrass [Wei]) algebraic geometry was just a chapter in analytic function theory. ... A new current appeared however (1870, [No1]) under the powerful influence of Max Noether who really put "geometry" and more "birational geometry" into "algebraic geometry."

It may also be noted that the algebracization of Max Noether's geometric ideas was furthered by the impetus given by his daughter, Emmy Noether [Noe]. So Max Noether, the father of algebraic geometry, may also be called the grandfather of modern algebra! Some years ago, I started writing up my "Calcutta Lectures" [A33] to do my bit in explaining what a wealth of results can follow if we keep deciphering Max Noether's ideas on infinitely near points and such. Perhaps some day I may have enough energy to bring these "Calcutta Lectures" into print. In the meantime, as always, it will be quite worth while to look into Max Noether's basic papers (such as [BNo] and [No2]), as well as into Severi's charming Vorlesungen [Se2], which is the German translation of his original Lezioni in Italian.

Parametrizing a Quartic with Three Double Points

Let us try to calculate an example today.

Example. Find a quartic (a plane curve of degree 4) with three double points and hence parametrize it rationally.

Recall the genus formula,

$$g = \frac{(n-1)(n-2)}{2} - \text{number of double points (counted properly)}.$$

So a quartic can have at most $\frac{(4-1)(4-2)}{2} = 3$ double points. If it does have three double points, then $g = 0$. We can parametrize it by rational functions. Let us actually try to do it. See Figure 20.1.

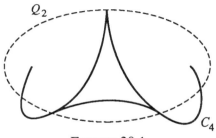

FIGURE 20.1

Now given three double points on a quartic C_4, we can fix any other point on it and pass conics Q_2 through these four points. Since $C_4 \cdot Q_2 = 4 \cdot 2 = 8$ points, exactly one $(= 8 - (2 + 2 + 2 + 1))$ point is left free. This would give the parametrization.

Now in the projective plane, we may assume that the three double points lie on the corners of the fundamental triangle. See Figure 20.2 on page 160.

In fact, in view of the fundamental theorem for projective plane, any four points can be given preassigned coordinates. Thus we can fix any other (simple) point on C_4 and assume it to be $(1, 1, 1)$. Now the equation of a quartic is, in general, given by a homogeneous equation in $\mathscr{X}, \mathscr{Y}, \mathscr{Z}$ of degree 4, i.e., an equation of the form

$$\bar{a}\mathscr{X}^4 + \bar{b}\mathscr{X}^3\mathscr{Y} + \bar{c}\mathscr{X}^2\mathscr{Y}^2 + \bar{d}\mathscr{X}\mathscr{Y}^3 + \bar{e}\mathscr{Y}^4 + \bar{f}\mathscr{X}^3\mathscr{Z} + \bar{g}\mathscr{X}^2\mathscr{Y}\mathscr{Z} + \cdots = 0.$$

Since we want the three corners of the fundamental triangle (the points $(0, 0, 1)$, $(0, 1, 0)$, and $(1, 0, 0)$) to be the double points of the quartic, (1) after putting $\mathscr{Z} = 1$, we should have no constant term and no linear

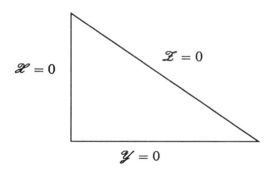

$\mathscr{X} = 0$

$\mathscr{Z} = 0$

$\mathscr{Y} = 0$

FIGURE 20.2

terms, (2) after putting $\mathscr{Y} = 1$, we should have no constant term and no linear terms, and (3) after putting $\mathscr{X} = 1$, we should have no constant term and no linear terms. In other words, without equating anybody to one, in the above equation there should be (1') no term in $\mathscr{Z}^4$ or $\mathscr{Z}^3\mathscr{X}$ or $\mathscr{Z}^3\mathscr{Y}$, (2') no term in $\mathscr{Y}^4$ or $\mathscr{Y}^3\mathscr{X}$ or $\mathscr{Y}^3\mathscr{Z}$, (3') no term in $\mathscr{X}^4$ or $\mathscr{X}^3\mathscr{Y}$ or $\mathscr{X}^3\mathscr{Z}$. Hence, the equation of the quartic must look like

$$a\mathscr{X}^2\mathscr{Y}^2 + b\mathscr{X}^2\mathscr{Y}\mathscr{Z} + c\mathscr{X}^2\mathscr{Z}^2 + d\mathscr{X}\mathscr{Y}^2\mathscr{Z} + e\mathscr{X}\mathscr{Y}\mathscr{Z}^2 + f\mathscr{Y}^2\mathscr{Z}^2 = 0.$$

For the point $(1, 1, 1)$ to be on it we must have

$$a + b + c + d + e + f = 0.$$

The process by which we have simplified the equation is similar to that of interpolation. Speaking of interpolation, we may make the following

REMARK. Lagrange Interpolation $\Rightarrow$ Dedekind Conductor Formula (i.e., derivative $=$ conductor $\times$ different).

Now for conics Q_2 to pass through the three points $(0, 0, 1)$, $(0, 1, 0)$, and $(1, 0, 0)$, we necessarily have no $\mathscr{X}^2$ term, no $\mathscr{Y}^2$ term, and no $\mathscr{Z}^2$ term. So the equation of such a conic would be

$$\alpha\mathscr{X}\mathscr{Y} + \beta\mathscr{Y}\mathscr{Z} + \gamma\mathscr{X}\mathscr{Z} = 0.$$

Since this is symmetric in $\mathscr{X}$, $\mathscr{Y}$, $\mathscr{Z}$, we may assume that $\beta = 1$. Now put $\gamma = t$. Since the conic passes through $(1, 1, 1)$, we get $\alpha = -1 - t$. Thus we may assume that our conic Q_2 is simply given by

$$(-1 - t)\mathscr{X}\mathscr{Y} + \mathscr{Y}\mathscr{Z} + t\mathscr{X}\mathscr{Z} = 0.$$

Dehomogenizing by putting $\mathscr{Z} = 1$ we get

$$(-1 - t)XY + Y + tX = 0, \quad \text{and hence,} \quad Y = \frac{-Xt}{(1 - X - Xt)}.$$

Thus we have a one-parameter family $Q_2 = Q_2(t)$ of conics. Now for the quartic

$$a\mathscr{X}^2\mathscr{Y}^2 + b\mathscr{X}^2\mathscr{Y}\mathscr{Z} + c\mathscr{X}^2\mathscr{Z}^2 + d\mathscr{X}\mathscr{Y}^2\mathscr{Z} + e\mathscr{X}\mathscr{Y}\mathscr{Z}^2 + f\mathscr{Y}^2\mathscr{Z}^2 = 0,$$

if we require that $\mathscr{Y} = \alpha\mathscr{X}$ and $\mathscr{Y} = -\alpha\mathscr{X}$ are the two tangents at $(0, 0, 1)$, then we get $e = 0$. Considering similar conditions at the other two double points, we get $b = 0$ and $d = 0$. The equation is reduced to

$$a\mathscr{X}^2\mathscr{Y}^2 + c\mathscr{X}^2\mathscr{Z}^2 + f\mathscr{Y}^2\mathscr{Z}^2 = 0.$$

To proceed further with this case, when all three double points are nodes, we note that $a + c + f = 0$. Hence we may take $f = 1$ and $c = \lambda$, and then we get $a = -1 - \lambda$. Thus we have

$$(-1 - \lambda)\mathscr{X}^2\mathscr{Y}^2 + \lambda\mathscr{X}^2\mathscr{Z}^2 + \mathscr{Y}^2\mathscr{Z}^2 = 0.$$

Dehomogenizing by putting $\mathscr{Z} = 1$ for the affine equation of C_4 we get $(-1 - \lambda)X^2Y^2 + \lambda X^2 + Y^2$. Hence $Y^2(1 - X^2 - \lambda X^2) = -\lambda X^2$. Therefore,

$$Y^2 = \frac{-\lambda X^2}{1 - X^2 - \lambda X^2}.$$

Squaring the affine equation $Y = \frac{-Xt}{1-X-Xt}$ of Q_2 we get

$$Y^2 = \frac{X^2 t^2}{(1 - X - Xt)^2}.$$

Comparing the two expressions for Y^2 we get

$$-\lambda(1 - X - Xt)^2 = t^2\left(1 - X^2 - \lambda X^2\right).$$

Simplifying the above, we have

$$-\lambda\left(1 + X^2 + X^2t^2 - 2X - 2Xt + 2X^2t\right) = t^2 - t^2X^2 - t^2\lambda X^2.$$

Hence,

$$\left(-\lambda - \lambda t^2 - 2t\lambda + t^2 + t^2\lambda\right)X^2 + (2\lambda + 2\lambda t)X + \left(-\lambda - t^2\right) = 0.$$

Therefore we get

$$\left(-\lambda - 2\lambda t + t^2\right)X^2 + 2\lambda(1 + t)X + \left(-\lambda - t^2\right) = 0.$$

To solve the above quadratic equation, consider the discriminant

$$\Delta = 4\lambda^2\left(1 + 2t + t^2\right) - 4\left(t^2 - \lambda - 2\lambda t\right)\left(-\lambda - t^2\right)$$
$$= 4\lambda^2 + 8\lambda^2 t + 4\lambda^2 t^2 + 4\left(\lambda t^2 - \lambda^2 - 2\lambda^2 t + t^4 - t^2\lambda - 2\lambda t^3\right)$$
$$= 4t^4 - 8\lambda t^3 + 4\lambda^2 t^2$$
$$= 4t^2(\lambda - t)^2.$$

Now $\sqrt{\Delta} = \pm 2(\lambda - t)t$. Therefore, the roots are

$$X = \frac{-2\lambda(1 + t) \pm 2\left(\lambda t - t^2\right)}{2\left(-\lambda - 2\lambda t + t^2\right)}.$$

That is,

$$X = \frac{-\lambda - \lambda t + \lambda t - t^2}{t^2 - 2\lambda t - \lambda} \quad \text{or} \quad X = \frac{-\lambda - \lambda t - \lambda t + t^2}{t^2 - 2\lambda t - \lambda} = 1.$$

Thus the only nontrivial solution is

$$X = \frac{-\lambda - t^2}{-\lambda - 2\lambda t + t^2}.$$

So we obtain a rational parametrization! (Note that the other solution corresponds to the common point $(1, 1, 1)$.)

REMARK. We can do the same thing for a quartic whose three double points are all cusps instead of nodes. For details, see Part V of my tame coverings paper [A05]. Lots of computations of the above type are made in this paper. These are facilitated by my earlier paper on cubic surfaces [A06]. The tame coverings paper arose out of the following. Among plane curves such as nodal cubic, nodal quartic etc., (see Figure 20.3), the cuspidal

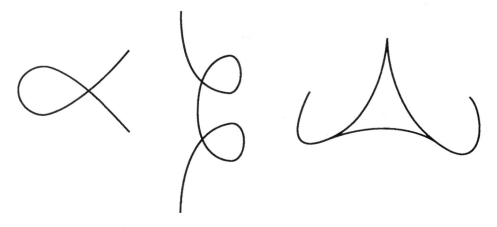

FIGURE 20.3

quartic is the simplest plane curve C for which the fundamental group $\pi_1(\mathbb{P}^2 \backslash C)$ is not abelian. In fact it is the unique nonabelian group of order 12 having a cyclic normal subgroup of order 3 with a cyclic group of order 4 as the factor group. Now the fundamental group $\pi_1(\mathbb{P}^2 \backslash C)$, with a base point $P \in \mathbb{P}^2 \backslash C$, consists of the equivalence classes of continuous closed paths (= continuous images of the unit circle) in $\mathbb{P}^2 \backslash C$ starting and ending at P, where two paths are considered equivalent if one can be continuously deformed into the other. For details see [STh]. See Figure 20.4.

Algebraists care about this fundamental group $\pi_1(\mathbb{P}^2 \backslash C)$ because it is intimately related to the Galois groups of the least Galois extensions of field extensions of the form $k(S)/k(\mathbb{P}^2)$, where $S : \phi(X, Y, Z) = 0$ is an algebraic surface whose discriminant locus $\mathrm{Res}_Z(\phi, \phi_Z) = 0$ is the plane curve

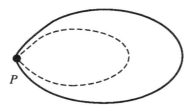

FIGURE 20.4

C in the projective (X, Y)-plane $\mathbb{P}^2$ over a field k. (Note that by a *normal extension L/K* we mean a finite algebraic field extension L/K such that the image of any bijective K-homomorphism of L is L itself. Then the group of all such K-homomorphisms of L is called the *Galois group* of L/K. Also note that by a *Galois extension L/K* we mean a normal extension L/K which is also *separable*, i.e., for every $z \in L$ we can find $H(Z) \in K[Z]$ such that $H(z) = 0 \neq H_Z(z)$. Finally, note that by the *least Galois extension* of a finite algebraic separable field extension L'/K we mean the Galois extension L/K such that L' is a subfield of L and such that there does not exist any Galois extension L^*/K for which L^* is a subfield of L with $L' \subset L^* \neq L$.) The three cuspidal quartic gives the smallest (degree) example of an irreducible plane curve C for which we can find a surface S such that the Galois group of the least Galois extension of $k(S)/k(\mathbb{P}^2)$ is nonabelian. Here we should really talk about the "branch locus" rather than the "discriminant locus." In this connection, we have the following

THEOREM OF ZARISKI. *If the branch locus C has only nodes as singularities then the fundamental group $\pi_1(\mathbb{P}^2 \setminus C)$ is abelian. Hence, the Galois groups of the corresponding Galois extensions are abelian.*

For details see [**Za1**], Chapter VIII of [**Za3**], and Parts I through VI of [**A05**].

NOTE. Just as the Galois group being solvable corresponds to the situation that the equation can be solved by successively taking roots (= square roots, cube roots, etc.), so the Galois group being abelian means that the equation can be solved by taking roots simultaneously and not successively.

LECTURE 21

Characteristic Pairs

Let P be a singular point of a plane curve $C : F(X, Y) = 0$ over an algebraically closed field k of characteristic zero. Assuming C to be analytically irreducible at P, we introduce a finite number of pairs of coprime positive integers $(\lambda_1, \mu_1), (\lambda_2, \mu_2), \ldots, (\lambda_h, \mu_h)$, which are called the *characteristic pairs* of C at P.

Geometric picture. In a moment, by using Newton's Theorem we shall define the characteristic pairs of C at P. But the characteristic pairs can actually be defined without assuming the characteristic of k to be zero. Here is a characteristic free geometric definition of the number h of characteristic pairs of C at P. Apply QDTs (quadratic transformations) until the total transform of C has only normal crossings. Then ultimately an exceptional line will meet at most three other exceptional lines. Moreover, the number of exceptional lines which do meet three others is exactly equal to the number h of characteristic pairs of C at P. See (the oversimplified) Figure 21.1 for $h = 2$.

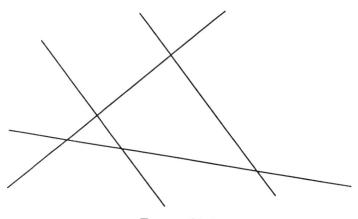

FIGURE 21.1

Parametrization. Going back to the assumption that k is an algebraically closed field of characteristic zero, we can define the characteristic pairs of C at P thus. Suppose that by translation we have brought P to the origin and then by rotation we have arranged matters so that $X = 0$ is not tangent to C at P. That is, the intersection multiplicity $I(C, X; P) = \text{mult}_P C = u$.

Now by Newton's Theorem, at P we can parametrize C as $X = T^u$ and

$$Y = \eta(T) = \cdots + b_{m_1} T^{m_1} + \cdots + b_{m_2} T^{m_2} + \cdots + b_{m_3} T^{m_3} + \cdots + b_{m_h} T^{m_h} + \cdots$$

where in the expansion, m_1 is the smallest exponent not divisible by u whose coefficient b_{m_1} is nonzero, m_2 is the smallest exponent not divisible by GCD (u, m_1) whose coefficient b_{m_2} is nonzero, m_3 is the smallest exponent not divisible by GCD (u, m_1, m_2) whose coefficient b_{m_3} is nonzero, ... , and m_h is the smallest exponent for which GCD $(u, m_1, \ldots, m_h) = 1$ whose coefficient b_{m_h} is nonzero. Let $m_0 = u$ and

$$d_{i+1} = \mathrm{GCD}(m_0, m_1, \ldots, m_i) \quad \text{for } 0 \le i \le h.$$

Note that the d_i's form a decreasing sequence of positive integers $d_1 > d_2 > d_3 > \cdots > d_h > d_{h+1} = 1$. Following Halphen [**Hal**] and Smith [**Smt**], we define the *characteristic pairs* $(\lambda_1, \mu_1), (\lambda_2, \mu_2), \ldots, (\lambda_h, \mu_h)$ of C at P by putting

$$\lambda_i = \frac{m_i}{d_{i+1}} \quad \text{and} \quad \mu_i = \frac{d_i}{d_{i+1}} \quad \text{for } 1 \le i \le h.$$

We note that the m-sequence $(m_0, m_1, \ldots, m_h)$ and these characteristic pairs determine each other. Together with the m-sequence, we define the q-sequence $(q_0, q_1, \ldots, q_h)$ by putting

$$q_0 = m_0, \quad q_1 = m_1, \quad q_2 = m_2 - m_1, \quad \ldots, \quad q_h = m_h - m_{h-1}.$$

Next, we define the s-sequence $(s_0, s_1, \ldots, s_h)$ by putting

$$s_0 = m_0 \quad \text{and} \quad s_i = \sum_{j=1}^{i} q_j d_j \quad \text{for } 1 \le i \le h.$$

Finally, we define the r-sequence $(r_0, r_1, \ldots, r_h)$ by putting

$$r_0 = m_0 \quad \text{and} \quad r_i = s_i/d_i \quad \text{for } 1 \le i \le h.$$

We note that each of the four sequences m, q, s, r determines the other. (Verify!) It turns out that the r-sequence is the best.

Intersection multiplicities. The r-sequence can be geometrically obtained thus. Consider the intersection multiplicity $I(C, D; P)$ as D varies over all curves, not having C as a component. These intersection multiplicities form a semigroup S of non-negative integers, i.e. a set of non-negative integers that contains 0 and is closed under addition. It can be shown that the semigroup S and the r-sequence $(r_0, r_1, \ldots, r_h)$ determine each other.

EQUIVALENCE OF THE r-SEQUENCE AND THE INTERSECTION SEMIGROUP S. The r-sequence $(r_0, r_1, \ldots, r_h)$ generates S as a semigroup, i.e., $S = r_0 \mathbb{N} + r_1 \mathbb{N} + \cdots + r_h \mathbb{N}$ where $\mathbb{N}$ denotes the set of all non-negative integers. Conversely, the intersection semigroup S determines the r-sequence

$(r_0, r_1, \ldots, r_h)$ as follows:

$$r_0 = \min\left(S \setminus \{0\}\right)$$
$$r_1 = \min\left(S \setminus r_0\mathbb{N}\right)$$
$$\vdots$$
$$r_i = \min\left(S \setminus r_0\mathbb{N} + \cdots + r_{i-1}\mathbb{N}\right)$$
$$\vdots$$
$$r_h = \min\left(S \setminus r_0\mathbb{N} + \cdots + r_{h-1}\mathbb{N}\right)$$
$$\varnothing = S \setminus r_0\mathbb{N} + \cdots + r_h\mathbb{N}.$$

Value semigroup. Let V be the discrete valuation of $k(C)$ having center P on C, and let R be the local ring of P on C. By the *value semigroup* of C at P we mean the semigroup of non-negative integers $\{V(z) : 0 \neq z \in R\}$. Now we have the

EQUIVALENCE OF VALUE SEMIGROUP AND INTERSECTION SEMIGROUP. Obviously, $S = $ *the value semigroup* $\{V(z) : 0 \neq z \in R\}$.

Multiplicity sequence. Since C is assumed to be analytically irreducible at P, by the Tangent Lemma it follows that the "multiplicity tree" of C at P is reduced to a sequence

$$m_0 = u = u_0 \geq u_1 \geq u_2 \geq \cdots \geq u_{j-1} \geq u_j \geq \cdots.$$

In other words, we have a unique infinite QDT sequence

$$(C_0, P_0) \to (C_1, P_1) \to (C_2, P_2) \to \cdots \to (C_{j-1}, P_{j-1}) \to (C_j, P_j) \to \cdots,$$

where $(C_0, P_0) = (C, P)$, (C_j, P_j) is the proper transform of (C_{j-1}, P_{j-1}) for all $j > 0$,

$$\text{mult}_{P_j} C_j = u_j \quad \text{for all } j \geq 0,$$

and $u_j = 1$ for all large enough j. We call $u_0, u_1, u_2, \ldots$ the *multiplicity sequence* of C at P. Again, the characteristic pairs and the multiplicity sequence determine each other.

EQUIVALENCE OF CHARACTERISTIC PAIRS AND MULTIPLICITY SEQUENCE. Let

$$\mu_{j,0} = \mu_j \mu_{j+1} \cdots \mu_h \quad \text{for } 1 \leq j \leq h$$

and take

$$\mu_{h+1,0} = 1 \quad \text{and} \quad \lambda_0 = 0.$$

For $1 \leq j \leq h$ we clearly have

$$\text{GCD}\left((\lambda_j - \lambda_{j-1}\mu_j)\mu_{j+1,0}, \, \mu_{j,0}\right) = \mu_{j+1,0}.$$

Applying Euclid's algorithm to find this GCD we get

$$(\lambda_j - \lambda_{j-1}\mu_j)\mu_{j+1,0} = q_{j,0}\mu_{j,0} + \mu_{j,1}$$

$$\mu_{j,0} = q_{j,1}\mu_{j,1} + \mu_{j,2}$$

$$\vdots$$

$$\mu_{j,i-1} = q_{j,i}\mu_{j,i} + \mu_{j,i+1}$$

$$\vdots$$

$$\mu_{j,h_j-1} = q_{j,h_j}\mu_{j,h_j}$$

where $q_{j,0}$ and h_j are positive integers and for $1 \leq i \leq h_j$ we have that $q_{j,i}$ and $\mu_{j,i}$ are positive integers with $\mu_{j,i} < \mu_{j,i-1}$. Now clearly

$$\mu_{j,0} > \mu_{j,1} > \mu_{j,2} > \cdots > \mu_{j,h_j} = \mu_{j+1,0}.$$

Let u_j' be the sequence $u_{j,0}', u_{j,1}', \ldots, u_{j,h_j}'$ of length $q_{j,0} + \cdots + q_{j,h_j}$, where $u_{j,i}'$ is the sequence of integers simply consisting of the integer $\mu_{j,i}$ repeated $q_{j,i}$ times. Then the sequence $u_1', u_2', \ldots, u_h', 1, 1, 1, \ldots$ coincides with the multiplicity sequence $u_0, u_1, u_2, \ldots$, i.e., $u_0 = u_{1,0}'$, $u_1 = u_{1,1}', \ldots, u_{h_j} = u_{1,h_j}'$, $u_{h_j+1} = u_{2,0}'$, $\ldots$. Conversely, by reversing this algorithm, we can express the characteristic pairs in terms of the multiplicity sequence.

REMARK. To the curve C at the point P, where it is assumed to be analytically irreducible, we have attached the three entities (1) characteristic pairs, (2) intersection semigroup (value semigroup), and (3) multiplicity sequence. We have indicated that these three are equivalent to each other. That is, any one of them determines the other two. The proof of this equivalence follows from the Inversion Theorem that I proved in [A16] and which was reproduced in [A20]. Also see [A27] and [A33]. To state the Inversion Theorem, we define the "reverse characteristic pairs" $(\lambda_1', \mu_1'), (\lambda_2', \mu_2'), \ldots, (\lambda_{h'}', \mu_{h'}')$ of C at P by expanding X in terms of Y thus. Let $u' = \operatorname{ord} \eta(T)$. Note that then u' is a positive integer with $u' \leq m_1$. Let $Y = T^{u'}$ and

$$X = \xi(T) = \cdots + c_{m_1'}T^{m_1'} + \cdots + c_{m_2'}T^{m_2'} + \cdots + c_{m_3'}T^{m_3'} + \cdots + c_{m_{h'}'}T^{m_{h'}'} + \cdots$$

be the parametrization of C at P obtained by Newton's Theorem. In the expansion, m_1' is the smallest exponent not divisible by u' whose coefficient $c_{m_1'}$ is nonzero, m_2' is the smallest exponent not divisible by $\operatorname{GCD}(u', m_1')$ whose coefficient $c_{m_2'}$ is nonzero, m_3' is the smallest exponent not divisible by $\operatorname{GCD}(u', m_1', m_2')$ whose coefficient $c_{m_3'}$ is nonzero, $\ldots$, and $m_{h'}'$ is the smallest exponent for which $\operatorname{GCD}(u', m_1', \ldots, m_{h'}') = 1$ whose coefficient $c_{m_{h'}'}$ is nonzero. Let $m_0' = u'$ and

$$d_{i+1}' = \operatorname{GCD}(m_0', m_1', \ldots, m_i') \quad \text{for } 0 \leq i \leq h'.$$

Note that the d_i''s form a decreasing sequence of positive integers $d_1' > d_2' > d_3' > \cdots > d_{h'}' > d_{h'+1}' = 1$. We define the *reverse characteristic pairs* $(\lambda_1', \mu_1'), (\lambda_2', \mu_2'), \ldots, (\lambda_{h'}', \mu_{h'}')$ of C at P by putting

$$\lambda_i' = \frac{m_i'}{d_{i+1}'} \quad \text{and} \quad \mu_i' = \frac{d_i'}{d_{i+1}'} \quad \text{for } 1 \le i \le h'.$$

Now we may state the

INVERSION THEOREM. *The relationship between the characteristic pairs and the reverse characteristic pairs is given by the following.*

(1) *If* $u' = u$, *then* $h' = h$, $\lambda_i' = \lambda_i$ *for* $1 \le i \le h$, *and* $\mu_i' = \mu_i$ *for* $1 \le i \le h$.

(2) *If* $u' = m_1$, *then* $h' = h$, $\lambda_1' = \mu_1$, $\mu_1' = \lambda_1$, $\lambda_i' = \lambda_i - (\lambda_1 - \mu_1)\mu_2\mu_3 \cdots \mu_i$ *for* $2 \le i \le h$, *and* $\mu_i' = \mu_i$ *for* $2 \le i \le h$.

(3) *If* $u \ne u' \ne m_1$, *then* $h' = h + 1$, $\lambda_1' = 1$, $\mu_1' = u'/u$, $\lambda_{i+1}' = \lambda_i - (u'/u - 1)\mu_1\mu_2 \cdots \mu_i$ *for* $1 \le i \le h$, *and* $\mu_{i+1}' = \mu_i$ *for* $1 \le i \le h$.

Genus formula. Recall that the genus g of the irreducible algebraic plane curve $C : F(X, Y) = 0$ of degree n is given by the formula

$$g = \frac{(n-1)(n-2)}{2} - \text{number of double points counted properly.}$$

To count properly, as a first approximation, if $\text{mult }_P C = \nu$, then it accounts for $\frac{\nu(\nu-1)}{2}$ double points. More precisely, it accounts for

$$\delta(P) = \sum^P \frac{\nu_i(\nu_i - 1)}{2}$$

double points where the sum $\sum^P$ is taken over multiplicities of points at P and infinitely near to P. Thus the genus formula is

$$g = \frac{(n-1)(n-2)}{2} - \sum_P \delta(P).$$

It can be shown that $\delta(P)$ can be expressed in terms of the r-sequence by the formula

$$\boxed{\delta(P) = \frac{1}{2} + \frac{1}{2}\sum_{i=0}^{h} r_i \left(\frac{d_i}{d_{i+1}} - 1\right)}$$

where $d_0 = 0$, provided C is analytically irreducible at P. If there is more than one branch at P, the formula is to be modified by adding the contributions of the various branches and increasing it by the sum of their mutual intersection multiplicities. This can be deduced from the theorems of Dedekind and Noether, which say that

$$F_Y(x, y)A' = \mathfrak{C}\mathfrak{D}' \quad \text{and} \quad \delta(P) = \text{length}_{R(P)}\mathfrak{C}R(P),$$

where x and y are the functions on C induced by X and Y, $\mathfrak{C}$ is the conductor of the affine coordinate ring $A = k[x, y]$ of C, A' is the integral

Discriminant locus = Branch locus + Projection of singular locus

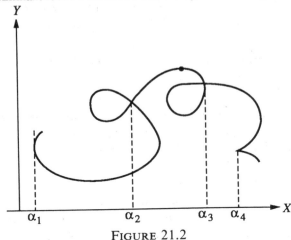

FIGURE 21.2

closure of A in the function field $K = k(x, y)$ of C, $\mathfrak{D}'$ is the ideal in A' given by the different $\mathfrak{D}$ of $K/k(x)$, and $R(P)$ is the local ring of P on C. Geometrically speaking, the theorems of Dedekind and Noether may be paraphrased into the *aphorism* shown in Figure 21.2. The Y-discriminant of F is $D(X) = \operatorname{Res}_Y(F, F_Y)$. By the discriminant locus we mean the set of points α such that $D(\alpha) = 0$. The projection of the singular locus consists of those points α for which $F(\alpha, \beta) = F_X(\alpha, \beta) = F_Y(\alpha, \beta) = 0$ for some β. In Figure 21.2, α_1 is a branch point but not the projection of a singular point, α_2 is the projection of a singular point but not a branch point, and α_3 and α_4 are branch points as well as the projections of singular points. So what is the branch locus? Here is the oldest description of the

Branch locus. Assume that $k = \mathbb{C}$ and

$$F(X, Y) = Y^n + a_1(X)Y^{n-1} + \cdots + a_n(X).$$

Then $F(X, Y) = 0$ defines Y as an n-valued function of X. For most values of X there are n distinct values of Y. Points with less than n corresponding values of Y are the discriminant points. Suppose for $X = \alpha$ the corresponding values of Y are $\beta_1, \ldots, \beta_l$. We want to decide whether α is a branch point or not. Take a nondiscriminant point α' so that we have n distinct values $\beta_1', \ldots, \beta_n'$ of Y corresponding to the value α' of X. Now for every value x of X in a small neighborhood of α' we get n distinct values $y_1, \ldots, y_n$ of Y in small neighborhoods of $\beta_1', \ldots, \beta_n'$ respectively. Thus we get n functions $y_1, \ldots, y_n$ of x in a neighborhood of α'. These n functions are continuous (where continuity can be suitably defined for multiple valued functions as well). In fact they are analytic. (This can be proved by Hensel's Lemma.) Let Γ be the path consisting of the segment from α' to a point α'' on a small circle around α, followed by the said small circle, and then followed by the segment from α'' to α'. By choosing α'' suitably, it can be arranged that Γ does not pass through any discriminant point. Now make analytic continuations along Γ. The functions $y_1, \ldots, y_n$ will undergo a permutation, and we define α is a *branch point* $\Leftrightarrow$ this permutation is not identity. Equivalently, the said permutation can be uniquely written as a product of disjoint cycles of lengths $e_1, \ldots, e_v$ (say), and we define: α is a branch point $\Leftrightarrow e_i > 1$ for some i. At any rate, every branch point is a discriminant point, but a discriminant point need not be a branch point. See Figure 21.3.

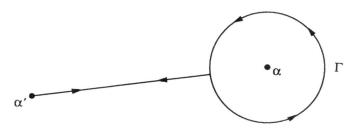

FIGURE 21.3

Let $\alpha_1^*, \ldots, \alpha_w^*$ be all the branch points, and let Γ_j^* be the path consisting of the segment from α' to a point α_j' on a small circle around α_j^*, followed by the said small circle, and then followed by the segment from α_j'' to α'. By choosing the nondiscriminant point α' suitably, it can be arranged that the paths $\Gamma_1^*, \ldots, \Gamma_w^*$ do not pass through any discriminant point and do not meet each other except at α'. Let θ_j be the permutation undergone by $y_1, \ldots, y_n$ as we make analytic continuations along Γ_j^*. Let G be the group of those permutations of $y_1, \ldots, y_n$ that can be obtained by making analytic continuations along various continuous closed paths starting and ending at α' and not passing through any discriminant point. It can easily be seen that then G is generated by the permutations $\theta_1, \ldots, \theta_w$ whose product $\theta_1 \cdots \theta_w$ is the identity permutation. Moreover, the permutation group G is *transitive*, i.e., given any $i \neq i'$, some member of G sends y_i to $y_{i'}$. See Figure 21.4 on page 000. G is called the *monodromy group* of F. Now the fundamental group $\pi_1(\mathbb{P}^1 \setminus \{\alpha_1^*, \ldots, \alpha_w^*\})$ of the complex projective line $\mathbb{P}^1$ punctured at the points $\alpha_1^*, \ldots, \alpha_w^*$, and with base point at α', consists of the equivalence classes of continuous closed paths in $\mathbb{P}^1 \setminus \{\alpha_1^*, \ldots, \alpha_w^*\}$ starting and ending at α', where two paths are equivalent if they can be continuously deformed into each other. The well-known *Monodromy Theorem* of complex analysis says that two equivalent paths give the same analytic continuation. Therefore, the monodromy group G is a homomorphic image of the fundamental group. Note that the fundamental group is a free group on $w - 1$ generators because it is generated by the equivalence classes of the paths $\Gamma_1^*, \ldots, \Gamma_w^*$, and the

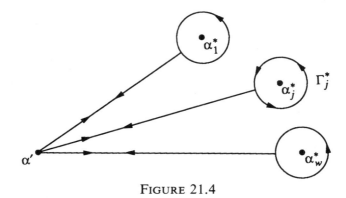

FIGURE 21.4

only relation between them is the fact that their product is the equivalence class of the "null" path. It can also be shown that the monodromy group G is isomorphic to the Galois group of the least Galois extension of $k(x, y)/k(x)$. Hence the said Galois group is generated by $w - 1$ generators. Because of the Riemann Existence Theorem every such group does occur. For details see [A03] and [Sie].

Problem. Find an algebraic proof of the fact that the said Galois group is generated by $w - 1$ generators and every such group does occur.

Ramification exponents. The above defined numbers $e_1, \ldots, e_v$ are called *ramification exponents.* There are at least three different methods of introducing them.

(1) *Riemann's method.* This is the above method in which $e_1, \ldots, e_v$ are defined as the lengths of the disjoint cycles.

(2) *Newton's method.* By Newton's Theorem we get $F(X + \alpha, Y) = \prod_{i=1}^{v^*} F_i^*(X, Y)$ where $F_i^*(X, Y)$ is a monic irreducible polynomial of positive degree e_i^* in Y with coefficients in $k[[X]]$. It can easily be seen that then we must have $v^* = v$ and, after a suitable relabelling, $e_i^* = e_i$ for $1 \le i \le v$.

(3) *Dedekind's method.* Let M^* be the maximal ideal in $A^* = k[x]$ generated by $x - \alpha$, and let A' be the integral closure of A^* in $k(x, y)$. It can be shown that then there are exactly v distinct maximal ideals $M_1, \ldots, M_v$ in A' whose intersection with A^* is M^*. By labelling them suitably we have $M^* A' = \prod_{i=1}^{v} M_i^{e_i}$.

Discriminant ideal. To recapitulate, α is a branch point $\Leftrightarrow e_i > 1$ for some i. The Krull-Dedekind procedure of finding all the branch points consists of introducing an ideal in the ring $A^* = k[x]$, which is called the discriminant of A^* in $K = k(x, y)$ and is denoted by $\mathrm{Disc}\,(A^*, K)$. For any $\alpha \in k$ we have $\mathrm{Disc}\,(A^*, K) \subset (x - \alpha)A^* \Leftrightarrow \alpha$ is a branch point. Quite generally, let B^* be any domain integrally closed in its quotient field L^*, let L be a finite algebraic field extension of L^*, and let $n' = [L : L^*]$. For any elements $\xi_1, \ldots, \xi_{n'}$ in L we consider the n' by n' matrix

$\left(\text{Trace}_{L/L^*} \left(\xi_i \xi_j \right) \right)$. We define

$$\text{Disc}_{L/L^*} (\xi_1, \ldots, \xi_{n'}) = \det \left(\text{Trace}_{L/L^*} \left(\xi_i \xi_j \right) \right).$$

We call $\text{Disc}_{L/L^*} (\xi_1, \ldots, \xi_{n'})$ the *discriminant* of $(\xi_1, \ldots, \xi_{n'})$ relative to L/L^*. Note that $\text{Trace}_{L/L^*}(\xi)$ is the sum of all conjugates of ξ. For any $\xi \in L$ we take the unique monic polynomial

$$H(Y) = Y^{n'} + c_1 Y^{n'-1} + \cdots + c_{n'}$$

of degree n' in Y with coefficients in L^* such that $H(\xi) = 0$ and $H(Y) = H^*(Y)^{n'/n^*}$ for some monic polynomial $H^*(Y)$ of degree $n^* = [L^*(\xi) : L^*]$ in Y with coefficients in L^*. Now $\text{Trace}_{L/L^*}(\xi) = -c_1$. The above definition of $\text{Disc}_{L/L^*} (\xi_1, \ldots, \xi_n)$ is motivated by the equation

$$\text{Disc}_{L/L^*} \left(1, \xi, \ldots, \xi^{n'-1} \right) = \text{Disc}_Y(H) = \text{Res}_Y(H, H_Y).$$

Upon letting B' to be the integral closure of B^* in L, we note that if $\xi_1, \ldots, \xi_{n'}$ are any elements in B' then $\text{Disc}_{L/L^*} (\xi_1, \ldots, \xi_{n'}) \in B^*$. We define

$\text{Disc}(B^*, L) = $ the ideal in B^* generated by all the discriminants

$$\text{Disc}_{L/L^*} (\xi_1, \ldots, \xi_{n'}) \text{ as } \xi_1, \ldots, \xi_{n'} \text{ vary over } B'.$$

We call $\text{Disc}(B^*, L)$ the *discriminant* of B^* in L. Given any prime ideal N^* in B^*, it can be seen that there are at most a finite number of prime ideals $N_1, \ldots, N_{v'}$ in B' whose intersection with B^* is N^*. Moreover, we have $\sum_{j=1}^{v'} [B/N_j : B^*/N^*]_s \leq n'$ where s denotes the separable part of the field degree; N^* is said to be unramified in L if $\sum_{j=1}^{v'} [B/N_j : B^*/N^*]_s = n'$. Referring to [A04] for details, we can now state the Dedekind-Krull

DISKRIMINANTENSATZ. *N^* is unramified in L iff $\text{Disc}(B^*, L) \not\subset N^*$.*

REMARK. In case $B^* = A^* = k[x]$, $L = K = k(x, y)$, and $N^* = M^* = (x - \alpha)A^*$, it can easily be seen that M^* is unramified in K iff α is not a branch point, i.e., iff $e_1 = \cdots = e_v = 1$.

Four incarnations of Bezout's Theorem. To put Dedekind's method in proper perspective, in the situation of the paragraph on discriminant ideal, assume that $\dim B^* = 1$ and L is separable over L^*. In other words, let B^* be a *Dedekind domain* (a one-dimensional normal Noetherian domain), let B' be the integral closure of B^* in a finite separable algebraic field extension L of the quotient field L^* of B^*, and let $n' = [L : L^*]$. Now the conclusion of Dedekind's method can be generalized to the following assertion, which may be regarded as the incarnation of Bezout's Theorem belonging to

(B1). COLLEGE ALGEBRA. *For any maximal ideal N^* in B^* we have $N^*B' = \prod_{i=1}^{v'} N_i^{e_i'}$ where $N_1, \ldots, N_{v'}$ are the distinct maximal ideals in B' containing N^*, and upon letting $f_i' = [B'/N_i : B^*/N^*]$ we have $\sum_{i=1}^{v'} e_i' f_i' = n'$.*

To see that the college algebra assertion is an incarnation of Bezout, we take $L = K =$ the function field $k(C)$ of the irreducible algebraic plane curve C, and $N^* =$ the ideal zB^* in the ring $B^* = k[z]$ where z is any (nonzero) rational function of C. Then (the number of zeros of z) $= n' =$ (the number of poles of z). Now here are the other three incarnations of Bezout.

(B2). COMPLEX ANALYSIS. *Number of zeros of a rational function on the irreducible algebraic plane curve C (or a meromorphic function on the Riemann surface of C) = number of its poles.*

(B3). HIGH SCHOOL ALGEBRA. *The degree of a univariate polynomial equals the number of its roots (counted with multiplicities).*

(B4). GEOMETRY. $C_m \cdot C_n = mn$ *points.*

Generalized tubular knots. Going back to the plane curve $C : F(X, Y) = 0$ over the complex field $k = \mathbb{C}$, where C is assumed to be analytically irreducible at $P = (0, 0)$, we shall now associate to C at P a *generalized tubular knot.* Let U be a small nice neighborhood of P in the complex (X, Y)-plane, say $U = \{(X, Y) : |X| < \epsilon, |Y| < \epsilon'\}$ where ϵ and ϵ' are small positive real numbers. Let U^* be the boundary of U. Now U^* is a 3-sphere, and the intersection of C with U^* can be shown to be a *knot*, i.e., a homeomorphic image of a circle. By a stereographic projection of the 3-sphere U^* (from a point not on C) we obtain a knot E in $\mathbb{R}^3$. We call E the *knot* of C (at P). To describe how E looks like, we define a certain class of knots. Let W be the surface of an ordinary torus, a torus with an ordinary (unknotted) circle Λ as axis. If J is any knot on W, then J makes a certain number of turns, say σ along the meridians of W (i.e., around Λ), and a certain number of turns, say τ, along the parallels of W. We call J a *torus knot* of type (σ, τ). More generally we can start with any knot Θ and consider the surface W_Θ of a torus with Θ as axis. If J_0 is any knot on W_Θ, then J_0 makes a certain number of turns, say σ_0, along the meridians of W_Θ, and a certain number of turns, say τ_0, along the parallels of W_Θ. We call J_0 a knot of type $(\Theta; \sigma_0, \tau_0)$. Given a knot J (in $\mathbb{R}^3$), we say that J is a tubular knot of type $(\sigma_1, \tau_1), \ldots, (\sigma_\gamma, \tau_\gamma)$, where $\sigma_1, \tau_1, \ldots, \sigma_\gamma, \tau_\gamma$ are integers greater than one with GCD$(\sigma_i, \tau_i) = 1$ for $1 \le i \le \gamma$, if there exist knots $J_1, \ldots, J_\gamma$ such that J_1 is a torus knot of type (σ_1, τ_1), and J_i is a knot of type $(J_{i-1}; \sigma_i, \tau_i)$ for $2 \le i \le \gamma$, and $J = J_\gamma$. It can be shown that the knot E of C is a tubular knot of type $(\lambda_1, \mu_1), \ldots, (\lambda_h, \mu_h)$. Given any other analytically irreducible plane curve

C^*, we say that the C is *topologically equivalent* to C^* if the knot E of C is *isotopic* to the knot E^* of C^*, i.e., if by a continuous deformation of $\mathbb{R}^3$ we can send E to E^*. Let us also say that C and C^* are *algebraically* (resp. *geometrically, arithmetically*) *equivalent* if they have the same characteristic pairs (resp. multiplicity sequence, value semigroup). It can be proved that *C and C^* are topologically equivalent iff they are algebraically equivalent.* Since we have already indicated that C and C^* are algebraically equivalent $\Leftrightarrow$ they are geometrically equivalent $\Leftrightarrow$ they are arithmetically equivalent, we conclude that *the four classifications, topological, algebraic, geometric, and arithmetical, of analytically irreducible plane curves lead to the same thing.* Note that if C is analytically reducible, then each branch of C gives rise to a knot. These knots are linked in some manner. For details about the topological aspects of singularities see [**Bra**] [**Bur**] [**Kah**] [**Mum**] [**Ree**] [**Za2**] and [**Za3**].

NOTE. As an example, for the cusp $Y^2 = X^3$, (1) there is only one characteristic pair, (2) the multiplicity sequence is $(2, 1, 1, \ldots)$, (3) the value semigroup consists of all integers greater than 1 (together with 0), and (4) the knot is the trefoil knot drawn in Figure 21.5.

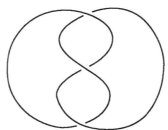

FIGURE 21.5

Criterion for One Place and Jacobian Problem

As we have said several times, a plane curve can be polynomially parametrized if and only if it can be rationally parametrized and has only one place at infinity. We also know that the curve is rational if and only if its genus is zero. We have given several methods of calculating the genus. We shall now complete the discussion of polynomial parametrization by giving an algorithmic criterion of having only one place at infinity. It turns out that in addition to polynomial parametrization, such a criterion also has applications to the

Plane Jacobian problem. If $F(X, Y)$ and $G(X, Y)$ are polynomials over a field k of characteristic 0 then:

$$\frac{J(F, G)}{J(X, Y)} = \text{ a nonzero constant}$$

$$\overset{?}{\Rightarrow} X \text{ and } Y \text{ are polynomials in } F \text{ and } G$$

where

$$\frac{J(F, G)}{J(X, Y)} = \text{ the Jacobian of } F \text{ and } G \text{ with respect to } X \text{ and } Y$$

$$= \det \begin{pmatrix} F_X & F_Y \\ G_X & G_Y \end{pmatrix}.$$

The above implication does not hold over fields of characteristic p; $(F, G) = (X + X^p, Y + Y^p)$ is a counterexample. Although we do not know the answer to the plane Jacobian problem, nevertheless we may also pose the

General Jacobian problem. If $f_1(X_1, \ldots, X_n), \ldots, f_n(X_1, \ldots, X_n)$ are n polynomials in n variables, over a field k of characteristic 0, then,

$$\frac{J(f_1, \ldots, f_n)}{J(X_1, \ldots, X_n)} = \text{ a nonzero constant}$$

$$\overset{?}{\Rightarrow} X_1, \ldots, X_n \text{ are polynomials in } f_1, \ldots, f_n,$$

where

$$\frac{J(f_1, \ldots, f_n)}{J(X_1, \ldots, X_n)} = \text{ the } Jacobian \text{ of } f_1, \ldots, f_n \text{ with respect to } X_1, \ldots, X_n$$

$$= \det \left(\frac{\partial(f_1, \ldots, f_n)}{\partial(X_1, \ldots, X_n)} \right)$$

and where

$$\frac{\partial(f_1, \ldots, f_q)}{\partial(X_1, \ldots, X_s)} = \text{ the } q \text{ by } s \text{ matrix } \left(\frac{\partial f_i}{\partial X_j}\right)_{1 \le j \le s}^{1 \le i \le q}.$$

REMARK. In my Purdue lectures of 1971–72 [A22], the general Jacobian problem is settled affirmatively in the Galois case. It is shown that if $\frac{J(f_1, \ldots, f_n)}{J(X_1, \ldots, X_n)}$ is a nonzero constant and $k(X_1, \ldots, X_n)/k(f_1, \ldots, f_n)$ is a Galois extension, then $X_1, \ldots, X_n$ are polynomials in $f_1, \ldots, f_n$. In my Tata lectures of 1977 [A26], the plane Jacobian problem is settled affirmatively in the one place at infinity case. It is shown that if $\frac{J(F, G)}{J(X, Y)}$ is a nonzero constant and the plane curve $F(X, Y) = 0$ has only one place at infinity, then X and Y are polynomials in F and G. So, as far as the plane Jacobian problem is concerned, it only remains to show that if $\frac{J(F, G)}{J(X, Y)}$ is a nonzero constant, then the plane curve $F(X, Y) = 0$ has only one place at infinity.

The Jacobian problem is clearly related to the

Inverse Function Theorem. For one variable, the Inverse Function Theorem says that if $X = $ a power series $\Theta(Y)$ with $\Theta(0) = 0$ then

$$\Theta_Y(0) \ne 0 \Rightarrow Y = \text{ a power series } \Lambda(X).$$

More generally, for n variables, the Inverse Function Theorem says that if $X_i = $ a power series $\Theta_i(Y_1, \ldots, Y_n)$ with $\Theta_i(0, \ldots, 0) = 0$ for $1 \le i \le n$ then

$$\frac{J(\Theta_1, \ldots, \Theta_n)}{J(Y_1, \ldots, Y_n)}(0, \ldots, 0) \ne 0 \Rightarrow Y_i$$
$$= \text{ a power series } \Lambda_i(X_1, \ldots, X_n) \quad \text{for } 1 \le i \le n.$$

In turn, the Inverse Function Theorem is closely related to the

Implicit Function Theorem. For one dependent variable, the Implicit Function Theorem says that if $F(X, Y)$ is a power series with $F(0, 0) = 0$ then:

$$F_Y(0, 0) \ne 0 \Rightarrow \text{ there exists a power series } \eta(X) \text{ with } \eta(0) = 0$$
$$\text{ such that } F(X, \eta(X)) = 0.$$

In general if $F(X_1, \ldots, X_m, Y)$ is a power series with $F(0, \ldots, 0, 0) = 0$ then

$$F_Y(0, \ldots, 0, 0) \ne 0 \Rightarrow \text{ there exists a power series } \eta(X_1, \ldots, X_m)$$
$$\text{ with } \eta(0, \ldots, 0) = 0$$
$$\text{ such that } F(X_1, \ldots, X_m, \eta(X_1, \ldots, X_m)) = 0.$$

Generally, for n dependent variables, the Implicit Function Theorem says if $f_i(X_1, \ldots, X_m, Y_1, \ldots, Y_n)$ is a power series with $f_i(0, \ldots, 0, 0, \ldots, 0) =$

0 for $1 \le i \le n$ then:

$$\frac{J(f_1, \ldots, f_n)}{J(Y_1, \ldots, Y_n)}(0, 0, \ldots, 0) \ne 0 \Rightarrow$$

there exist power series $\eta_i(X_1, \ldots, X_m)$ with

$\eta_i(0, \ldots, 0) = 0$ for $1 \le i \le n$ such that

$f_i(X_1, \ldots, X_m, \eta_1(X_1, \ldots, X_m), \ldots, \eta_n(X_1, \ldots, X_m)) = 0$
for $1 \le i \le n$.

Exercise. Show that the Implicit Function Theorem for two variables ($2n$ variables) implies, and is implied by, the Inverse Function Theorem for 1 variable (n variables). See [A07].

Irreducible varieties. Recall that an irreducible algebraic variety W in n-space over a field k is the locus of a finite number of polynomial equations $h_1(X_1, \ldots, X_n) = \cdots = h_q(X_1, \ldots, X_n) = 0$, such that $h_1, \ldots, h_q$ generate a prime ideal in the polynomial ring $k[X_1, \ldots, X_n]$. Upon letting $x_1, \ldots, x_n$ to be the functions on W induced by $X_1, \ldots, X_n$ we have

$$\dim W = \operatorname{trdeg}_k k(x_1, \ldots, x_n).$$

For any point $P = (\alpha_1, \ldots, \alpha_n)$ of W upon letting $R(P)$ to be the local ring of P on W we have

$$P \text{ is a simple point of } W \Leftrightarrow R(P) \text{ is regular.}$$

Exercise. Show that

$$\dim W = n - \operatorname{rank} \left. \frac{\partial(h_1, \ldots, h_q)}{\partial(X_1, \ldots, X_n)} \right|_{X_1 = x_1, \ldots, X_n = x_n}.$$

Show that

P is a simple point of $W \Leftrightarrow$

$$\operatorname{rank} \left. \frac{\partial(h_1, \ldots, h_q)}{\partial(X_1, \ldots, X_n)} \right|_{X_1 = \alpha_1, \ldots, X_n = \alpha_n} = n - \dim W,$$

whereas

P is a singular point of $W \Leftrightarrow$

$$\operatorname{rank} \left. \frac{\partial(h_1, \ldots, h_q)}{\partial(X_1, \ldots, X_n)} \right|_{X_1 = \alpha_1, \ldots, X_n = \alpha_n} < n - \dim W.$$

General varieties. Let W' be any algebraic variety in n-space over the field k and let $h'_1, \ldots, h'_{q'}$ be generators of the ideal of W' in $k[X_1, \ldots, X_n]$. Note that the said ideal is its own radical. Let $W_1, \ldots, W_l$ be the irreducible components of W'. Note that then

$$\dim W' = \max_{1 \le i \le l} \dim W_i.$$

For any point $P' = (\alpha'_1, \ldots, \alpha'_n)$ of W' we put

$$\dim_{P'} W' = \max_{P' \in W_i} \dim W_i,$$

where the max is taken over all i for which $P' \in W_i$. We recall that P' is a simple point of W iff $P' \in W_i$ for only one i and P' is a simple point of W_i for that i.

Exercise. In case of $k = \mathbb{C}$, show that there exists a neighborhood U' of P' in $\mathbb{C}^n$ such that for every neighborhood U of P' in U' we have

$$\dim_{P'} W' = n - \max_{(\beta_1, \ldots, \beta_n) \in U \cap W'} \operatorname{rank} \left. \frac{\partial(h'_1, \ldots, h'_{q'})}{\partial(X_1, \ldots, X_n)} \right|_{X_1 = \beta_1, \ldots, X_n = \beta_n}.$$

Show that

$$P' \text{ is a simple point of } W' \Leftrightarrow \operatorname{rank} \left. \frac{\partial(h'_1, \ldots, h'_{q'})}{\partial(X_1, \ldots, X_n)} \right|_{X_1 = \alpha'_1, \ldots, X_n = \alpha'_n} = n - \dim_{P'} W',$$

whereas

$$P' \text{ is a singular point of } W' \Leftrightarrow \operatorname{rank} \left. \frac{\partial(h'_1, \ldots, h'_{q'})}{\partial(X_1, \ldots, X_n)} \right|_{X_1 = \alpha'_1, \ldots, X_n = \alpha'_n} < n - \dim_{P'} W'.$$

Moreover, upon letting $\operatorname{Sing}(W')$ to be the singular locus of W', show that the irreducible components $W_1, \ldots, W_l$ are the closures of the connected components of $W' \setminus \operatorname{Sing}(W')$. Thus (in the complex case), the irreducible components of an algebraic variety can be found topologically. Hence, in particular, for a hypersurface given by one polynomial equation this gives a topological method of finding the irreducible factors of that polynomial. Extend all this to finding, topologically, the local irreducible components of analytic varieties and the irreducible factors of a convergent power series. See [A07].

Plane Jacobian problem and characteristic pairs. In the Purdue lectures [A22] and Tata lectures [A26], a connection is made between the plane Jacobian problem and characteristic pairs at infinity in the following sense. Let n^* and m^* be the (total) degrees of F and G respectively. By a linear transformation we can arrange matters so that F is a monic polynomial of degree n^* in Y with coefficients in $k[X]$, and G is a monic polynomial of degree m^* in Y with coefficients in $k[X]$. Let k^* be an algebraic closure of $k((X))$. That is, let k^* be an algebraically closed overfield of $k((X))$ such that k^* is an algebraic extension of $k((X))$. By Newton's Theorem we can find $T \in k^*((1/Y))$ with $T^{-n^*} = F$. Then we can expand G in $k^*((T))$ by writing

$$G = \cdots + b^*_{m^*_1} T^{m^*_1} + \cdots + b^*_{m^*_2} T^{m^*_2} + \cdots + b^*_{m^*_3} T^{m^*_3} + \cdots + b^*_{m^*_{h^*}} T^{m^*_{h^*}} + \cdots,$$

where in the expansion, m_1^* is the smallest exponent not divisible by n^* whose coefficient $b_{m_1^*}^*$ is nonzero, m_2^* is the smallest exponent not divisible by GCD (n^*, m_1^*) whose coefficient $b_{m_2^*}^*$ is nonzero, m_3^* is the smallest exponent not divisible by GCD (n^*, m_1^*, m_2^*) whose coefficient $b_{m_3^*}^*$ is nonzero, ..., and $m_{h^*}^*$ is the smallest exponent whose coefficient $b_{m_{h^*}^*}^*$ is nonzero and which is such that GCD $(n^*, m_1^*, \ldots, m_{h^*}^*)$ divides every integer m^{**} for which the coefficient of $T^{m^{**}}$ is nonzero. Let $m_0^* = n^*$ and

$$d_{i+1}^* = \mathrm{GCD}(m_0^*, m_1^*, \ldots, m_i^*) \quad \text{for } 0 \le i \le h^*.$$

Now $(m_i^*/d_{i+1}^*, d_i^*/d_{i+1}^*)_{1 \le i \le h^*}$ may be called the characteristic pairs of (F, G) at infinity. In my Purdue lectures of 1971, I showed that the plane Jacobian problem has an affirmative answer if either (a) $h^* \le 2$ or (b) $h^* = 3$ and $d_3^* \le 4$. Note that if GCD (n^*, m^*) is a prime number, then $h^* = 1$. If GCD (n^*, m^*) is a product of two prime numbers then $h^* \le 2$. From this I deduced an affirmative answer when $\min(m^*, n^*) \le 52$. Some of this can be found in [A22] and [A26].

REMARK. At any rate, not much seems to be known about the Jacobian problem. For a survey see [BCW].

One place at infinity. We are now ready to give the criterion for one place at infinity. Referring to [A31] and [A32] for details, here we shall present things only descriptively. So consider the plane curve $C : F(X, Y) = 0$ where

$$F(X, Y) = Y^N + a_1(X)Y^{N-1} + \ldots + a_N(X)$$

with polynomials $a_1(X), \ldots, a_N(X)$ over an algebraically closed field k of characteristic 0. Let

$d_1' = r_0' = N$

$g_1' = Y$ and $r_1' = \deg_X \mathrm{Res}_Y(F, g_1')$

$d_2' = \mathrm{GCD}(r_0', r_1')$

$g_2' = \sqrt[d_2']{F}$ and $r_2' = \deg_X \mathrm{Res}_Y(F, g_2')$

$d_3' = \mathrm{GCD}(r_0', r_1', r_2')$

$g_3' = \sqrt[d_3']{F}$ and $r_3' = \deg_X \mathrm{Res}_Y(F, g_3')$

$d_4' = \mathrm{GCD}(r_0', r_1', r_2', r_3')$, and so on.

We obtain sequences

$r' = (r_0', r_1', r_2', \ldots, r_{h'}')$

$d_2' > d_3' > d_4' > \cdots > d_{h'+1}' = d_{h'+2}' = \cdots$

$g' = (g_1', g_2', \ldots, g_{h'+1}')$.

Now we have the

CRITERION FOR ONE PLACE AT INFINITY. *The curve C has only one place at infinity (i.e., only one point at infinity and only one place at that point) if and only if $d_{h'+1}' = 1$ and $r_1'd_1' > r_2'd_2' > \cdots > r_h'd_h'$ and g_{j+1}' is degreewise* *straight relative to* (r', g', g'_j) *for* $1 \le j \le h'$ (*in the sense we shall describe in a moment*).

Approximate roots. What is the meaning of $g'_2 = \sqrt[d']{F}$ or $g'_3 = \sqrt[d']{F}$? More generally, for a monic polynomial $F = Y^N + a_1 Y^{N-1} + \cdots + a_N$ and a positive integer D, what does it mean to compute $\Gamma = \sqrt[D]{F}$? We had better assume that D divides N, because otherwise we cannot even start. Now $\Gamma = Y^{N/D} + \hat{b}_1 Y^{N/D-1} + \cdots + \hat{b}_{N/D}$ such that .... If things could be exact, then $F - \Gamma^D = 0$. Otherwise we try to minimize $\deg(F - \Gamma^D)$ by making it $< N - \frac{N}{D}$. There is a unique monic polynomial Γ which satisfies this degree bound. It is called the *approximate* D^{th} *root* of F. This polynomial can easily be obtained by considering the inequality $\deg(F - \Gamma^D) < N - \frac{N}{D}$ and successively solving for $\hat{b}_1, \ldots, \hat{b}_{N/D}$ in terms of $a_1, \ldots, a_N$. Another way of obtaining Γ is by the Γ-adic expansion of F, by generalizing the method of completing the square. So, starting with any monic polynomial Γ of degree $\frac{N}{D}$, consider the Γ-adic expansion of F by writing

$$F = \Gamma^D + B_1 \Gamma^{D-1} + \cdots + B_D,$$

where the B_i are unique polynomials of degree $<$ the degree of Γ. Now $F - \Gamma^D = B_1 \Gamma^{D-1} + \cdots + B_D$. Hence, $\deg(F - \Gamma^D) < N - \frac{N}{D}$ iff $B_1 = 0$. To make $B_1 = 0$ we use the completing the Dth power method by writing

$$\begin{aligned} F &= \Gamma^D + B_1 \Gamma^{D-1} + \cdots + B_D \\ &= \left(\Gamma + \frac{B_1}{D}\right)^D + B_1^* \left(\Gamma + \frac{B_1}{D}\right)^{D-1} + \cdots + B_D^* \end{aligned}$$

to get $\deg B_1^* < \deg B_1$ provided $B_1 \ne 0$, and so on. In the definitions of $g'_2, g'_3, \ldots, g'_{h'+1}$, the roots are meant to be approximate roots.

One place at a point at finite distance. By a slight rewording of the criterion, we get a criterion for one place at any given point of C at finite distance. Namely, by a translation we bring it to the origin. By a rotation we arrange matters so that $F(0, Y) \ne 0$. Upon letting $N^* = \operatorname{ord} F(0, Y)$, by the Weierstrass Preparation Theorem $F(X, Y) = \delta(X, Y)F^*(X, Y)$ where $\delta(X, Y)$ is a power series with $\delta(0, 0) \ne 0$ and

$$F^*(X, Y) = Y^{N^*} + a_1^*(X)Y^{N^*-1} + \cdots + a_{N^*}^*(X)$$

with power series $a_1^*(X), \ldots, a_{N^*}^*(X)$ for which $a_1^*(0) = \cdots = a_{N^*}^*(0) = 0$. Again letting the root signs stand for approximate roots, let

$d_1 = r_0 = N^*$

$g_1 = Y$ and $r_1 = \operatorname{ord}_X \operatorname{Res}_Y(F, g_1)$

$d_2 = \operatorname{GCD}(r_0, r_1)$

$g_2 = \sqrt[d_2]{F}$ and $r_2 = \operatorname{ord}_X \operatorname{Res}_Y(F, g_2)$

$d_3 = \operatorname{GCD}(r_0, r_1, r_2)$

$g_3 = \sqrt[d_3]{F}$ and $r_3 = \operatorname{ord}_X \operatorname{Res}_Y(F, g_3)$
$d_4 = \operatorname{GCD}(r_0, r_1, r_2, r_3)$, and so on.

Thus we obtain sequences

$$r = (r_0, r_1, r_2, \ldots, r_h)$$
$$d_2 > d_3 > d_4 > \cdots > d_{h+1} = d_{h+2} = \cdots$$
$$g = (g_1, g_2, \ldots, g_{h+1}).$$

Now we have the

CRITERION FOR ONE PLACE AT THE ORIGIN. *The curve* C *has only one place at the origin if and only if* $d_{h+1} = 1$ *and* $r_1 d_1 < r_2 d_2 < \cdots < r_h d_h$ *and* g_{j+1} *is straight relative to* (r, g, g_j) *for* $1 \le j \le h$ *(in the sense we shall describe in a moment).*

Polynomial expansion. To describe the idea of straightness, first we discuss polynomial expansion. Now a polynomial

$$H = c_0 Y^m + c_1 Y^{m-1} + \cdots + c_i Y^{m-i} + \cdots + c_m$$

can be considered as the Y-adic expansion of itself. Given any monic polynomial Γ in Y with coefficients functions of X, instead of the Y-adic expansion of H we can consider its Γ-adic expansion

$$H = \sum \tilde{B}_i \Gamma^i \quad \text{with } \deg \tilde{B}_i < \deg \Gamma$$

where deg is the Y-degree. For example, the Γ-adic expansion of $Y^5 + 2Y^3 + 3Y + 3$ is $Y(Y^2 + 1)^2 + 2Y + 3$ where $\Gamma = Y^2 + 1$. This is analogous to the usual decimal, or 10-adic, expansion. For example $4375 = 5 \times 10^0 + 7 \times 10^1 + 3 \times 10^2 + 4 \times 10^3$. Note that, in connection with approximate roots, we have already used the idea of Γ-adic expansion. Just as in the case of 10-adic expansion, to obtain the Γ-adic expansion of H, by the division algorithm we successively write

$$H = \tilde{B}_0 + H_0 \Gamma = \tilde{B}_0 + (\tilde{B}_1 + H_1 \Gamma)\Gamma$$

and so on.

Generalizing the idea of 10-adic expansion, we get the idea of n-adic expansion with $n = (n_1, n_2, \ldots, n_{h+1})$ where $n_1, \ldots, n_{h+1}$ are positive integers with $1 = n_1 | n_2 | n_3 | \ldots | n_{h+1}$. (That is, n_1 divides n_2, n_2 divides n_3, and so on.) Now any integer b has a unique n-adic expansion $b = e_1 n_1 + e_2 n_2 + \cdots + e_{h+1} n_{h+1}$ where $0 \le e_i < \frac{n_{i+1}}{n_i}$ for $1 \le i \le h$, and e_{h+1} is free. For example, $4375 = 5 \times 10^0 + 7 \times 10^1 + 43 \times 10^2$. If only three digits are allowed then the last digit is free. Here, 43 is not constrained.

Next, consider a polynomial vector $g = (g_1, g_2, \ldots, g_{h+1})$ where g_i is a monic polynomial of degree n_i in Y for $1 \le i \le h + 1$, and $1 = n_1 | n_2 | n_3 | \ldots | n_{h+1}$. Now we have the g-adic expansion

$$H = \sum \gamma_{e_1 e_2 \cdots e_{h+1}} g_1^{e_1} g_2^{e_2} \cdots g_{h+1}^{e_{h+1}}$$

where the summation is over all non-negative integers $e_1, \ldots, e_{h+1}$ with $0 \le e_i < \frac{n_{i+1}}{n_i}$ for $1 \le i \le h$, and where $\gamma_{e_1 e_2 \cdots e_{h+1}}$ are obtained by additions and multiplications from the coefficients of H and $g_1, \ldots, g_{h+1}$. Thus if the coefficients of H and $g_1, \ldots, g_{h+1}$ belong to a ring R, then the coefficients $\gamma_{e_1 e_2 \cdots e_{h+1}}$ also belong to R. (Note that if $h = 1$ and $g = (g_1, g_2)$ with $g_1 = Y$ and $g_2 = \Gamma$, then the g-adic expansion is reduced to the Γ-adic expansion.)

To describe straightness, let us assume that the integers $n_1, \ldots, n_{h+1}$ are obtained by putting

$$d_i = \mathrm{GCD}(r_0, r_1, \ldots, r_{i-1}) \quad \text{and} \quad n_i = \frac{d_1}{d_i} \quad \text{for } 1 \le i \le h + 1$$

where $r = (r_0, r_1, \ldots, r_h)$ is any given sequence of integers. Let us also assume that the coefficients of H and $g_1, \ldots, g_{h+1}$ are power series in X. Now we define

$$\mathrm{fint}(r, g, H) = \min \left(\sum_{j=0}^{h} e_j r_j \right) \quad \text{with } e_0 = \mathrm{ord}\, \gamma_{e_1 e_2 \cdots e_{h+1}},$$

where the min is taken over all non-negative integers $e_1, \ldots, e_{h+1}$ for which $\gamma_{e_1 e_2 \cdots e_{h+1}} \ne 0 = e_{h+1}$.

Here *fint* is supposed to be an abbreviation of the phrase "formal intersection multiplicity," which in turn is meant to suggest some sort of analogy with intersection multiplicity of plane curves.

For $1 \le j \le h$, let $u(j) = n_{j+1}/n_j$ and consider the g_j-adic expansion

$$g_{j+1} = g_j^{u(j)} + \sum_{i=1}^{u(j)} g_{ji} g_j^{u(j)-i},$$

where we note that g_{ji} is a polynomial of degree less than n_j in Y whose coefficients are power series in X. We say that g_{j+1} is *straight* relative to (r, g, g_j) if

$$(u(j)/i)\mathrm{fint}(r, g, g_{ji}) \ge \mathrm{fint}(r, g, g_{ju(j)}) = u(j)\mathrm{fint}(r, g, g_j)$$

for $1 \le i \le u(j)$.

Again let

$$d_i' = \mathrm{GCD}(r_0', r_1', \ldots, r_{i-1}') \quad \text{and} \quad n_i' = \frac{d_1'}{d_i'} \quad \text{for } 1 \le i \le h' + 1,$$

where $r' = (r_0', r_1', \ldots, r_{h'}')$ is any sequence of integers. Consider a polynomial vector $g' = (g_1', g_2', \ldots, g_{h+1}')$ where g_i' is a monic polynomial of degree n_i' in Y with coefficients polynomials in X for $1 \le i \le h' + 1$, and $1 = n_1' | n_2' | n_3' | \ldots | n_{h'+1}'$. Now, assuming the coefficients of H to be polynomials in X, we have the g'-adic expansion

$$H = \sum \gamma_{e_1 e_2 \cdots e_{h'+1}}' g_1'^{e_1} g_2'^{e_2} \cdots g_{h'+1}'^{e_{h'+1}},$$

where the summation is over all non-negative integers $e_1, \ldots, e_{h'+1}$ with $0 \le e_i < n'_{i+1}/n'_i$ for $1 \le i \le h'$, and where the coefficients $\gamma'_{e_1 e_2 \cdots e_{h'+1}}$ are polynomials in X. Now we define

$$\text{fing}(r', g', H) = \max \left(\sum_{j=0}^{h'} e_j r'_j \right) \quad \text{with } e_0 = \deg \gamma'_{e_1 e_2 \cdots e_{h'+1}},$$

where the max is taken over all non-negative integers $e_1, \ldots, e_{h'+1}$ for which $\gamma'_{e_1 e_2 \cdots e_{h'+1}} \ne 0 = e_{h'+1}$.

Here *fing* is supposed to be an abbreviation of the phrase "degreewise formal intersection multiplicity," which is meant to suggest some sort of analogy with intersection multiplicity of plane curves.

For $1 \le j \le h$, let $u'(j) = n'_{j+1}/n'_j$ and consider the g'_j-adic expansion

$$g'_{j+1} = g'^{u'(j)}_j + \sum_{i=1}^{u'(j)} g'_{ji} g'^{u'(j)-i}_j$$

where we note that g'_{ji} is a polynomial of degree less than n'_j in Y whose coefficients are polynomials in X; we say that g'_{j+1} is *degreewise straight* relative to (r', g', g'_j) if

$$(u'(j)/i)\text{fing}(r', g', g'_{ji}) \le \text{fing}(r', g', g'_{ju'(j)}) = u'(j)\text{fing}(r', g', g'_j)$$

for $1 \le i \le u'(j)$.

REMARK. The above criterion for one place at the origin, remains valid for any monic polynomial

$$F^*(X, Y) = Y^{N^*} + a_1^*(X)Y^{N^*-1} + \cdots + a_{N^*}^*(X)$$

in Y whose coefficients $a_1^*(X), \ldots, a_{N^*}^*(X)$ are power series in X with $a_1^*(0) = \cdots = a_{N^*}^*(0) = 0$. Namely, for any such F^* we have the

IRREDUCIBILITY CRITERION. *$F^*(X, Y)$ is an irreducible power series if and only if $d_{h+1} = 1$ and $r_1 d_1 < r_2 d_2 < \cdots < r_h d_h$ and g_{j+1} is straight relative to (r, g, g_j) for $1 \le j \le h$.*

Problem. This is an algorithm for determining if F^* has only one power series factor. Generalize this to an algorithm to find the number of power series factors of F^*.

Newton polygon. The adjectives straight and degreewise straight are meant to suggest that we are considering some kind of generalization of Newton polygon. Namely, given any F^*, we plot all points (p, q) such that $X^p Y^q$ occurs in the expansion of F^* with a nonzero coefficient. Now the Newton polygon of F^* is the convex hull of all these points. See Figure 22.1 on page 186. If F^* is irreducible then the Newton polygon is a straight line. However, the converse may not be true. Namely, if $\text{ord } a_i^*(X) \ge i$ for $1 \le$

$i \leq N^*$, then upon letting $Y = Y' + \lambda X$ and $F'(X, Y') = F^*(X, Y)$, for all except a finite number of values of λ we have that the Newton polygon F' is the 45° line, regardless of whether F^* (and hence F') is irreducible or not.

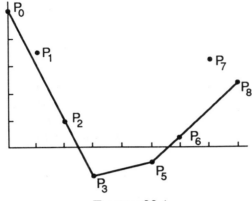

FIGURE 22.1

Concordance with characteristic pairs. The fact that we used the same notation $r = (r_0, r_1, \ldots, r_h)$ for the sequence obtained above and the r-sequence introduced in the lecture on characteristic pairs was not an accident. Indeed, they are the same! In greater detail, assume that $F^*(X, Y)$ is an irreducible power series and N^* does not divide $\operatorname{ord} a_{N^*}^*(X)$. Now by Newton's Theorem, we can parametrize the curve $F^*(X, Y) = 0$ as $X = T^{N^*}$ and

$$Y = b_{m_1} T^{m_1} + \cdots + b_{m_2} T^{m_2} + \cdots + b_{m_3} T^{m_3} + \cdots + b_{m_h} T^{m_h} + \cdots$$

where in the expansion, m_1 is the smallest exponent whose coefficient b_{m_1} is nonzero, m_2 is the smallest exponent not divisible by $\operatorname{GCD}(N^*, m_1)$ whose coefficient b_{m_2} is nonzero, m_3 is the smallest exponent not divisible by $\operatorname{GCD}(N^*, m_1, m_2)$ whose coefficient b_{m_3} is nonzero, $\ldots$, and m_h is the smallest exponent for which $\operatorname{GCD}(N^*, m_1, \ldots, m_h) = 1$ and whose coefficient b_{m_h} is nonzero. Let $m_0 = N^*$ and

$$d_{i+1} = \operatorname{GCD}(m_0, m_1, \ldots, m_i) \quad \text{for } 0 \leq i \leq h.$$

Also let

$$q_0 = m_0, \quad q_1 = m_1, \quad q_2 = m_2 - m_1, \quad \ldots, \quad q_h = m_h - m_{h-1}$$

and

$$s_0 = m_0 \quad \text{and} \quad s_i = \sum_{j=1}^{i} q_j d_j \quad \text{for } 1 \leq i \leq h.$$

Finally, let

$$r_0 = m_0 \quad \text{and} \quad r_i = \frac{s_i}{d_i} \quad \text{for } 1 \leq i \leq h.$$

Now the quantities $h, r_0, r_1, \ldots, r_h, d_1, \ldots, d_{h+1}$ coincide with their values defined in terms of approximate roots. For proof see [A32].

NOTE. For an irreducible algebraic plane curve $C : F(X, Y) = 0$ over a field k, polynomial parametrization is a property of the ring $k[X, Y]/(F(X, Y))$ while rational parametrization is a property of the field $k(X, Y)/(F(X, Y))$. This is how we would like to have the latter field. But unfortunately (!) we must write it as the quotient field of $K[X, Y]/(F(X, Y))$.

Inversion Formula and Jacobian Problem

In the lecture on characteristic pairs we met with an inversion formula for one variable fractional power series. In connection with parametrizations, we had occasion to invert one or several power series in the usual sense, with integral exponents. An explicit formula for inverting power series in several variables was obtained by me in 1973–74 in connection with the Jacobian problem. I included it in my Purdue seminar on 9 July 1974. The notes were written up by C. Christensen. The Christensen notes [A24] were privately circulated. As a result the formula became known as Abhyankar's Inversion Formula. Now let us reproduce the discussion of this formula from the Christensen notes.

Definitions/notations. Given any field k of characteristic zero and any positive integer n, let

$$A = k[[X]] \quad \text{with } X = (X_1, \ldots, X_n)$$
$$B = A^n = \{f = (f_1, \ldots, f_n) : f_i \in A \text{ for } 1 \le i \le n\}$$
$$M(B) = M(A)^n = \{f = (f_1, \ldots, f_n) : f_i \in M(A) \text{ for } 1 \le i \le n\}$$
$$M(B)^e = \{f = (f_1, \ldots, f_n) : f_i \in A \text{ with ord } f_i \ge e \text{ for } 1 \le i \le n\}$$
$$Z = \mathbb{N}^n = \{p = (p_1, \ldots, p_n) : p_i \in \mathbb{N} \text{ for } 1 \le i \le n\}$$

where $\mathbb{N}$ is the set of all non-negative integers. For $f = (f_1, \ldots, f_n)$ in B, let

$$J(f) = \text{ the Jacobian of } f = \det \left(\frac{\partial f_i}{\partial X_j} \right)_{1 \le j \le n}^{1 \le i \le n}.$$

For any $\gamma \in A$ and $f \in M(B)$, we define

$$\gamma(f) = \gamma(f_1, \ldots, f_n) \in A.$$

For any $g \in B$ and $f \in M(B)$, we define

$$g(f) = (g_1(f), \ldots, g_n(f)) \in B.$$

For any $u \in B$ and $p \in Z$, we define

$$u^{[p]} = \frac{u_1^{p_1} \cdots u_n^{p_n}}{p_1! \cdots p_n!}.$$

189

For any $p \in Z$, we define D^p by

$$D^p = \frac{\partial^{p_1 + \cdots + p_n}}{\partial^{p_1} X_1 \cdots \partial^{p_n} X_n}.$$

Let $I = \{f \in B : X - f \in M(B)^2\}$. If $f \in I$, then we say that f is a *near identity* (automorphism).

Inner product. For each $h \in A$ and $f \in I$, we define

(1) $$\langle h, f \rangle = \sum_{p \in Z} D^p \left[h(f) J(f) (X - f)^{[p]} \right] \in A.$$

Concerning this inner product $\langle h, f \rangle$, we shall prove the

THEOREM. *For each $h \in A$ and $f \in I$, we have $\langle h, f \rangle = h$.*

Inversion. For any $f \in I$, clearly there exists a unique $\zeta \in I$ such that $\zeta(f(X)) = X$. ζ is called the *inverse* of f. More generally, given any $\eta \in B$ and $f \in I$, there exists a unique $\lambda \in B$ such that $\lambda(f(X)) = \eta(X)$. We claim that λ is given by the formula

(2) $$\sum_{p \in Z} D^p \left[\eta(X) J(f) (X - f)^{[p]} \right] = \lambda(X).$$

Hence, in particular, by taking $\eta = X$, the inverse ζ of f is given by the formula

(2′) $$\sum_{p \in Z} D^p \left[X J(f) (X - f)^{[p]} \right] = \zeta(X).$$

Expression (2) follows from (1) and the theorem as follows. Given any $\eta \in B$ and $f \in I$, we let $\lambda \in B$ be such that $\lambda(f(X)) = \eta(X)$. Now, for any i, by letting $h = \lambda_i$ in (1), we get

$$\sum_{p \in Z} D^p \left[\eta_i(X) J(f) (X - f)^{[p]} \right] = \sum_{p \in Z} D^p \left[\lambda_i(f) J(f) (X - f)^{[p]} \right]$$
$$= \langle \lambda_i, f \rangle$$
$$= \lambda_i(X)$$

where the last equality follows by putting $h = \lambda_i$ in the theorem. Similarly, (2) can be converted to (1), i.e., the theorem can be deduced from formula (2).

Binomial theorem. $(u + v)^{[p]} = \sum_{i+j=p} u^{[i]} v^{[j]}$.

True for f and g, then true for the composite. As a preparation for the proof of the theorem, let $f \in I$ and $g \in I$ be such that for all $h \in A$, we have $\langle h, f \rangle = h$ and $\langle h, g \rangle = h$. We shall show that then $\langle h, g(f) \rangle = h$

for all $h \in A$. The proof is as follows.

$$\langle h, g(f) \rangle$$

$$= \sum_{p \in Z} D^p \left[h(g(f)) J(g(f)) (X - g(f))^{[p]} \right]$$

$$= \sum_{p \in Z} D^p \left[h(g(f)) J(f) (J(g)(f)) (X - f + f - g(f))^{[p]} \right]$$

$$= \sum_{p \in Z} D^p \left[h(g(f)) J(f) (J(g)(f)) \sum_{i+j=p} (X - f)^{[i]} (f - g(f))^{[j]} \right]$$

$$= \sum_{j \in Z} D^j \left[\sum_{i \in Z} D^i \left[\gamma_j(f) J(f) (X - f)^{[i]} \right] \right]$$

where $\gamma_j(X) = h(g) J(g) (X - g)^{[j]} \in A$. But

$$\sum_{j \in Z} D^j \left[\sum_{i \in Z} D^i \left[\gamma_j(f) J(f) (X - f)^{[i]} \right] \right]$$

$$= \sum_{j \in Z} D^j \left[[h(g) J(g) (X - g)^{[j]}] \right]$$

$$= h.$$

Therefore $\langle h, g(f) \rangle = h$.

f is a nice composite. The automorphism $f \in I$ can be written as the composition of automorphisms that change only one variable. Namely, for $i = 1, \ldots, n$, let $g^{(i)} \in I$ be defined by putting

$$g_j^{(i)}(X) = X_j \quad \text{for } j = 1, \ldots, i-1, i+1, \ldots, n$$

and

$$g_i^{(i)}(X) = \text{ the unique function satisfying}$$

$$g_i^{(i)}(X_1, \ldots, X_i, f_{i+1}, \ldots, f_n) = f_i.$$

Now clearly $f = g^{(1)}(g^{(2)}(\cdots(g^{(n)})\cdots))$.

Monomial transforms. An alternate argument uses the following result. Let S denote the sub-semigroup of I generated by the automorphisms that change only one variable by a constant times a monomial of degree greater than 1. Now S is dense in I, i.e., for each $f \in I$ and positive integer δ, there exists a $g \in S$ such that $f - g \in M(B)^\delta$.

Convention. Note that by convention

$$0^i = \begin{cases} 1 & \text{if } i = 0 \\ 0 & \text{if } i > 0. \end{cases}$$

PROOF OF THEOREM. Since f can be written as the composition of automorphisms which change only one variable, it suffices to prove the theorem when f changes only one variable. By relabeling the variables, we may assume that f changes only X_1, and then we have $h(f) = h(f_1, X_2, \ldots, X_n)$. Furthermore, since

$$h(X_1, \ldots, X_n) = \sum h_i(X_2, \ldots, X_n)X_1^i$$

where $h_i(X_2, \ldots, X_n)$ is constant with respect to X_1, and since $\langle h, f \rangle$ is additive on h, it suffices to prove the theorem when h is X_1^m times a constant where m is a non-negative integer. In fact, it is enough to prove it for $h = X_1^m$. Let $u = X - f$. Now clearly,

$$\sum_{p \in Z} D^p \left[h(f)J(f)(X-f)^{[p]} \right] = \sum_{p=0}^{\infty} \frac{1}{p!} \frac{\partial^p}{\partial X_1^p} \left[(X_1 - u_1)^m (1 - u_1')u_1^p \right],$$

where u' denotes the X_1-partial derivative of u. Hence, we are reduced to showing that

$$\sum_{p=0}^{\infty} \frac{1}{p!} \frac{\partial^p}{\partial X_1^p} \left[(X_1 - u_1)^m (1 - u_1')u_1^p \right] = X_1^m.$$

Now

$$\sum_{p=0}^{\infty} \frac{1}{p!} \frac{\partial^p}{\partial X_1^p} \left[(X_1 - u_1)^m (1 - u_1')u_1^p \right]$$

$$= \sum_{p=0}^{\infty} \frac{\partial^p}{\partial X_1^p} \left[\sum_{i=0}^{m} \frac{(-1)^i}{p!} \binom{m}{i} X_1^{m-i} \left(u_1^{p+i} - u_1^{p+i}u_1' \right) \right]$$

$$= \sum_{p=0}^{\infty} \sum_{i=0}^{m} \sum_{j=0}^{p} \frac{(-1)^i}{p!} \binom{m}{i}\binom{p}{j} \left(\frac{\partial^j X_1^{m-i}}{\partial X_1^j} \right) \left(\frac{\partial^{p-j} u_1^{p+i}}{\partial X_1^{p-j}} - \frac{1}{p+i+1} \frac{\partial^{p-j+1} u_1^{p+i+1}}{\partial X_1^{p-j+1}} \right)$$

$$= \sum_{p=0}^{\infty} \sum_{i=0}^{m} \sum_{j=0}^{\min(p,m-i)} \frac{(-1)^i m! X_1^{m-i-j}}{i!(p-j)!j!(m-i-j)!} \left(\frac{\partial^{p-j} u_1^{p+i}}{\partial X_1^{p-j}} - \frac{1}{p+i+1} \frac{\partial^{p-j+1} u_1^{p+i+1}}{\partial X_1^{p-j+1}} \right)$$

$$= \sum_{t=0}^{m} \sum_{i=0}^{t} \sum_{q=0}^{\infty} \frac{(-1)^i X_1^{m-t} m!}{i!q!(t-i)!(m-t)!} \left(\frac{\partial^q u_1^{q+t}}{\partial X_1^q} - \frac{1}{q+t+1} \frac{\partial^{q+1} u_1^{q+t+1}}{\partial X_1^{q+1}} \right),$$

where $i + j = t$ and $p - j = q$. Now the last expression equals

$$\sum_{t=0}^{m} \binom{m}{t} X_1^{m-t} \rho(t)\theta(t),$$

where

$$\rho(t) = \sum_{i=0}^{t} (-1)^i \binom{t}{i} = \begin{cases} 1 & \text{if } t = 0 \\ 0 & \text{if } t > 0 \end{cases}$$

and

$$\theta(t) = \sum_{q=0}^{\infty} \left[\frac{1}{q!} \frac{\partial^q u_1^{q+t}}{\partial X_1^q} - \frac{1}{q!(q+t+1)} \frac{\partial^{q+1} u_1^{q+t+1}}{\partial X_1^{q+1}} \right]$$

$$= u_1^t + \sum_{q=0}^{\infty} \frac{\partial^{q+1} u_1^{q+t+1}}{\partial X_1^{q+1}} \left(\frac{1}{(q+1)!} - \frac{1}{q!(q+t+1)} \right)$$

$$= u_1^t + t \left[\sum_{q=0}^{\infty} \frac{1}{(q+1)!(q+t+1)} \frac{\partial^{q+1} u_1^{q+t+1}}{\partial X_1^{q+1}} \right].$$

Note that $\theta(0) = 1$. Therefore $\langle h, f \rangle = h$.

Condensed formula. The theorem says that

$$\sum_{p \in Z} D^p \left[h(f)J(f)(X - f)^{[p]} \right] = h.$$

Long formula. Taylor expansion and the condensed formula yield

$$\sum_{p, r \in Z} D^{p+r} \left(h(f)J(f)(X - f)^{[p]} \right) \Big|_{X=0} X^{[r]} = h.$$

Algebraic formula. Taylor expansion and the condensed formula also yield that if

$$\sum_{p \in Z} \frac{h(f)J(f)(Y + X - f)^p}{X^p} = \sum_{l \in \mathbb{Z}^n} \Lambda_l(Y) X^l \quad \text{with } \Lambda_l(X) \in A$$

(where $Y = (Y_1, \ldots, Y_n)$ and $X^l = X_1^{l_1} \cdots X_n^{l_n}$ and so on), then $\Lambda_l(X) = h(X)\zeta(X)^{-l}$ (or equivalently $\Lambda_l(f) = h(f)(X)^{-l}$) for all $l \in Z$. (Hint: First check the special case when $l = 0 = (0, \ldots, 0)$ and then deduce the general case by substituting $h\zeta^{-l}$ for h).

Jacobian problem. Let $u = X - f$. For any $p \in Z$, by the Binomial Theorem we have $(Y + u)^p = \sum_{i+j=p} Y^i u^j \binom{i+j}{i}$ where $\binom{i+j}{i} = \frac{(i_1+j_1)! \cdots (i_n+j_n)!}{i_1! j_1! \cdots i_n! j_n!}$. Let

$$\Delta_{li} = \sum_{j \in Z} \Delta_{li}^{(j)} \quad \text{where} \quad u^j \binom{i+j}{i} = \sum_{l \in \mathbb{Z}^n} \Delta_{li}^{(j)} X^{i+j+l} \quad \text{with } \Delta_{li}^{(j)} \in k$$

and define $\widehat{\Delta}_{li} = ((\widehat{\Delta}_{li})_e)_{1 \le e \le n} \in k^n$ by putting $(\widehat{\Delta}_{li})_e = \Delta_{l(e)i}$ where $l(e)_q = l_q$ if $q \ne e$ and $l(e)_q = l_q - 1$ if $q = e$. Letting h range over the components of ζ, by the algebraic formula we see that if $J(f) \in k$ then for all $i \in Z$ we have $\widehat{\Delta}_{0i} = J(f)$ times the coefficient of X^i in ζ. Hence, $\zeta \in k[X]^n \Leftrightarrow \widehat{\Delta}_{0i} = 0$ for all $i >> 0$. Similarly, taking $h(X) = \theta(X) = \sum_{i \in Z} \theta_i X^i$ with $\theta_i \in k$ such that $\theta(f) = \frac{1}{J(f)}$, by the algebraic formula we see that $\Delta_{0i} = \theta_i$ for all $i \in Z$. Hence, $J(f) \in k \Leftrightarrow \Delta_{0i} = 0$ for all $i > 0$. Therefore, the Jacobian problem is equivalent to asking if $f \in k[X]^n$ and $\Delta_{0i} = 0$ for all $i > 0$ implies that $\widehat{\Delta}_{0i} = 0$ for all $i >> 0$.

NOTE. On the importance of *inversion*, here is a quotation from the chapter on Abel in Bell's *Men of mathematics* [**Bee**]:

> Instead of assuming that people are depraved because they drink to excess, Galton *inverted* this hypothesis...For the moment we need note only that Galton, like Abel, *inverted* his problem — turned it upside-down and inside-out, back-end-to and foremostend-backward...'You must always *invert*,' as Jacobi said when asked the secret of his mathematical discoveries.

He was recalling what Abel and he had done.

Surfaces

So far our stress was on curves. Now it will be on surfaces and higher dimensional varieties. Let us start with some examples of surfaces.

(1) *The (X, Y)-plane.* $Z = 0$. See Figure 24.1.

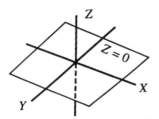

FIGURE 24.1

(2) *General plane.* $aX + bY + cZ + d = 0$. See Figure 24.2.

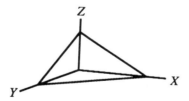

FIGURE 24.2

(3) *Sphere.* $X^2 + Y^2 + Z^2 - 1 = 0$. See Figure 24.3.

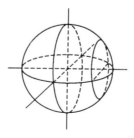

FIGURE 24.3

(4) *Ellipsoid.* $\frac{X^2}{a^2} + \frac{Y^2}{b^2} + \frac{Z^2}{c^2} - 1 = 0$. See Figure 24.4.

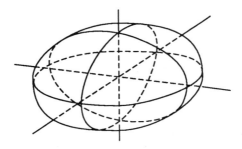

FIGURE 24.4

(5) *Hyperboloid of one sheet.* $\frac{X^2}{a^2} + \frac{Y^2}{b^2} - \frac{Z^2}{c^2} - 1 = 0$. See Figure 24.5.

FIGURE 24.5

(6) *Hyperboloid of two sheets.* $\frac{X^2}{a^2} - \frac{Y^2}{b^2} - \frac{Z^2}{c^2} - 1 = 0$. See Figure 24.6.

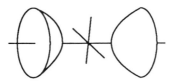

FIGURE 24.6

The last three surfaces are examples of a *conicoid*, a degree 2 surface. Further, we can consider some examples of cubic surfaces:

(7) *Nonsingular cubic surface.* $X^3 + Y^3 + Z^3 - 1 = 0$.

(8) *Cubic surface with isolated singularity.* $X^3 + Y^3 + Z^3 = 0$. The origin is the only singular point.

(9) *Cubic surface with double line.* The equation of the surface is $Z^2 - XY^2 = 0$. The double line is the X-axis, i.e., the line $Y = Z = 0$. The section of the surface by the plane $X = \alpha$ is $Z^2 - \alpha Y^2 = (Z - \sqrt{\alpha}Y)(Z + \sqrt{\alpha}Y) = 0$. That is, it consists of a pair of lines except when $\alpha = 0$. For

$\alpha \neq 0$, the two lines are given by $X - \alpha = Z - \sqrt{\alpha}Y = 0$ and $X - \alpha = Z + \sqrt{\alpha}Y = 0$. These lines get closer to the Y-axis as $\alpha \to 0$. At $\alpha = 0$ they merge into the line $X = Z = 0$. Let us draw these lines on our surface schematically. See Figure 24.7.

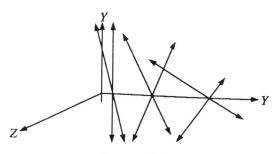

FIGURE 24.7

In general, a surface S in 3-space is the locus of a 3-variable polynomial equation $S : f(X, Y, Z) = 0$. The degree n of f is the *degree* of S. A surface of degree 1 is a plane. A surface of degree 2 is called a *conicoid* or a *quadric*. A surface of degree 3 is called a *cubic surface*, and so on.

Given a point $P = (\alpha, \beta, \gamma)$ of S, we can expand around P to get

$$f(X, Y, Z) = \sum_{i+j+l \leq n} a_{ijl} X^i Y^j Z^l = \sum b_{ijl}(X - \alpha)^i (Y - \beta)^j (Z - \gamma)^l.$$

By letting $f^*(X^*, Y^*, Z^*) = \sum b_{ijl} X^{*i} Y^{*j} Z^{*l}$ we have

$$d = \text{mult }_P S = \text{ord}_P f = \text{ord } f^* = \min i + j + l \quad \text{such that } b_{ijl} \neq 0.$$

Then P is a simple point, double point, triple point, ... , singular point, according as $d = 1$, $d = 2$, $d = 3$, ... , $d > 1$.

Note that we always have $d \leq n = \deg f$. Let us consider the case of a cone. We define a cone as follows: S is a *cone* with *vertex* at P means $d = \text{mult}_P S = n$. We clearly have that: S *is a cone with vertex at* P $\Leftrightarrow f^*(X^*, Y^*, Z^*)$ *is homogeneous of degree* n. A cone of degree 2 is called a *quadric cone*, a cone of degree 3 is called a *cubic cone*, ... , and so on. The cubic surface with isolated singularity, listed in the above item (8), is an example of a cubic cone. A cone of degree 1 with vertex P is simply a plane passing through P. An irreducible quadric cone is simply the usual *circular cone*. Figure 24.8 on page 198 shows a picture of the *horizontal circular cone* $X^2 - Y^2 - Z^2 = 0$.

By translation we may assume that $P = (0, 0, 0)$. Now S is a cone with vertex P iff

$$f(X, Y, Z) = \sum_{i+j+l=n} a_{ijl} X^i Y^j Z^l.$$

If S is a cone with vertex P then we may denote it by C or C_n because the word cone starts with the letter c. However you may say that the notation

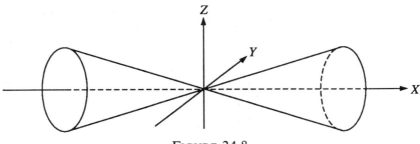

FIGURE 24.8

C_n is used for a plane curve of degree n (affine or projective). But this does not lead to confusion, because we maintain that

cone and projective plane curve are equivalent notions.

Equivalent doesn't mean same. The difference is, when we think of a cone, we don't identify proportional points. Proportional points define a line $L_{(a,b,c)}$ through the origin in 3-space, which is given by the parametric equations

$$L_{(a,b,c)} : \begin{cases} X = at \\ Y = bt \\ Z = ct. \end{cases}$$

We can also define a cone geometrically. S is a cone with vertex P means, for any point Q of S with $P \neq Q$, the entire line PQ is on S.

Given a cone we can take its section by a plane. Then we get a curve in the (affine) plane. The best section is the one by the projective plane (set of lines through the origin in 3-space). This "best section" may be called the *intrinsic section*. In taking that, we don't have to change the equation.

So let $S : f(X, Y, Z) = 0$ be a cone of degree n with vertex at $P = (0, 0, 0)$, and let C_n denote its section by the projective plane. In other words, the homogeneous equation of C_n is $f(X, Y, Z) = 0$ or $f(\mathscr{X}, \mathscr{Y}, \mathscr{Z}) = 0$ where $(\mathscr{X}, \mathscr{Y}, \mathscr{Z})$ are homogeneous coordinates in the projective plane ($(\mathscr{X}, \mathscr{Y}, \mathscr{Z})$ are merely calligraphic incarnations of the roman (X, Y, Z)). Thus the cone S consists of exactly those lines $L_{(a,b,c)}$ for which (a, b, c) are the homogeneous coordinates of a point of C_n, i.e., $f(a, b, c) = 0$. Let $P_1, P_2, \ldots$ be the singularities of C_n with multiplicities $d_1, d_2, \ldots$. Now, assuming $n > 1$, the singular locus $\mathrm{Sing}(S)$ of S consists of the vertex P together with the lines $PP_1, PP_2, \ldots$, which are the *multiple lines* of S. In fact PP_i is a d_i-*fold line* of S, and upon denoting $\mathrm{Sing}_d(S)$ to be the set of all points of S of multiplicity at least d, we have

$$\mathrm{Sing}_d(S) = \bigcup_{d_i \geq d} PP_i \quad \text{for } 2 \leq d < n$$

and

$$\text{Sing}_n(S) = \begin{cases} \{P\} & \text{if } C_n \text{ has no } n\text{-fold points} \\ PQ & \text{if } C_n \text{ consists of lines (at least 2) through a point } Q \\ S & \text{if } C_n \text{ consists of one line counted } n \text{ times.} \end{cases}$$

Here are some examples of cones:

(10) *Cuspidal cubic cone.* $\widehat{S}_3 : ZY^2 - X^3 = 0$. We call this the cuspidal cubic cone because the intrinsic section $\widehat{C}_3 : \mathscr{Z}\mathscr{Y}^2 - \mathscr{X}^3 = 0$ is the cuspidal cubic, i.e., the cubic plane curve having a cusp at the origin as its only singularity. Now the vertex of $\widehat{S}_3$ is the origin $P = (0, 0, 0)$, and we have $\text{Sing}_3(\widehat{S}_3) = \{P\}$. The *cuspidal line* $L : X = Y = 0$ is a double line of $\widehat{S}_3$, and we have $\text{Sing}_2(\widehat{S}_3) = L$ and $\text{Sing}(\widehat{S}_3) = \{P\} \cup L$.

(11) *Nodal cubic cone.* $\widetilde{S}_3 : ZY^2 - ZX^2 - X^3 = 0$. We call this the nodal cubic cone because the intrinsic section $\widetilde{C}_3 : \mathscr{Z}\mathscr{Y}^2 - \mathscr{Z}\mathscr{X}^2 - \mathscr{X}^3 = 0$ is the nodal cubic, i.e., the cubic plane curve having a node at the origin as its only singularity. Now the vertex of $\widetilde{S}_3$ is the origin $P = (0, 0, 0)$, and we have $\text{Sing}_3(\widetilde{S}_3) = \{P\}$. The *nodal line* $L : X = Y = 0$ is a double line of $\widetilde{S}_3$, and we have $\text{Sing}_2(\widetilde{S}_3) = L$, and $\text{Sing}(\widetilde{S}_3) = \{P\} \cup L$.

(12) *Fermat cone.* $\dot{S}_n : X^n \pm Y^n \pm Z^n = 0$ of any degree $n > 1$. Here we have four choices of the signs, but the vertex is always at the origin $P = (0, 0, 0)$. The intrinsic section is the nonsingular projective plane curve $\mathscr{X}^n \pm \mathscr{Y}^n \pm \mathscr{Z}^n = 0$ of degree n. So $\text{Sing}(\dot{S}_n) = \text{Sing}_n(\dot{S}_n) = \{P\}$.

Now *reverting to the case of a general surface* $S : f(X, Y, Z) = 0$ of degree n (without assuming S to be a cone), we put

$$\text{Sing}(S) = \text{ the singular locus of } S$$

$$= \text{ the set of all singular points of } S$$

and

$$\text{Sing}_d(S) = \text{ the } d\text{-fold locus of } S$$

$$= \text{ the set of all points of } S \text{ of multiplicity at least } d.$$

We note that

$$\text{Sing}(S) = \bigcup_{2 \le d \le n} \text{Sing}_d(S).$$

Now the singular locus can also be obtained by setting the partials equal to zero provided $\text{char } k = 0$ or $\text{mult} < \text{char } k \ne 0$ where, as usual, k is the field of coefficients of f. (That is why we have been saying $X^i Y^j Z^l$ instead of $X^i Y^j Z^k$. So in case $\text{char } k = p \ne 0$, in (12) we assume $p \nmid n$, and in (10) and (11) we assume $p \ne 2, 3$, and in (3) to (6) we assume $p \ne 2$.) Assuming $\text{char } k = 0$ or $\text{mult} < \text{char } k \ne 0$ we have

$$\text{mult}_P S = \min \ i + j + l \quad \text{such that} \quad \left. \frac{\partial^{i+j+l} f}{\partial X^i \partial Y^j \partial Z^l} \right|_P \ne 0.$$

Assuming P to be the origin, and expanding f around P we have

$$f(X, Y, Z) = f_d(X, Y, Z) + f_{d+1}(X, Y, Z) + \cdots + f_j(X, Y, Z) + \cdots$$
$$\text{with } d = \text{mult}_P C$$

where f_j is a homogeneous polynomial of degree j. Now substituting the equations of the line $L_{(a,b,c)}$ into the equation of S we get

$$f(at, bt, ct) = f_d(a, b, c)t^d + f_{d+1}(a, b, c)t^{d+1} + \cdots + f_j(a, b, c)t^j + \cdots.$$

So, the intersection multiplicity of $L_{(a,b,c)}$ with S at $P = (0, 0, 0)$ is greater than d iff $f_d(a, b, c) = 0$. In other words, the tangent lines to S at $P = (0, 0, 0)$ constitute the *tangent cone* $f_d(X, Y, Z) = 0$ of S at P. Here $f_d(X, Y, Z)$ is called the *initial form* of $f(X, Y, Z)$.

Let us work out some examples in greater detail.

Cuspidal cubic cone. Consider the cuspidal cubic cone $\widehat{S}_3 : f(X, Y, Z) = 0$ where $f(X, Y, Z) = ZY^2 - X^3$. The origin $X = Y = Z = 0$ is clearly a triple point. The curve $\widehat{C}_3$ (a cuspidal cubic) has a double point at the origin $X = Y = 0$ where, strictly speaking, $X = \frac{\mathscr{X}}{\mathscr{Z}}$ and $Y = \frac{\mathscr{Y}}{\mathscr{Z}}$. So this should give a double line of $\widehat{S}_3$. The first partials of $f(X, Y, Z)$ are given by $f_X = -3X^2$, $f_Y = 2ZY$, $f_Z = Y^2$. So $X = Y = 0$ is a double line of $\widehat{S}_3$. Note that the multiplicity is exactly 2, because some of the second partials are $f_{X^2} = -6X$, $f_{Y^2} = 2Z$, $f_{YZ} = 2Y$. These are all zero only if $X = Y = Z = 0$. Moreover, if we make the plane QDT

$$\left\{ \begin{array}{l} X = X' \\ Y = X'Y' \end{array} \right\} \quad \text{with reverse equations} \quad \left\{ \begin{array}{l} X' = X \\ Y' = Y/X \end{array} \right\}$$

to resolve the singularity of $\widehat{C}_3$ at the origin, then that would, by treating Z as a "dummy variable," resolve the multiple line of $\widehat{S}_3$.

Monoidal transformation of 3-space. By treating Z as a dummy variable we mean making the MDT (*monoidal transformation*)

$$\left\{ \begin{array}{l} X = X' \\ Y = X'Y' \\ Z = Z' \end{array} \right\} \quad \text{with reverse equations} \quad \left\{ \begin{array}{l} X' = X \\ Y' = Y/X \\ Z' = Z \end{array} \right\}.$$

Now the plane QDT blows up the origin in the (X, Y)-plane into the Y'-axis $X' = 0$ in the (X', Y')-plane. So the MDT blows up the Z-axis in the (X, Y, Z)-space into the (Y', Z')-plane $X' = 0$ in the (X', Y', Z')-space. The Z-axis is called the *center* of the MDT. The (Y', Z')-plane is called the *exceptional plane*. By applying the MDT to the cuspidal cubic cone $\widehat{S}_3$ we get

$$Z'X'^2Y'^2 - X'^3 = X'^2 \left[Z'Y'^2 - X' \right].$$

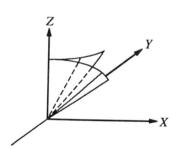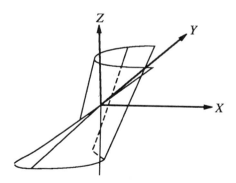

FIGURE 24.9

So the *total transform* $Z'X'^2Y'^2 - X'^3 = 0$ of $\widehat{S}_3$ consists of the exceptional plane $X' = 0$ and the *proper transform* $Z'Y'^2 - X' = 0$, which is nonsingular because the X'-derivative equals the nonzero constant -1. See Figure 24.9.

Similar things would happen in the general case. Thus, consider a general surface $S : f(X, Y, Z) = 0$ that need not be a cone. Its singular locus would consist of points as well as curves so that

$$\text{Sing}(S) = \underset{d_1}{\{P_1\}} \cup \underset{d_2}{\{P_2\}} \cup \ldots \cup \underset{d_h}{\{P_h\}} \cup \underset{\delta_1}{\Gamma_1} \cup \underset{\delta_2}{\Gamma_2} \cup \ldots \cup \underset{\delta_e}{\Gamma_e}$$

where P_i is a d_i-fold point and Γ_j is a δ_j-fold curve. How do we define the multiplicity δ of a singular curve Γ of S? One way is $f = f_X = f_Y = f_Z = 0$ at every point of Γ. If we keep taking partials, then after some steps we are at a stage when the partials of a certain order are zero at only finitely many points of Γ. All the partials of lower order are zero at all the points of Γ. This would then define the multiplicity of Γ to be the number δ such that all the partials of order $< \delta$ are zero at every point of Γ, and some partial of order δ is nonzero at most points of Γ. (In algebraic geometry, "most" means the points of nonvalidity are contained in a proper subvariety. So in our case of a curve Γ, at "most" points of Γ means at all but finitely many points of Γ. In analysis, "most" means outside a set of measure zero. In every category in mathematics, "most" has an appropriate meaning.)

This definition is not computationally effective since one has to check the value of the partials at every point of Γ. Here the power of college algebra helps. It shows that the pair (curve, point) is the same as the pair (surface, curve). Thus, defining the multiplicity of a curve on a surface is no more work than defining the multiplicity of a point on a curve. As we shall see in a moment, algebraically speaking, a nonsingular curve is "like" a line.

Now if we have a double line through a triple point then blowing up this double line may worsen the multiplicity of the triple point. However, if we blow up a triple line through a triple point that doesn't happen. What should we do if we have the former situation? In this case one should blow up this point. This brings us to a

Quadratic transformation of 3-*space.* This is given by

$$\left.\begin{cases} X = X' \\ Y = X'Y' \\ Z = X'Z' \end{cases}\right\} \quad \text{or by the reverse equations} \quad \left.\begin{cases} X' = X \\ Y' = Y/X \\ Z' = Z/X \end{cases}\right\}.$$

By making this *quadratic transformation*, the origin in the (X, Y, Z)-space blows up into the (Y', Z')-plane $X' = 0$ in the (X', Y', Z')-space. The origin $(0, 0, 0)$ is called the *center* of this QDT (quadratic transformation) and the (Y', Z')-plane is again called the *exceptional plane.* (So an MDT blows up a line into a plane, whereas a QDT blows up a point into a plane.) Now for example consider a

Fermat cone. $X^n \pm Y^n \pm Z^n = 0$ (with $n > 1$ and nondivisible by the characteristic). The partial X, Y, Z derivatives respectively are nX^{n-1}, $\pm nY^{n-1}$, $\pm nZ^{n-1}$. Hence, the origin $(0, 0, 0)$ is the only singular point. The above QDT changes the Fermat cone to

$$X'^n \pm X'^n Y'^n \pm X'^n Z'^n = X'^n \left[1 \pm Y'^n \pm Z'^n\right] = 0,$$

where the *total transform* $X'^n \pm X'^n Y'^n \pm X'^n Z'^n = 0$ consists of the exceptional plane $X' = 0$ and the proper transform $1 \pm Y'^n \pm Z'^n = 0$, which is a nonsingular "cylinder."

General cone. What we just said for a Fermat cone $X^n \pm Y^n \pm Z^n = 0$ applies equally well to a cone over any nonsingular curve, that is, a cone $S : f(X, Y, Z) = 0$ where $f(X, Y, Z)$ is a homogeneous polynomial that when regarded as defining a curve C in the projective plane is such that C is nonsingular. In other words, if we apply a QDT to such a cone S over a nonsingular curve C, taking the center of the QDT to be at the vertex of the cone, then the proper transform S' of S is nonsingular. Moreover, by taking the intersection $E \cap S'$ of the proper transform S' with the exceptional plane E we get back our curve C. Thus there is an alternative way to look at the intrinsic section of a cone which we talked about earlier. Indeed, as we just said, it is $E \cap S'$. This alternative way of looking at the intrinsic section remains valid even when C is allowed to have singularities.

Tangent cone. Now, if we have a point P on a general surface $S :$ $f(X, Y, Z) = 0$ which need not be a cone, and we make a QDT with center at P to get the exceptional plane E and the proper transform S' of S, then what is $E \cap S'$? In brief, $E \cap S' =$ a section of the tangent cone T of S at P. In greater detail, translating P to the origin and then expanding around P we have

$$f(X, Y, Z) = f_d(X, Y, Z) + f_{d+1}(X, Y, Z) + \cdots + f_j(X, Y, Z) + \cdots$$
$$\text{with } d = \text{mult}_P C,$$

where f_j is a homogeneous polynomial of degree j. By applying the above QDT with center at $P = (0, 0, 0)$ we get

$$f(X', X'Y', X'Z') = X'^d f'(X', Y', Z'),$$

where

$$f'(X', Y', Z')$$
$$= f_d(1, Y', Z') + X' f_{d+1}(1, Y', Z') + \cdots + X'^{j-d} f_j(1, Y', Z') + \cdots.$$

So the total transform $f(X', X'Y', X'Z') = 0$ of S consists of the exceptional plane $E : X' = 0$ and the proper transform $S' : f'(X', Y', Z') = 0$. Hence, $E \cap S'$ is the curve $C'_d : f_d(1, Y', Z') = 0$. The tangent cone T of S at P is given by $f_d(X, Y, Z) = 0$. Its intrinsic section is the curve $C_d : f_d(\mathcal{X}, \mathcal{Y}, \mathcal{Z}) = 0$ in the projective plane $\mathbb{P}^2$. We can cover $\mathbb{P}^2$ by three affine planes:

$$\mathbb{P}^2 = \mathbb{A}^2_{(Y', Z')} \cup \mathbb{A}^2_{(X'', Z'')} \cup \mathbb{A}^2_{(X^*, Z^*)},$$

where

$$\left\{ \begin{array}{l} Y' = \mathcal{Y}/\mathcal{X} \\ Z' = \mathcal{Z}/\mathcal{X} \end{array} \right\} \text{ and } \left\{ \begin{array}{l} X'' = \mathcal{X}/\mathcal{Y} \\ Z'' = \mathcal{Z}/\mathcal{Y} \end{array} \right\} \text{ and } \left\{ \begin{array}{l} X^* = \mathcal{X}/\mathcal{Z} \\ Y^* = \mathcal{Y}/\mathcal{Z} \end{array} \right\}.$$

Correspondingly, C_d is covered by the three affine plane curves:

$$\left\{ \begin{array}{l} C'_d : f_d(1, Y', Z') = 0 \\ C''_d : f_d(X'', 1, Z'') = 0 \\ C^*_d : f_d(X^*, Y^*, 1) = 0. \end{array} \right.$$

This brings us to the fact that, strictly speaking, the QDT with center $(0, 0, 0)$ is also given by the three "coordinate patches":

$$\left\{ \begin{array}{l} X' = X \\ Y' = Y/X \\ Z' = Z/X \end{array} \right\} \text{ and } \left\{ \begin{array}{l} X'' = X/Y \\ Y'' = Y \\ Z'' = Z/Y \end{array} \right\} \text{ and } \left\{ \begin{array}{l} X^* = X/Z \\ Y^* = Y/Z \\ Z^* = Z \end{array} \right\}.$$

Likewise, the MDT centered at a line is really given by two coordinate patches which come from the two coordinate patches required by the corresponding plane QDT. As a matter of fact, while talking about resolution of plane curve singularities, it was this phenomenon that a plane QDT requires two coordinate patches, which made it desirable to avoid "$X = 0$ being tangent." At any rate, we have the

DICTUM. *When we make a QDT on 3-space, the exceptional plane is a projective plane. Namely, it is the projective plane consisting of lines in the 3-space passing through the center of the QDT. Likewise, when we make a QDT on 2-space, the exceptional line is a projective line. Namely, it is the projective line consisting of all lines in the 2-space passing through the center of the QDT. This then is the most natural definition of, and the real raison d'être for, projective lines and projective planes.*

Nodal cubic cone. The equation of the surface is $ZY^2 - ZX^2 - X^3 = 0$. The singular locus consists of the triple point $(0, 0, 0)$ and the double line $X = Y = 0$. As in the above example of a cuspidal cubic cone, by applying the MDT with center $X = Y = 0$ we get

$$Z'X'^2Y'^2 - Z'X'^2 - X'^3 = X'^2 \left[Z'Y'^2 - Z' - X' \right].$$

So the total transform consists of the exceptional plane $X' = 0$ and the proper transform $Z'Y'^2 - Z' - X' = 0$, which is nonsingular because the X'-derivative equals the nonzero constant -1.

Quadratic or monoidal? In introducing QDTs in 3-space we said that if we blow up a double line through a triple point it may worsen the triple point. The examples of cuspidal cubic cone and nodal cubic cone did not exhibit this phenomenon, because there we did blow up a double line through a triple point, and the triple point actually got resolved. But this should be regarded as an accident. For a nonaccidental illustration of worsening a singularity, consider the

(13) *Surface with double line through triple point.* $Y^3 - X^2Z^{n-2} = 0$ with $n > 5$. This has a triple point at the origin, and $X = Y = 0$ is a double line. By making the MDT with $X = Y = 0$ as center, we get

$$Y^3 - X^2Z^{n-2} = X'^2[X'Y'^3 - Z'^{n-2}] = 0$$

as the total transform. The proper transform is $X'Y'^3 - Z'^{n-2} = 0$, which has the origin as a 4–fold point.

For an example of the simplifying action on a double point, when we blow up a double line passing through it, let us turn to the surface considered in item (9). This will also give us an opportunity to describe what a "pinch point" is.

Nonconical cubic surface with double line. The surface considered in item (9) is given by $f(X, Y, Z) = 0$, where $f(X, Y, Z) = Z^2 - XY^2$. The singular locus is given by $\{f = f_X = f_Y = f_Z = 0\}$. The first partials are given by $f_X = Y^2$, $f_Y = 2XY$, $f_Z = 2Z$. So the singular locus consists of $Y = Z = 0$, i.e., the X-axis, which is a double line because the second partial f_{Z^2} equals the nonzero constant 2. Consider the variable point $Q_\alpha = (\alpha, 0, 0)$ of the double line $Y = Z = 0$. Around Q_α the equation of the surface looks like

$$Z^2 - (X - \alpha)Y^2 - \alpha Y^2 = 0,$$

so the tangent cone at Q_α is $Z^2 - \alpha Y^2 = 0$, which factors into the two tangent planes $Z = \sqrt{\alpha}Y$ and $Z = -\sqrt{\alpha}Y$, except when $\alpha = 0$. If $\alpha = 0$, the tangent plane is $Z = 0$ *counted twice*. Thus at the origin the two tangent planes get "pinched" into one. Such a point of the double line is called a *pinch point* of the surface. If $\alpha \neq 0$, then around Q_α the above equation can be written as $Z^2 = Y^2 [\alpha + (X - \alpha)]$, and factoring this into power series factors we get

$$Z = \pm\sqrt{\alpha}\, Y \left[1 + \frac{(X - \alpha)}{\alpha}\right]^{1/2}$$

$$= \pm\sqrt{\alpha}\, Y \left[1 + \frac{1}{2}\left(\frac{X - \alpha}{\alpha}\right) - \frac{1}{8}\left(\frac{X - \alpha}{\alpha}\right)^2 + \cdots\right].$$

But if $\alpha = 0$ then the equation of the surface is $Z^2 = XY^2$. This cannot be factored into power series around $Q_0 = (0, 0, 0)$.

REMARK. The singular locus of any surface $f(X, Y, Z) = 0$ is always "closed." That is, it can be given by polynomial equations, namely, $f = f_X = f_Y = f_Z = 0$. However, the above example shows that the "analytic reducibility locus" (i.e., the locus of points where the surface is analytically reducible) is *not* necessarily closed. In general we can consider some special property $\mathscr{P}$ of a point P, and ask if the set of points with this property is closed, i.e., if it can be given by polynomial equations. For example, we may ask whether the locus $\mathrm{Sing}_d(S)$ of "d or more fold" points of S is closed. The criterion of multiplicity in terms of partial derivatives shows that this is so if either $\mathrm{char}\, k = 0$ or $d < \mathrm{char}\, k \neq 0$. Actually it turns out that $\mathrm{Sing}_d(S)$ is closed without any such assumption.

By adding an extra term to the example of a nonconical cubic surface with double line, we get the more general

(14) *Surface with double line through double point.* $Z^2 - XY^2 + aY^n = 0$ with $n > 2$. This has a double point at the origin, and $Y = Z = 0$ is a double line. To resolve this double line, by applying the MDT with the line $Y = Z = 0$ as center, by making the substitution

$$\begin{cases} X = X' \\ Y = Y' \\ Z = Y'Z', \end{cases}$$

we get

$$Y'^2 Z'^2 - X'Y'^2 - aY'^n = Y'^2 \left[Z'^2 - X' - aY'^{n-2}\right].$$

The total transform consists of the exceptional plane $Y' = 0$, and the proper transform $Z'^2 - X' - aY'^{n-2} = 0$, which is nonsingular because its X'-derivative equals the nonzero constant -1. For $a = 0$, i.e., for the nonconical surface, the proper transform is the "parabolic cylinder" $Z'^2 = X'$.

Now the simplest examples of the resolving action of an MDT are provided.

(15) *Cuspidal cubic cylinder.* $Y^2 - X^3 = 0$. See Figure 24.10 on page 206.

(16) *Nodal cubic cylinder.* $Y^2 - X^2 - X^3 = 0$. See Figure 24.11 on page 206.

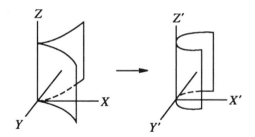

FIGURE 24.10

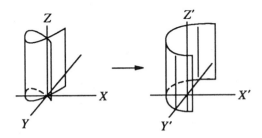

FIGURE 24.11

These are to be thought of as loci in the 3-space of the variables (X, Y, Z). We didn't give these examples to begin with because, as they don't involve the third variable explicitly, they may not look very convincing. At any rate, $X = Y = 0$, i.e., the Z-axis, is a double line for each of them. By applying the MDT centered at the line $X = Y = 0$, as the respective total transforms we get

$$Y^2 - X^3 = X'^2 Y'^2 - X'^3 = X'^2[Y'^2 - X'] = 0$$

and

$$Y^2 - X^2 - X^3 = X'^2 Y'^2 - X'^2 - X'^3 = X'^2[Y'^2 - 1 - X'] = 0.$$

So the respective proper transforms are the nonsingular *parabolic cylinders* $Y'^2 = X'$ and $Y'^2 = 1 + X'$.

Thus we have studied the simplifying effect of QDTs and MDTs on the singularities of some special surfaces. Here the center of a QDT was a point, and the center of an MDT was a line. Eventually, we are aiming at the following

PROPOSITION. *By a finite succession of QDTs centered at points and MDTs centered at nonsingular curves, the singularities of any surface can be resolved.*

So how do we blow up a curve? We have said that a nonsingular curve is "like" a line. However, for this we must make an analytic coordinate change. Let us explain

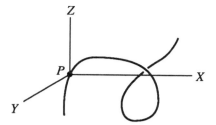

FIGURE 24.12

How to think of a space curve as a line. Consider an irreducible space curve C. See Figure 24.12. Take a simple point P of C. We may assume that $P = (0, 0, 0)$. At P we can find a local parametrization

$$\begin{cases} X = \lambda(t) \\ Y = \mu(t) \\ Z = \nu(t), \end{cases}$$

where $\lambda(t), \mu(t), \nu(t)$ are power series in t with

$$\min\{\operatorname{ord}\lambda(t), \operatorname{ord}\mu(t), \operatorname{ord}\nu(t)\} = 1.$$

Say $\operatorname{ord}\nu(t) = 1$. Then $\lambda(t) = \lambda^*(\nu(t))$ and $\mu(t) = \mu^*(\nu(t))$, where $\lambda^*(t)$ and $\mu^*(t)$ are power series in t devoid of constant term. By the analytic coordinate change

$$\begin{cases} \widehat{X} = X - \lambda^*(Z) \\ \widehat{Y} = Y - \mu^*(Z) \\ \widehat{Z} = Z \end{cases}$$

the curve C becomes the line $\widehat{X} = \widehat{Y} = 0$. In some sense, we can do this without assuming P to be a simple point of C. Still assuming P to be the origin $(0, 0, 0)$, C is given by a height 2 prime ideal N in $R = k[X, Y, Z]$. It can be shown that R_N is a 2-dimensional regular local ring. Hence $NR_N = (\hat{x}, \hat{y})R_N$ for some $\hat{x}, \hat{y}$ in R_N. Actually, we can choose $\hat{x}, \hat{y}$ in R. In other words, although C may not be complete intersection in 3-space, it is so in its own neighborhood. We can find $\hat{x}, \hat{y}$ in R such that

$$(\hat{x}, \hat{y})R = N \cap N_1^* \cap \cdots \cap N_h^*,$$

where N_i^* is a primary ideal belonging to a height 2 prime ideal N_i different from N. In fact, $\hat{x}, \hat{y}$ can be chosen so that $N_i^* = N_i$ for all i. In case P is a simple point of C, the argument shows how to find $\widehat{X}, \widehat{Y}, \widehat{Z}$ in $\widehat{R} = k[[X, Y, Z]] =$ the completion of R_J where $J = (X, Y, Z)R$, such that $(\widehat{X}, \widehat{Y}, \widehat{Z})\widehat{R} = M(\widehat{R})$ and $(\widehat{X}, \widehat{Y})\widehat{R} = N\widehat{R}$. By slightly modifying this argument we can show that, in case P is a simple point of C, we can actually find $\widehat{X}, \widehat{Y}, \widehat{Z}$ in R such that $(\widehat{X}, \widehat{Y}, \widehat{Z})R_J = M(R_J)$ and $(\widehat{X}, \widehat{Y})R_J = NR_J$.

Remark. The experience with curves shows that the higher the singularity, the easier it is to parametrize rationally. To some extent this remains true for surfaces and hypersurfaces. But not quite all the way. Namely, a surface of degree n, if it is a monoid (if it has an $(n-1)$-fold point), then we can rationally parametrize it by taking lines through that point. But not so if it is a cone, (if it has a point of multiplicity n), because then it can be rationally parametrized iff a section of it can be so parametrized. Thus, the cone $X^3 + Y^3 + Z^3 = 0$ cannot be rationally parametrized because its section is a curve of genus 1. There is this slight discrepancy between curves and surfaces because a plane curve of degree n having an n-fold point simply reduces to n lines through that point.

NOTE. My father, Professor Shankar Keshav Abhyankar, was my first teacher. I have the following fond memory of him. I think I was in high school. Having browsed through a book on plane coordinate geometry, for three days I was continuously asking him questions about 3-dimensional coordinate geometry and he was continuously answering them. On the fourth day he became very angry with me because during his college period he found that he had forgotten the material of the subject he was supposed to be teaching, because his mind was completely occupied with my questions. His anger did not matter to me because my intensive course of 3-dimensional coordinate geometry, covering much material from R.J.T. Bell [**Bel**] or C. Smith [**Smi**], was already over. Looking back at these books written around the turn of the century, indeed I find them to be an excellent initiation into the algebraic geometry of surfaces and curves in 3-space. Also, I believe that the $AF + BG$ type argument of finding a circle passing through the common points of two given circles, and the fact that such nice things are contained in Salmon's books [**Sa1**], [**Sa2**], was imparted to me by him. I suppose Salmon's books must have been in one or both of the colleges he studied in, Holkar College at Indore and Science College at Nagpur. But Salmon was not in the Victoria College at Gwalior where I studied. I saw Salmon (his books) first time at Harvard. Pitaji (Father) also taught me the following two aphorisms: (1) Series is the beginning and the end of mathematics. (2) Inequalities are the backbone of analysis. Also he said that Euler purified trigonometry.

Hypersurfaces

Having discussed singularities of curves and surfaces, let us now consider a hypersurface $S : f(X_1, X_2, \ldots, X_n) = 0$ in n-space for any n, and let us study the singularity of S at any point P. To do this, by a translation we bring P to the origin $(0, \ldots, 0)$, and then expand f around P to get

$$f(X_1, \ldots, X_n)$$
$$= f_d(X_1, \ldots, X_n) + f_{d+1}(X_1, \ldots, X_n) + \cdots + f_j(X_1, \ldots, X_n) + \cdots$$

where

$$d = \text{mult}\,_P S = \text{ord}\, f(X_1, \ldots, X_n)$$

and f_j is a homogeneous polynomial of degree j. Here f_d, the sum of the least degree terms of f, is called the *initial form* of f. Having brought P to the origin, we could just as well let S be an analytic hypersurface; we could assume that f is a power series. To find where a line

$$L_{(a_1, \ldots, a_n)} : \begin{cases} X_1 = a_1 t \\ X_2 = a_2 t \\ \vdots \\ X_n = a_n t \end{cases}$$

through the origin meets S, by substituting the equations of the line in the equation of S, we get

$$f(a_1 t, \ldots, a_n t)$$
$$= f_d(a_1, \ldots, a_n)t^d + f_{d+1}(a_1, \ldots, a_n)t^{d+1} + \cdots + f_j(a_1, \ldots, a_n)t^j + \cdots.$$

Hence, the intersection multiplicity of $L_{(a_1, \ldots, a_n)}$ with S at P is greater than d iff $f_d(a_1, \ldots, a_n) = 0$. Thus the tangent lines to S at P fill up the cone $T : f_d(X_1, \ldots, X_n) = 0$ of degree d with vertex at the origin, and so we call T the *tangent cone* or the *tangent hypercone* of S at P. Note that by a *cone*, or a *hypercone*, with vertex at the origin we mean an algebraic hypersurface whose defining equation is homogeneous.

By generalizing the idea of plane QDT, we get the QDT (*quadratic transformation*) of n-space with center at the origin. This sends the $(X_1, \ldots, X_n)$-space to the $(X'_1, \ldots, X'_n)$-space (see Figure 25.1, page 210) by means of

FIGURE 25.1

the equations $\begin{cases} X_1 = X_1' \\ X_2 = X_1' X_2' \\ \vdots \\ X_n = X_1' X_n' \end{cases}$ or the reverse equations $\begin{cases} X_1' = X_1 \\ X_2' = X_2/X_1 \\ \vdots \\ X_n' = X_n/X_1 \end{cases}$.

It blows up the origin in the $(X_1, \ldots, X_n)$-space into the *exceptional hyperplane* $E' : X_1' = 0$ in the $(X_1', \ldots, X_n')$-space. Moreover, it blows up the line $L_{(a_1, \ldots, a_n)}$ into the exceptional hyperplane E' and the line $L_{(a_1, \ldots, a_n)}'$: $X_2' - (a_2/a_1) = \cdots = X_n' - (a_n/a_1) = 0$, provided $a_1 \neq 0$. This line $L_{(a_1, \ldots, a_n)}'$, which is parallel to the X_1'-axis, may be called the *proper transform* of $L_{(a_1, \ldots, a_n)}$. Why did we have to make the proviso of $a_1 \neq 0$? That is because, strictly speaking, the "image space" of the QDT is covered by the n "affine charts" $\mathscr{A}_1^n, \ldots, \mathscr{A}_n^n$, where $\mathscr{A}_i^n$ is the space of the variables $(X_1^{(i)}, \ldots, X_n^{(i)})$ that are given by

the equations: $\begin{cases} X_1 = X_1^{(i)} X_i^{(i)} \\ X_2 = X_2^{(i)} X_i^{(i)} \\ \vdots \\ X_{i-1} = X_{i-1}^{(i)} X_i^{(i)} \\ X_i = X_i^{(i)} \\ X_{i+1} = X_{i+1}^{(i)} X_i^{(i)} \\ \vdots \\ X_n = X_n^{(i)} X_i^{(i)} \end{cases}$ or the reverse equations: $\begin{cases} X_1^{(i)} = X_1/X_i \\ X_2^{(i)} = X_2/X_i \\ \vdots \\ X_{i-1}^{(i)} = X_{i-1}/X_i \\ X_i^{(i)} = X_i \\ X_{i+1}^{(i)} = X_{i+1}/X_i \\ \vdots \\ X_n^{(i)} = X_n/X_i \end{cases}$.

So really the QDT blows up the origin into projective $(n-1)$-space, which consists of lines $L_{(a_1, \ldots, a_n)}$ through the origin in n-space, where $(a_1, \ldots, a_n)$ is the corresponding homogeneous coordinate n-tuple.

To formally define the *projective* $(n-1)$-*space* $\mathbb{P}_k^{n-1}$ over a field k (the coefficient field of f), consider the set of all nonzero n-tuples $(a_1, \ldots, a_n)$ of elements of k (nonzero means $a_i \neq 0$ for some i), and declare two nonzero tuples $(a_1, \ldots, a_n)$ and $(b_1, \ldots, b_n)$ to be equivalent if they are proportional, that is, if there exists $c \neq 0$ in k such that $a_i = cb_i$ for $1 \leq i \leq n$. Now $\mathbb{P}^{n-1}_k$ may be defined to be the set of all equivalence classes of all nonzero n-tuples of elements of k. Clearly then the points of $\mathbb{P}^{n-1}_k$ bijectively correspond to the lines $L_{(a_1, \ldots, a_n)}$ in n-space through the origin. If we represent the homogeneous coordinates in $\mathbb{P}^{n-1}_k$ by $(\mathscr{X}_1, \ldots, \mathscr{X}_n)$ then

$$\mathbb{P}^{n-1}_k = \bigcup_{i=1}^{n} \mathbb{A}^{n-1}_i$$

where $\mathbb{A}^{n-1}_i$ is the affine $(n-1)$-space of the variables

$$\left(\frac{\mathscr{X}_1}{\mathscr{X}_i}, \ldots, \frac{\mathscr{X}_{i-1}}{\mathscr{X}_i}, \frac{\mathscr{X}_{i+1}}{\mathscr{X}_i}, \ldots, \frac{\mathscr{X}_n}{\mathscr{X}_i} \right).$$

By applying the QDT to S and looking at the chart $\mathscr{A}^n_1 = \mathscr{A}'^n$, we get

$$f(X'_1, X'_1 X'_2, \ldots, X'_1 X'_n) = X'^d_1 f'(X'_1, X'_2, \ldots, X'_n),$$

where

$$f'(X'_1, X'_2, \ldots, X'_n) = f_d(1, X'_2, \ldots, X'_n) + X'_1 f_{d+1}(1, X'_2, \ldots, X'_n) + \cdots$$
$$+ X'^{j-d}_1 f_j(1, X'_2, \ldots, X'_n) + \cdots.$$

So $f(X'_1, X'_1 X'_2, \ldots, X'_1 X'_n) = 0$ is the *total transform* of S. It consists of the exceptional hyperplane $E' : X'_1 = 0$ and the *proper transform* S' : $f'(X'_1, X'_2, \ldots, X'_n) = 0$. Now $E' \cap S' : X'_1 = f_d(1, X'_2, \ldots, X'_n) = 0$.

Let $(a_1, \ldots, a_n)$ be the homogeneous coordinates of a point Q of the intrinsic section C_d of the tangent hypercone T. By the *intrinsic section* of a hypercone $\phi(X_1, \ldots, X_n) = 0$ in n-dimensional space we mean the hypersurface in $\mathbb{P}^{n-1}$ having $\phi(\mathscr{X}_1, \ldots, \mathscr{X}_n) = 0$ as its homogeneous equation. Now $C_d = C^{(1)}_d \cup \cdots \cup C^{(n)}_d$ where $C^{(i)}_d = C_d \cap \mathbb{A}^{n-1}_i$ = the set of points Q of C_d with $a_i \neq 0$. Clearly $Q \mapsto Q' = Q^{(1)} = (0, b_2, \ldots, b_n)$, with $b_i = \frac{a_i}{a_1}$, gives a bijection of $C'_d = C^{(1)}_d$ onto $E' \cap S'$. Note that if $E^{(i)}$ and $S^{(i)}$ are the exceptional hyperplane and the proper transform of S in $\mathscr{A}^n_i$ with $E^{(1)} = E'$ and $S^{(1)} = S'$, then $\hat{E} = E^{(1)} \cup \cdots \cup E^{(n)}$ is the *exceptional* projective $(n-1)$-space and $\hat{S} = S^{(1)} \cup \cdots \cup S^{(n)}$ is the *complete* proper transform of S.

For a moment assume Q is a point of C'_d, i.e., $a_1 \neq 0$. Let us translate Q to the origin in $\mathbb{A}^{n-1}_1 = \mathbb{A}'^{n-1}$ by putting

$$X^*_2 = X'_2 - b_2, \ldots, X^*_n = X'_n - b_n \text{ and } F^*_d(X^*_2, \ldots, X^*_n)$$
$$= f_d\left(1, X^*_2 + b_2, \ldots, X^*_n + b_n\right).$$

Now $F^*_d(0, \ldots, 0) = 0$ and $\text{mult}_Q C_d = \text{ord} F^*_d$. Let $X^*_1 = X'_1$. Now $(X^*_1, \ldots, X^*_n)$ is a coordinate system in $\mathbb{A}^n_1$ with respect to which the point

Q' is at the origin. Let

$$f^*(X_1^*, \ldots, X_n^*) = f'\left(X_1^*, X_2^* + b_2, \ldots, X_n^* + b_n\right).$$

Then $f^*(X_1^*, \ldots, X_n^*) = 0$ is the equation of S' in the coordinate system $(X_1^*, \ldots, X_n^*)$. We have

$$f^*(X_1^*, \ldots, X_n^*) = F_d^*(X_2^*, \ldots, X_n^*) + X_1^* \theta(X_1^*, \ldots, X_n^*),$$

where

$$\theta(X_1^*, \ldots, X_n^*) = f_{d+1}\left(1, X_2^* + b_2, \ldots, X_n^* + b_n\right) + \cdots$$
$$+ X_1^{*j-d-1} f_j\left(1, X_2^* + b_2, \ldots, X_n^* + b_n\right) + \cdots.$$

Hence,

$$\operatorname{mult}_{Q'} S' = \operatorname{ord} f^*(X_1^*, \ldots, X_n^*) \le \operatorname{ord} F_d^*(X_2^*, \ldots, X_n^*) = \operatorname{mult}_Q C_d \le d.$$

Now by definition $\operatorname{mult}_{Q'} \widehat{S} = \operatorname{mult}_{Q'} S'$. Therefore,

$$\operatorname{mult}_{Q'} \widehat{S} \le \operatorname{mult}_Q C_d \le d.$$

Hence, in particular,

$$\operatorname{mult}_Q C_d = 1 \Rightarrow \operatorname{mult}_{Q'} \widehat{S} = 1.$$

We may summarize the discussion in the following

PROPOSITION. *The QDT of n-space with center at a point P explodes P into an exceptional hyperplane which is a projective $(n-1)$-space $\mathbb{P}^{n-1}$. In effect, this $\mathbb{P}^{n-1}$ corresponds to the lines in n-space passing through P. Applying this QDT to a hypersurface (algebraic or analytic) S having P as a d-fold point transforms S into the proper transform $\widehat{S}$ together with the exceptional $\mathbb{P}^{n-1}$. (In effect, $\widehat{S}$ is obtained by puncturing S at P and filling the hole by a section of the tangent hypercone of S at P.) Let C_d be the intrinsic section of the tangent hypercone T of S at P. Then we have a natural bijection from C_d to $\widehat{S} \cap \mathbb{P}^{n-1}$. If under this bijection a point Q of C_d corresponds to a point Q' of $\widehat{S} \cap \mathbb{P}^{n-1}$, then $\operatorname{mult}_{Q'} \widehat{S} \le \operatorname{mult}_Q C_d \le d$. (The said bijection of C_d onto $\widehat{S} \cap \mathbb{P}^{n-1}$ is actually a "biregular map," a correspondence of algebraic varieties that preserves function fields and local rings.) Hence, in particular, if the tangent hypercone T is almost nonsingular, i.e., if it is nonsingular except at its vertex P (equivalently, if its intrinsic section C_d is nonsingular), then $\widehat{S}$ has no singular point in $\widehat{S} \cap \mathbb{P}^{n-1}$. Thus if the tangent hypercone of S at P is almost nonsingular, then one QDT resolves the singularities of S in the neighborhood of P. For $n = 2$, this reduces to the old observation that a d-fold point of a plane curve is resolved by one quadratic transformation if the curve has d distinct tangents at that point. For $d = 2$, this means a node.*

As we have said, the "image space" of the QDT is covered by the affine charts $\mathscr{A}_1^n, \ldots, \mathscr{A}_n^n$. It may be noted that these charts are highly overlapping. Likewise, the exceptional hyperplane $\widehat{E} = \mathbb{P}^{n-1}$ is covered by the charts $\mathbb{A}_1^{n-1}, \ldots, \mathbb{A}_n^{n-1}$, which are again highly overlapping. In the bijective correspondence between $\widehat{E} = \mathbb{P}^{n-1}$ and the set of all lines through the origin in the $(X_1, \ldots, X_n)$-space, the line $L_{(a_1, \ldots, a_n)}$ corresponds to a point in all those $\mathbb{A}_i^{n-1}$'s for which $a_i \neq 0$. In this correspondence, points of $\widehat{S} \cap \mathbb{P}^{n-1}$ come from lines on the tangent hypercone T. So if $L_{(0, \ldots, 0, 1)}$ does not lie on T, then $\widehat{S} \cap \mathbb{P}^{n-1}$ is completely covered by the charts $\mathscr{A}_1^n, \ldots, \mathscr{A}_{n-1}^n$. Hence, while studying the singularity of S at P, we need not look at the chart $\mathscr{A}_n^n$. We summarize this observation in the

DON'T LEMMA. *If the X_n-axis, i.e., the line $X_1 = \cdots = X_{n-1} = 0$, is not tangent to the hypersurface S at $P = (0, \ldots, 0)$, then in making the QDT of the $(X_1, \ldots, X_n)$-space with center at P we don't have to divide by X_n, at least as far as the effect of the QDT on S is concerned.*

REMARK. While tackling a resolution problem in $(X_1, \ldots, X_n)$-space, this lemma helps reduce it to a resolution problem in $(X_1, \ldots, X_{n-1})$-space, thus leading to an inductive argument. For such a reduction, the lemma needs to be supplemented by completing the square procedures.

To explain how the Don't Lemma leads to an inductive argument in the notation, assume that the X_n-axis is not tangent to S at P, and $\mathrm{mult}_{Q'} S' = d$. Since $f_d(X_1, \ldots, X_n)$ is a homogeneous polynomial of degree d, we can write

$$f_d(X_1, \ldots, X_n) = \sum_{i=0}^{d} \psi_i(X_2, \ldots, X_n) X_1^{d-i},$$

where $\psi_i(X_2, \ldots, X_n)$ is a homogeneous polynomial of degree i. Since

$$F_d^*(X_2^*, \ldots, X_n^*) = f_d(1, X_2^* + b_2, \ldots, X_n^* + b_n)$$

and

$$d = \mathrm{mult}_{Q'} S' \leq \mathrm{mult}_Q C_d = \mathrm{ord}\, F_d^*(X_2^*, \ldots, X_n^*) \leq d,$$

we must have

$$F_d^*(X_2^*, \ldots, X_n^*) = \psi_d(X_2^*, \ldots, X_n^*)$$
$$= \text{a homogeneous polynomial of degree } d.$$

Now we are assuming that the X_n-axis is not tangent to S at P. We are assuming that $f_d(0, \ldots, 0, 1) \neq 0$. Hence, we must have

$$F_d^*(0, \ldots, 0, X_n^*) = \psi_d(0, \ldots, 0, X_n^*) = c X_n^{*d} \quad \text{with } 0 \neq c \in k.$$

Let

$$f^*(X_1^*, \ldots, X_n^*)$$
$$= f_d^*(X_1^*, \ldots, X_n^*) + f_{d+1}^*(X_1^*, \ldots, X_n^*) + \cdots + f_j^*(X_1^*, \ldots, X_n^*) + \cdots$$

be the expansion of f^*, where f_j^* is a homogeneous polynomial of degree j. Since

$$f^*(X_1^*, \ldots, X_n^*) = F_d^*(X_2^*, \ldots, X_n^*) + X_1^* \theta(X_1^*, \ldots, X_n^*),$$

we get

$$F_d^*(0, \ldots, 0, X_n^*) = f_d^*(0, \ldots, 0, X_n^*) = cX_n^{*d} \quad \text{with } 0 \neq c \in k.$$

Hence, in particular,

$$f_d^*(0, \ldots, 0, 1) = c \neq 0.$$

Thus, as an extension of the Don't Lemma, we have proved the

DIDN'T LEMMA. *If we didn't have to divide by* X_n, *and if there is no drop in multiplicity after a QDT, then for the next QDT we don't have to divide by* X_n, *or rather by the translate of its transform.*

Details of Didn't Lemma. In greater detail, if the X_n-axis is not tangent to the hypersurface $S : f(X_1, \ldots, X_n) = 0$ at the d-fold point $P = (0, \ldots, 0)$ of S, with $d > 0$, and if after making the QDT $(X_1', X_2', \ldots, X_n') = (X_1, X_2/X_1, \ldots, X_n/X_1)$, we have a d-fold point $Q' = (0, b_2, \ldots, b_n)$ of $E' \cap S'$ where $E' : X_1' = 0$ is the exceptional hyperplane and $S' : f'(X_1', X_2', \ldots, X_n') = 0$ is the proper transform, then upon letting $(X_1^*, X_2^*, \ldots, X_n^*) = (X_1', X_2' - b_2, \ldots, X_n' - b_n)$ to bring Q' to the origin, we have that the X_n^*-axis, i.e., the line $X_1^* = \cdots = X_{n-1}^* = 0$, is not tangent to S' at Q'.

Inductive argument. Suppose we want to see whether a d-fold singular point P of the (algebraic or analytic) hypersurface $S : f(X_1, \ldots, X_n) = 0$ can be simplified by QDTs. Then first by translation we bring P to the origin $(0, \ldots, 0)$. Afterwards by rotation we arrange matters so that the X_n-axis is not tangent to S at P. Now we make the QDT centered at P and look at the chart $(X_1/X_i, \ldots, X_{i-1}/X_i, X_i, X_{i+1}/X_i, \ldots, X_n/X_i)$ for some $i < n$. By relabeling $X_1, \ldots, X_{n-1}$, we may assume that $i = 1$. We are looking at the chart $(X_1', X_2', \ldots, X_n') = (X_1, X_2/X_1, \ldots, X_n/X_1)$. Let $S' : f'(X_1', X_2', \ldots, X_n') = 0$ be the corresponding proper transform of S. Suppose we find a d-fold point $Q' = (0, b_2, \ldots, b_n)$ of S' that corresponds to P (which lies on the exceptional hyperplane E'). Again we make the translation $(X_1^*, X_2^*, \ldots, X_n^*) = (X_1', X_2' - b_2, \ldots, X_n' - b_n)$ to bring Q' to the origin. Because of the Didn't Lemma, the X_n^*-axis is not tangent to S' at Q'. Hence, we *don't* have to make a rotation any more. Now make the QDT centered at Q' and look at the proper transform S'' of S' relative to the chart $(X_1^*/X_j^*, \ldots, X_{j-1}^*/X_j^*, X_j^*, X_{j+1}^*/X_j^*, \ldots, X_n^*/X_j^*)$ for some $j < n$. Suppose we again find a d-fold point $Q'' = (c_1, \ldots, c_{j-1}, 0, c_{j+1}, \ldots, c_n)$ of S' that corresponds to Q'. Again we bring Q'' to the origin by the translation

$$(X_1^{**}, X_2^{**}, \ldots, X_n^{**})$$
$$= \left(\frac{X_1^*}{X_j^*} - c_1, \ldots, \frac{X_{j-1}^*}{X_j^*} - c_{j-1}, X_j^*, \frac{X_{j+1}^*}{X_j^*} - c_{j+1}, \ldots, \frac{X_n^*}{X_j^*} - c_n \right).$$

Once again, by the Didn't Lemma, the X_n^{**}-axis is not tangent to S'' at Q'', and so on. Assuming we have an infinite QDT sequence of n-spaces of d-fold points emanating from P, that seems to give rise to a certain problem in the $(n-1)$-space of the variables $(X_1, \ldots, X_{n-1})$. At any rate, as an iterative version of the Didn't Lemma we have the

NEVER LEMMA. *In making successive QDTs, because we started with a convenient rotation, we never have to rotate again until the multiplicity drops.*

Case of plane curves. In accordance with the inductive argument, in case of a plane curve (when $n = 2$), we noted that $f_d(1, X_2') = b^*(X_2' - b_2)^d$ with $b^* \neq 0$. We proceeded to make the preliminary coordinate change $X_2 \to X_2 - b_2 X_1$, which resulted in making b_2 to be zero and insuring that no translation was involved after the QDT. No drop in multiplicity after another QDT indicated a further preliminary coordinate change $X_2 \to X_2 - b_2 X_1 - c_2 X_1^2$. If the multiplicity never drops as we keep making QDT, then we get an infinite sequence of constants $b_2, c_2, d_2, \ldots$. Then by making the ultimate coordinate change $X_2 \to Y^* = X_2 - b_2 X_1 - c_2 X_1^2 - d_2 X_1^3 - \cdots$, we conclude that $F = \delta^* Y^{*d}$, where δ^* is a unit in $k[[X_1, X_2]]$. However, as explained in the alternative way of finding the number of steps in the lecture on resolution of singularities of plane curves, if we use WPT and make an SDT (Shreedharacharya Transformation, completing the square operation), then a translation is not necessary until the multiplicity drops.

Higher dimensional case. The above cited WPT + SDT method can be generalized to the n-dimensional case. Since we are assuming $f_d(0, \ldots, 0, 1) \neq 0$, by WPT we can write

$$f(X_1, \ldots, X_n) = \hat{\delta}(X_1, \ldots, X_n)\hat{f}(X_1, \ldots, X_n),$$

where

$$\hat{\delta}(X_1, \ldots, X_n) \in k[[X_1, \ldots, X_n]] \quad \text{with } \hat{\delta}(0, , \ldots, 0) \neq 0$$

and where

$$\hat{f}(X_1, \ldots, X_n) = X_n^d + \sum_{i=1}^{d} \hat{f}_i(X_1, \ldots, X_{n-1})X_n^{d-i}$$

with $\hat{f}_i(X_1, \ldots, X_{n-1}) \in k[[X_1, \ldots, X_{n-1}]]$ and $\hat{f}_i(0, \ldots, 0) = 0$

is a distinguished polynomial. Now, in case d is not divisible by the characteristic of k, by making the SDT $X_n \longrightarrow X_n + \frac{\hat{f}_1}{d}$ we can arrange $\hat{f}_1$ to be zero. With the proviso that $\hat{f}_1$ is zero to begin with, we must have $b_n = 0$, and likewise, $c_n = 0$, and so on. Thus having made the SDT, there is no translation in the n-th variable until the multiplicity drops. Now the successive proper transforms of $\hat{f}$ are going to be

$$(X_n/\Theta)^d + \sum_{i=2}^{d} \left(\hat{f}_i(X_1, \ldots, X_{n-1})/\Theta^i \right) (X_n/\Theta)^{d-i},$$

with $\Theta = Y_1 Y_2 \cdots Y_j$ where $Y_1 = 0$, $Y_2 = 0$, $\ldots$, $Y_j = 0$ are the successive exceptional hyperplanes. So the coefficient $(\hat{f}_i / \Theta^i)$ is neither the proper transform of $\hat{f}_i = 0$ nor its total transform, but something in between. Here $\hat{f}_i(X_1, \ldots, X_{n-1}) = 0$ is a hypersurface in the $(n-1)$-space of the variables $(X_1, \ldots, X_{n-1})$. We are more convincingly reduced to a problem in lower dimension.

Monoidal transformations. For any integer m with $1 < m \leq n$, we can make a QDT on the $(X_1, \ldots, X_m)$-space and "product" it with the $(X_{m+1}, \ldots, X_n)$-space to get what we call the MDT (= *monoidal transformation*) of the *n*-space *centered* at the $(n-m)$-dimensional linear subspace $L : X_1 = \cdots = X_m = 0$. In greater detail, the MDT with center L sends the $(X_1, \ldots, X_n)$-space into the $(X'_1, \ldots, X'_n)$-space by means of

the equations:
$$\begin{cases} X_1 = X'_1 \\ X_2 = X'_1 X'_2 \\ \vdots \\ X_m = X'_1 X'_m \\ X_{m+1} = X'_{m+1} \\ \vdots \\ X_n = X'_n \end{cases}$$
or the reverse equations:
$$\begin{cases} X'_1 = X_1 \\ X'_2 = X_2 / X_1 \\ \vdots \\ X'_m = X_m / X_1 \\ X'_{m+1} = X_{m+1} \\ \vdots \\ X'_n = X_n \end{cases}.$$

The origin in the $(X_1, \ldots, X_n)$-space is blown up into the linear $(m-1)$-dimensional subspace of the $(X'_1, \ldots, X'_n)$-space given by $L' : X'_1 = X'_{m+1} = \cdots = X'_n = 0$. More generally, a point $(0, \ldots, 0, b_{m+1}, \ldots, b_n)$ of L is blown up into the linear $(m-1)$-dimensional subspace of the $(X'_1, \ldots, X'_n)$-space given by $L' : X'_1 = X'_{m+1} - b_{m+1} = \cdots = X'_n - b_n = 0$. Thus the entire center L is blown up into the *exceptional hyperplane* $E' : X'_1 = 0$. Again, strictly speaking, the "image space" of the MDT is covered by m "affine charts," and so on. Now we can uniquely write

$$f(X_1, \ldots, X_n) = \sum_{j=e}^{\infty} g_j(X_1, \ldots, X_m) \quad \text{with } g_e(X_1, \ldots, X_m) \neq 0$$

where $g_j(X_1, \ldots, X_m)$ is a homogeneous polynomial of degree j in $X_1, \ldots, X_m$ with coefficients in $k[[X_{m+1}, \ldots, X_n]]$. We define the *multiplicity* of S at L by putting $\text{mult}_L S = e$. By applying the MDT with center L, we get $f(X'_1, \ldots, X'_n) = X'^e_1 g'(X'_1, \ldots, X'_n)$, where

$$g'(X'_1, \ldots, X'_n) = \sum_{j=e}^{\infty} X'^{j-e}_1 g_j(1, X'_2, \ldots, X'_m).$$

So, assuming $e > 0$, S goes over into the exceptional hyperplane $E' : X'_1 = 0$ and the *proper transform* $S' : g'(X'_1, \ldots, X'_n) = 0$. (Note that now $X_1', \ldots, X_n', E', S'$ have a different meaning than in the discussion of QDT.)

Clearly $e \leq d$, and $e = d \Leftrightarrow f_d(X_1, \ldots, X_n) = h(X_1, \ldots, X_m, 0, \ldots, 0)$ where $g_e(X_1, \ldots, X_m) = h(X_1, \ldots, X_n) \in k[[X_1, \ldots, X_n]]$. That is $h(X_1, \ldots, X_n)$ is the expression of $g_e(X_1, \ldots, X_m)$ as a power series in $X_1, \ldots, X_n$ with coefficients in k. In view of this remark, and by suitably modifying the above calculations made for QDT, we can establish the following

SPECIAL PERMISSIBILITY CRITERION. *If L is an $(n-m)$-dimensional linear subspace passing through the point P of the hypersurface S in n-space, with $1 < m \leq n$, and if L is equimultiple for S at P, i.e., $\text{mult}_L S = e = d = \text{mult}_P S$, then for every $P' \in L' \cap S'$ we have $\text{mult}_{P'} S' \leq \text{mult}_P S$ where S' is the proper transform of S under the MDT centered at L, and L' is the linear subspace into which the origin blows up.*

Now let U be any $(n-m)$-dimensional irreducible subvariety of S passing through P and having a simple point at P, and suppose that $1 < m \leq n$. To make an MDT *centered* at U, by making a suitable coordinate change, we may assume that $U = L$, and then we may follow the procedure. The resulting version of the above lemma may now be stated as the general

PERMISSIBILITY CRITERION. *If U is an $(n - m)$-dimensional irreducible subvariety passing through the point P of the hypersurface S in n-space, with $1 < m \leq n$, and if U has a simple point at P and U is equimultiple for S at P, then for every $P' \in L' \cap S'$ we have $\text{mult}_{P'} S' \leq \text{mult}_P S$ where S' is the proper transform of S under the MDT centered at U, and L' is the linear subspace into which the origin blows up.*

Simple point of a subvariety. In the above context, we note that an $(n - m)$-dimensional irreducible subvariety U in the $(X_1, \ldots, X_n)$-space passing through the origin $P = (0, \ldots, 0)$, has a simple point at P iff it has a local parametrization

$$\begin{cases} X_1 = \lambda_1(t_1, \ldots, t_{n-m}) \\ \quad \vdots \\ X_n = \lambda_n(t_1, \ldots, t_{n-m}) \end{cases}$$

at P, where $\lambda_1(t_1, \ldots, t_{n-m}), \ldots, \lambda_n(t_1, \ldots, t_{n-m})$ are power series in $t_1, \ldots, t_{n-m}$ such that $\lambda_1(0, \ldots, 0) = \cdots = \lambda_n(0, \ldots, 0) = 0$, and such that

the rank of the Jacobian matrix $\left(\dfrac{\partial \lambda_i}{\partial t_j} \Big|_{t_1 = \cdots = t_{n-m} = 0} \right)_{1 \leq i \leq n, \ 1 \leq j \leq n-m} = n - m.$

By relabeling $X_1, \ldots, X_n$, we can arrange matters such that

$$\det \left(\frac{\partial \lambda_i}{\partial t_j} \bigg|_{t_1 = \cdots = t_{n-m} = 0} \right)_{m+1 \leq i \leq n, \ 1 \leq j \leq n-m} \neq 0.$$

Then for $1 \leq i \leq m$ we have

$$\lambda_i(t_1, \ldots, t_{n-m}) = \lambda_i^* \left(\lambda_{m+1}(t_1, \ldots, t_{n-m}), \ldots, \lambda_n(t_1, \ldots, t_{n-m}) \right)$$

where $\lambda_i^*(t_1, \ldots, t_{n-m})$ are power series in $t_1, \ldots, t_{n-m}$ which are zero at $(0, \ldots, 0)$. Now by the analytic coordinate change

$$\begin{cases} \hat{X}_1 = X_1 - \lambda_1^*(X_{m+1}, \ldots, X_n) \\ \vdots \\ \hat{X}_m = X_m - \lambda_m^*(X_{m+1}, \ldots, X_n) \\ \hat{X}_{m+1} = X_{m+1} \\ \vdots \\ \hat{X}_n = X_n \end{cases}$$

the subvariety U becomes the $(n-m)$-dimensional linear subspace $\hat{L} : \hat{X}_1 = \cdots = \hat{X}_m = 0$. This explains the phrase "we may assume that $U = L$."

  Order along a subvariety. Let U be an irreducible $(n-m)$-dimensional subvariety in the affine n-space $\mathscr{A}$. Here $\mathscr{A}$ could be the affine n-space or the "local n-space." In the first case, U is given by a prime ideal N in $R = k[X_1, \ldots, X_n]$, and we let $A = R_N$. In the second case, U is given by a prime ideal N^* in $R^* = k[[X_1, \ldots, X_n]]$, and we let $A = R_{N^*}^*$. Now in both the cases, A is a regular local ring. Let P be a point of the hypersurface $S : f(X_1, \ldots, X_n) = 0$ in $\mathscr{A}$ where in the first case, $f \in R$ and in the second case, $f \in R^*$. In the second case, P is required to be the origin, whereas in the first case, we bring P to the origin by translation. The *multiplicity* of S at U is defined by

$$\mathrm{mult}\,_U S = \min e \text{ such that } f \in M(A)^e.$$

In the discussion we have "proved" that if U has a simple point at P, then

$$\mathrm{mult}\,_U S \leq \mathrm{mult}\,_P S.$$

It can be shown that the inequality remains true without the assumption of U having a simple point at P. We say that U is *equimultiple* for S at P if equality holds. Moreover, it can be shown that U is equimultiple for S at most points of U.

  Question. Then why not let an MDT have a center that has a singular point at P?

  Answer. Because then the singularity at P may worsen.

  Semicontinuity of order. The above inequality $\mathrm{mult}\,_U S \leq \mathrm{mult}\,_P S$ is a special case of a result in local algebra called semicontinuity of order. For this, see Nagata's *Local rings* [Na2].

Task. Having defined QDTs and MDTs of n-space, i.e., quadratic and monoidal transformations of the $(X_1, \ldots, X_n)$-space, where the centers of the QDTs are points and the centers of the MDTs are nonsingular subvarieties, we are ready to attack the problem of resolution of singularities. We have the necessary tools. Consider a hypersurface $S : f(X_1, \ldots, X_n) = 0$ that may be algebraic or analytic, and let P be a singularity of S that needs to be resolved. We hope to resolve it by successively blowing up suitable centers.

Question. How do you know which center to blow up?

Very brief answer. It is, in general, a delicate matter.

Cryptic answer. Complete a succession of SDTs until the power series f is expressed in a neat and tight form from which it becomes obvious what center to choose. If we blow up that center , then the singularity improves. As in the case of plane curves, such a neat and tight form should enable us to read off the "first characteristic data" or the "supermultiplicity" of S at P. For details, you may see my Springer monograph [A28] on canonical desingularization.

REMARK. Now a QDT is simply an MDT with zero dimensional center. But for convenience of talking, usually we shall say MDT when the center is positive dimensional.

NOTE. Desingularization (resolution of singularities) of curves was obtained in the nineteenth century by Riemann, Noether, and Dedekind. Riemann [Rie] did it by function theory, Noether [No1] by geometry ($\sim$ high school algebra), and Dedekind [DWe] by college algebra. Desingularization of surfaces over a field k was achieved by Walker [Wal] in 1935 for $k = \mathbb{C}$, by Zariski [Za4] in 1939 for char $k = 0$, and by Abhyankar [A02] in 1956 for char $k \neq 0$. Desingularization of algebraic varieties over a field of characteristic zero was accomplished by Zariski [Za5] in 1944 for dimension 3, and by Hironaka [Hir] in 1964 for any dimension. In 1965, Abhyankar [A10] carried out desingularization of "arithmetical" surfaces. Subsequently, in 1966, Abhyankar [A12] proved desingularization of three-dimensional algebraic varieties for fields of characteristic different from $2, 3, 5$.

We talked about neat and tight forms of power series. Neat and tight is meant to be a friendly substitute for canonical. The reference to the Springer monograph [A28] on Canonical Desingularization is mostly for characteristic zero. You may also like to read my various algorithmic papers on characteristic p resolution such as [A08], [A13], [A15], where you will encounter numerous semitight and pretight and prototight, ... , expressions. The algorithms in the arithmetic case [A09], [A11], are even more enjoyable. I hope you will then proceed to resolve singularities of arithmetic solids, i.e., surfaces defined over integers, thought of as 3-dimensional objects. Such a 3-dimensional object can be viewed as a family of surfaces, one for each characteristic.

Resolution of Singularities of Algebraic Surfaces

To solve a quadratic equation

$$aZ^2 + bZ + c = 0 \quad \text{with} \ a \neq 0,$$

following Shreedharacharya, we complete the square

$$a(Z^2 + \frac{b}{a}Z + \frac{b^2}{4a^2} - \frac{b^2}{4a^2}) + c = a(Z + \frac{b}{2a})^2 + (c - \frac{b^2}{4a}) = 0$$

to get

$$Z = \frac{-b \pm \sqrt{b^2 - 4ac}}{2a}$$

provided the characteristic is not 2. We call this an SDT (Shreedharacharya transformation = completing the square operation).

We shall now use SDTs to resolve singularities of a surface $S : f(X, Y, Z) = 0$ in 3-space over a field k. The general plan is thus. Let d be the maximum multiplicity of points of S. Blow up all the isolated d-fold points and singularities of d-fold curves. These are "bad" d-fold points. Some simple points of the d-fold curves may also be "bad". At any rate, the bad d-fold points are finite in number; blow them up by QDTs (quadratic transformations). On the resulting surface there may still be a finite number of bad d-fold points. Again blow them up by QDTs, and so on. After a finite number of steps of QDTs there will be no bad d-fold points left. This is where we use SDTs. Now blow up d-fold curves by MDTs (monoidal transformations). On the blown-up surface there will not be any bad d-fold points. Again blow up the new d-fold curves by MDTs, and so on. After a finite number of steps of MDTs there will be no more d-fold curves left. Now the multiplicity everywhere will be less than d. Thus, a finite number of QDTs followed by a finite number of MDTs decrease d. So we are done by induction on d.

To carry out this plan, we must define good points (bad = not good) and, assuming $d > 1$, prove the following three lemmas.

(1) Bad d-fold points are finite in number.

(2) There is no infinite quadratic sequence of bad d-fold points.

(3) There is no infinite MDT sequence of good d-fold curves. Moreover, we should also take into account the following three auxiliary results.

(i) $d < \infty$.

(ii) If we either blow up a d-fold point by a QDT, or blow up an irreducible nonsingular d-fold curve by an MDT, then the multiplicity of any resulting point is at most d.

(iii) If there are no bad d-fold points, then any irreducible d-fold curve is nonsingular and the surface, obtained by blowing it up by an MDT, has no bad d-fold point.

We get a heuristic proof of (i), at least when char $k = 0$, by noting that d is the maximum such that the partials

$$\left\{ \frac{\partial^{i+j+\kappa} f}{\partial X^i \partial Y^j \partial Z^\kappa} \right\}_{i+j+\kappa < d}$$

have a common solution. Projectively, we have to work with a finite number of affine patches. Thus d is reasonably computable. It may be remarked that (i) need not be true for an "analytic" surface in $\mathbb{C}^3$ because it may require infinitely many power series in infinitely many neighborhoods to define it.

A more precise proof of (i), in case of an algebraic surface in any characteristic, can be obtained thus. The singular locus of S, being a proper closed subset of S, consists of a finite number of curves and a finite number of points. Each of these curves has only a finite number of points at which the multiplicity is greater than the multiplicity of the curve. Adding these to the previous finite set of points, we still get only a finite set of points which need to be looked at. Considering the multiplicities at the finite number of irreducible curves and the finite number of points which "needed to be looked at," we get a finite set of positive integers. d is the largest among them.

Now (ii) is a special case of the general fact that if we blow up an equidimensional nonsingular center, then the multiplicity does not increase. (iii) follows from the precise definition of good points, which we shall describe in a moment. Let

$$\Delta = \text{ the } d\text{-fold locus of } S = \text{ the set of all } d\text{-fold points of } S.$$

Note that then Δ consists of a finite number of d-fold irreducible curves and a finite number of isolated d-fold points, i.e., d-fold points which are not on any of these curves.

Let P be a d-fold point of S.

Philosophical definition of good point. " P is a *good* point" means that it takes the same effort to resolve P as it does to resolve a singularity of a curve.

In greater detail, the isolated d-fold points, as well as the singular points of the various irreducible d-fold curves are certainly declared to be bad points. Moreover, points which are common to more than two irreducible d-fold curves are also regarded as bad. Finally, a point where two irreducible d-fold curves meet is also considered bad unless the two curves meet there normally, i.e., like coordinate axes. Actually, what we have so far described

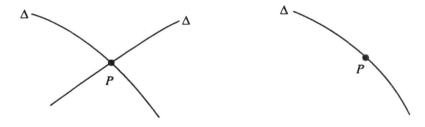

FIGURE 26.1

as bad points may be called *very bad*. A point which is not very bad is called *pregood*. See Figure 26.1.

The d-fold point P is said to be *good* if it is pregood, and every d-fold point P_1^*, which is on the surface S_1^* obtained by applying an MDT to S centered at any irreducible component of Δ passing through P, and which corresponds to P, is again pregood, and any d-fold point P_2^*, which lies on the surface S_2^* obtained by applying an MDT to S_1^* centered at any irreducible component of the d-fold locus of S_1^* passing through P_1^*, and which corresponds to P_1^*, is again pregood, and so on.

Now lemmas (1) and (3) follow from the resolution of singularities of plane curves. By a *special* d-fold point let us mean a d-fold point that is either very bad or a point which lies on two irreducible d-fold curves. Then the special d-fold locus Δ' of S, i.e., the set of all special d-fold points of S, is obviously a finite set. Now $\Delta \setminus \Delta'$ is a nonsingular curve on $S \setminus \Delta'$. Let S_1' be the surface obtained by monoidally blowing up $\Delta \setminus \Delta'$ on $S \setminus \Delta'$. Let Δ_1 be the d-fold locus of S_1', and let Δ_1' be the special d-fold locus S_1'. Again $\Delta_1 \setminus \Delta_1'$ is a nonsingular curve on $S_1' \setminus \Delta_1'$. Let S_2' be the surface obtained from $S_1' \setminus \Delta_1'$ by monoidally blowing up $\Delta_1 \setminus \Delta_1'$, and so on. By resolution of singularities of plane curves (viewed in the college algebra way), this must stop after a finite number of steps. That is, we shall reach a surface S_j' for which $\Delta_j = \Delta_j'$. Let Δ_i^* be the image of Δ_i' under the natural map of S_i' onto S. Let $\Delta^* = \Delta' \cup \Delta_1^* \cup \cdots \cup \Delta_j^*$. Then Δ^* is a finite subset of Δ, and every point in $\Delta \setminus \Delta^*$ is good. Therefore, bad d-fold points are finite in number, which proves (1). (3) follows from the resolution of singularities of plane curves in a similar but somewhat easier manner.

As we shall see in a moment, (2) also follows from resolution of singularities of plane curves, but in a somewhat subtle manner. This time we have to use total resolution of curves rather than proper resolution. This is where we also need SDT, WPT, and whatnot. Now (2) can be paraphrased positively by saying that if (S_0, P_0), (S_1, P_1), (S_2, P_2), $\ldots$ is any infinite sequence, where (2.1) P_j is a point on a surface S_j for all $j \geq 0$, such that (2.2) S_j is obtained by applying a QDT to S_{j-1} centered at P_{j-1}, and P_j corresponds to P_{j-1}, and

$$\max_{Q \in S_j} \text{mult}_Q S_j = \text{mult}_{P_j} S_j = \text{mult}_{P_{j-1}} S_{j-1} = \max_{Q' \in S_{j-1}} \text{mult}_{Q'} S_{j-1}$$

for all $j > 0$, then (2.3) P_m is a good point of S_m for some m (and infinitely many m).

To prove (2), let there be given any infinite sequence

$$(S_0, P_0), (S_1, P_1), (S_2, P_2), \ldots$$

satisfying (2.1) and (2.2). Without loss of generality, we may suppose that $S = S_0$ and $P = P_0$. Our aim is to show (2.3).

By translation, we can bring P to the origin $(0, 0, 0)$. Then by rotation we can arrange matters so that the Z-axis is not tangent to S at P. Now by WPT we can write

$$f(X, Y, Z) = \hat{\delta}(X, Y, Z)\hat{f}(X, Y, Z),$$

where

$$\hat{\delta}(X, Y, Z) \in k[[X, Y, Z]] \quad \text{with } \hat{\delta}(0, 0, 0) \neq 0$$

and where

$$\hat{f}(X, Y, Z) = Z^d + \sum_{i=1}^{d} \hat{f}_i(X, Y)Z^{d-i}$$

$$\text{with} \quad \hat{f}_i(X, Y) \in k[[X, Y]] \quad \text{and} \quad \hat{f}_i(0, 0) = 0$$

is a distinguished polynomial.

Assuming that d is not divisible by the characteristic of k, we can make the SDT

$$Z^* = Z + \frac{\hat{f}_1(X, Y)}{d} \quad \text{and} \quad f^*(X, Y, Z^*) = \hat{f}\left(X, Y, Z^* - \frac{\hat{f}_1(X, Y)}{d}\right)$$

to get

$$f^*(X, Y, Z^*) = Z^{*d} + \sum_{i=2}^{d} f_i^*(X, Y)Z^{*d-i}$$

$$\text{with} \quad f_i^*(X, Y) \in k[[X, Y]] \quad \text{and} \quad f_i^*(0, 0) = 0.$$

Since $\text{ord} f^*(X, Y, Z^*) = \text{mult}_P S = d$, we must have

$$\text{ord} f_i^*(X, Y) \geq i \quad \text{for } 2 \leq i \leq d.$$

Now let us be bold and put

$$G(X, Y) = \text{GCD}\{f_i^*(X, Y)^{1/i} : 2 \leq i \leq d\}.$$

More precisely, $G(X, Y)$ is the nonzero unique (up to unit factor) member of $k[[X, Y]]$ such that $G(X, Y)^i$ divides $f_i^*(X, Y)$ in $k[[X, Y]]$ for $2 \leq$

$i \leq d$, and such that $G(X, Y)$ is divisible by every $\widehat{G}(X, Y)$ whose i^{th} power $\widehat{G}(X, Y)^i$ divides $f_i^*(X, Y)$ in $k[[X, Y]]$ for $2 \leq i \leq d$. Let

$$\theta = \left[\min_{2 \leq i \leq d} \frac{\operatorname{ord} f_i^*(X, Y)}{i} \right] - \operatorname{ord} G(X, Y).$$

Let us note that by definition, the curve $G(X, Y)$ has a *normal crossing* at the origin $\Leftrightarrow G(X, Y) = \epsilon X^{*A} Y^{*B}$ where ϵ is a unit in $k[[X, Y]]$, A and B are non-negative integers, and X^* and Y^* are elements of order one in $k[[X, Y]]$ whose tangents are perpendicular, i.e., for which we have $M(k[[X, Y]]) = (X^*, Y^*)k[[X, Y]]$.

To check whether a point is good, here is a nice

ALGEBRAIC CRITERION. *P is good $\Leftrightarrow \theta < 1$ and $G(X, Y)$ has a normal crossing.*

Now S_1 and P_1 are respectively given by $f_1(X_1, Y_1, Z_1) = 0$ and $P_1 = (0, 0, 0)$, where either

(1)
$$(X_1, Y_1, Z_1) = \left(X, \frac{Y}{X} - b_1, \frac{Z^*}{X} - c_1 \right)$$

and

$$f_1(X_1, Y_1, Z_1) = f^*(X_1, X_1[Y_1 + b_1], X_1[Z_1 + c_1])/X_1^d,$$

or (2)
$$(X_1, Y_1, Z_1) = \left(\frac{X}{Y} - a_1, Y, \frac{Z^*}{Y} - c_1 \right)$$

and

$$f_1(X_1, Y_1, Z_1) = f^*(Y_1[X_1 + a_1], Y_1, Y_1[Z_1 + c_1])/Y_1^d,$$

or (3)
$$(X_1, Y_1, Z_1) = \left(\frac{X}{Z^*}, \frac{Y}{Z^*}, Z^* \right)$$

and

$$f_1(X_1, Y_1, Z_1) = f^*(X_1 Z_1, Y_1 Z_1, Z_1)/Z_1^d.$$

In case (3) we get

$$f_1(X_1, Y_1, Z_1) = 1 + \sum_{i=2}^{d} \left(f_i^*(X_1 Z_1, Y_1 Z_1)/Z_1^i \right).$$

Since $\operatorname{ord} f_i^*(X, Y) \geq i$ for $2 \leq i \leq d$, we see that

$$\left(f_i^*(X_1 Z_1, Y_1 Z_1)/Z_1^i \right) \in k[[X_1, Y_1, Z_1]]$$

$$\text{and } \operatorname{ord} \left(f_i^*(X_1 Z_1, Y_1 Z_1)/Z_1^i \right) \geq i.$$

Hence, $\operatorname{ord} f_1(X_1, Y_1, Z_1) = 0$, which is a contradiction because by assumption we have $\operatorname{ord} f_1(X_1, Y_1, Z_1) = \operatorname{mult}_{P_1} S_1 = d > 0$. *Thus, case (3) is ruled out.*

In case (1) we have

$$f_1(X_1, Y_1, Z_1) = f^*(X_1, X_1[Y_1 + b_1], X_1[Z_1 + c_1])/X_1^d$$

$$= (Z_1 + c_1)^d + \sum_{i=2}^{d} f_{1i}(X_1, Y_1)(Z_1 + c_1)^{d-i}$$

$$= \left[(Z_1 + c_1)^d + \sum_{i=2}^{d} f_{1i}(0, 0)(Z_1 + c_1)^{d-i} \right] + \alpha(X_1, Y_1, Z_1),$$

where

$$f_{1i}(X_1, Y_1) = \left(f_i^*(X_1, X_1[Y_1 + b_1])/X_1^i \right) \in k[[X_1, Y_1]]$$

and

$$\alpha(X_1, Y_1, Z_1) = \sum_{i=2}^{d} [f_{1i}(X_1, Y_1) - f_{1i}(0, 0)](Z_1 + c_1)^{d-i} \in k[[X_1, Y_1, Z_1]]$$

$$\text{with } \alpha(0, 0, Z_1) = 0.$$

Hence, because of the assumption that $\text{mult}_{P_1} S_1 = \text{ord} f_1(X_1, Y_1, Z_1) = d$, we get

$$\left[(Z_1 + c_1)^d + \sum_{i=2}^{d} f_{1i}(0, 0)(Z_1 + c_1)^{d-i} \right] = Z_1^d.$$

Now we have the

BINOMIAL ARGUMENT. *By the binomial theorem, the coefficient of* Z_1^{d-1} *in the above LHS is* dc_1, *but the coefficient of* Z_1^{d-1} *in the above RHS is* 0. *Hence,* $dc_1 = 0$. *Therefore, because* d *is nondivisible by the characteristic of* k, *we must have* $c_1 = 0$.

Similarly, in case (2) we must have $c_1 = 0$.

Thus, case (3) *never occurs, and in cases* (1) *and* (2) *there is no translation in* $\frac{Z^*}{X}$ *and* $\frac{Z^*}{Y}$, *respectively, i.e.,* $c_1 = 0$.

Now iterating this procedure, we conclude that the infinite QDT sequence

$$(S_1, P_1) \to (S_2, P_2) \to \cdots \to (S_{j-1}, P_{j-1}) \to (S_j, P_j) \to \cdots$$

is given by $S_j : f_j(X_j, Y_j, Z_j) = 0$ and $P_j = (0, 0, 0)$. For all $j > 1$ we have either

$$(1_j) \qquad (X_j, Y_j, Z_j) = \left(X_{j-1}, \frac{Y_{j-1}}{X_{j-1}} - b_j, \frac{Z_{j-1}}{X_{j-1}} \right)$$

and

$$f_j(X_j, Y_j, Z_j) = f_{j-1}(H_j, H_j[Y_j + b_j], H_j Z_j)/H_j^d \text{ with } H_j = X_j,$$

or

$$(2_j) \qquad (X_j, Y_j, Z_j) = \left(\frac{X_{j-1}}{Y_{j-1}} - a_j, Y_{j-1}, \frac{Z_{j-1}}{Y_{j-1}} \right)$$

and

$$f_j(X_j, Y_j, Z_j) = f_{j-1}(H_j[X_j + a_j], H_j, H_j Z_j)/H_j^d \text{ with } H_j = Y_j.$$

Let $(X_0, Y_0) = (X, Y)$, and let $H_1 = X_1$ in case (1), and $H_1 = Y_1$ in case (2). Now, for all $j > 0$ we have

$$f_j(X_j, Y_j, Z_j) = \left[Z_j^d + \sum_{i=2}^{d} f_{ji}(X_j, Y_j) Z_j^{d-i} \right],$$

where

$$f_{ji}(X_j, Y_j) = \frac{f_i^*(X, Y)}{(H_1 H_2 \cdots H_j)^i} \in k[[X_j, Y_j]]$$

with $\operatorname{ord} f_{ji}(X_j, Y_j) \geq i$ for $2 \leq i \leq d$.

We have the infinite plane QDT sequence

$$(X, Y) = (X_0, Y_0) \to (X_1, Y_1) \to \cdots \to (X_{j-1}, Y_{j-1}) \to (X_j, Y_j) \to \cdots.$$

For all $j > 0$ we have either

$(1_j')$ $(X_{j-1}, Y_{j-1}) = (H_j, H_j[Y_j + b_j])$ with $H_j = X_j$,

or $(2_j')$ $(X_{j-1}, Y_{j-1}) = (H_j[X_j + a_j], H_j)$ with $H_j = Y_j$.

If $f^*(X, Y, Z^*) = Z^{*d}$, then we have nothing more to say. In other words, we should assume that S is free from multiple components. Then, since $d > 1$, this can not occur anyway. Now, assuming that $f^*(X, Y, Z^*) \neq Z^{*d}$, by the principalized total resolution of plane curves, there exist positive integers e and l, with $2 \leq e \leq d$, such that

$$f_{li}(X_l, Y_l)^e \in f_{le}(X_l, Y_l)^i k[[X_l, Y_l]] \quad \text{for } 2 \leq i \leq d$$

and

$$f_{le}(X_l, Y_l) = \delta X_l^r Y_l^s,$$

where δ is a unit in $k[[X_l, Y_l]]$, and r and s are positive integers.

Concerning the infinite plane QDT sequence, we have used the

PRINCIPALIZED TOTAL RESOLUTION OF PLANE CURVES. *Given any finite number of elements* $\phi_1, \phi_2, \ldots, \phi_{d'}$ *in* $k[[X, Y]]$ *at least one of which is not zero, there exist positive integers* e' *and* l', *with* $1 \leq e' \leq d'$, *such that* $\phi_i \in \phi_{e'} k[[X_{l'}, Y_{l'}]]$ *for* $1 \leq i \leq d'$, *and* $\phi_{e'} = \delta' X_{l'}^{r'} Y_{l'}^{s'}$, *where* δ' *is a unit in* $k[[X_{l'}, Y_{l'}]]$, *and* r' *and* s' *are non-negative integers.*

The claim follows from the principalized total resolution of plane curves by taking $d' = d$, $\phi_1 = 0$, and $\phi_i = f_i^*(X, Y)^{d!/i}$ for $2 \leq i \leq d$. To obtain the principalized total resolution of plane curves, we note that by refining the proof of the proper resolution of plane curves, we get the

TOTAL RESOLUTION OF PLANE CURVES. *Given any* $0 \neq \phi \in k[[X, Y]]$, *there exists a positive integer* $\hat{l}$ *such that for every* $l'' \geq \hat{l}$ *we have* $\phi = \delta'' X_{l''}^{r''} Y_{l''}^{s''}$, *where* δ'' *is a unit in* $k[[X_{l''}, Y_{l''}]]$, *and* r'' *and* s'' *are non-negative integers.*

It is quite easy to see that given any monomials ψ_1 and ψ_2 in X_j, Y_j, for any positive integer j, we either have $\psi_1 \in \psi_2 k[[X_l, Y_l]]$ for all large enough l, or $\psi_2 \in \psi_1 k[[X_l, Y_l]]$ for all large enough l. Therefore, the principalized total resolution of plane curves follows from the total resolution of plane curves.

Looking at the two dominant terms of f_l we have

$$Z_l^d + f_{le}(X_l, Y_l)Z_l^{d-e} = Z_l^{d-e}\left[Z_l^e + \delta X_l^r Y_l^s\right].$$

Hence we may state the

FIRST APPROXIMATION. *At this stage we can say that the surface* S_l *can be "approximated" by the "canonical form"* $Z_l^e + \delta X_l^r Y_l^s = 0$.

Now clearly

$$f_{l+1,e}(X_{l+1}, Y_{l+1}) = \frac{f_{le}(X_l, Y_l)}{H_{l+1}^e} = \frac{\delta X_l^r Y_l^s}{H_{l+1}^e} = \tilde{\delta} X_{l+1}^{\tilde{r}} Y_{l+1}^{\tilde{s}},$$

where $\tilde{\delta}$ is a unit in $k[[X_{l+1}, Y_{l+1}]]$, and $\tilde{r}$ and $\tilde{s}$ are non-negative integers. Upon letting r_0, s_0, $\tilde{r}_0$, $\tilde{s}_0$ be the unique integers such that $0 \leq r_0 < e$, $0 \leq s_0 < e$, $0 \leq \tilde{r}_0 < e$, $0 \leq \tilde{s}_0 < e$, and $r - r_0 \in e\mathbb{Z}$, $s - s_0 \in e\mathbb{Z}$, $r - \tilde{r}_0 \in e\mathbb{Z}$, $s - \tilde{s}_0 \in e\mathbb{Z}$, we easily see that if $r_0 + s_0 \geq e$ then $\tilde{r}_0 + \tilde{s}_0 < r_0 + s_0$. Therefore, for some integer $m \geq l$ we have

$$f_{mi}(X_m, Y_m)^e \in f_{me}(X_m, Y_m)^i k[[X_m, Y_m]] \quad \text{for } 2 \leq i \leq d$$

and

$$f_{me}(X_m, Y_m) = \delta^* X_m^{r^*} Y_m^{s^*},$$

where δ^* is a unit in $k[[X_m, Y_m]]$, and r^* and s^* are non-negative integers such that, upon letting r_0^* and s_0^* be the unique integers such that $0 \leq r_0^* < e$, $0 \leq s_0^* < e$, and $r^* - r_0^* \in e\mathbb{Z}$, $s^* - s_0^* \in e\mathbb{Z}$, we have $r_0^* + s_0^* < e$. By applying the "$\Leftarrow$" part of the above algebraic criterion to the surface S_m, we immediately see that P_m is a good point of S_m.

Nonstability. In the argument, $\tilde{r}_0 + \tilde{s}_0 < r_0 + s_0$ *provided* $r_0 + s_0 \geq e$. This shows that goodness is not stable. For example, by one QDT the good double point $Z^2 = X^2 Y$ becomes the bad double point $Z_1^2 = X_1 Y_1$. As another example, by one QDT the good triple point $Z^3 = X^4 Y^4$ becomes the bad triple point $Z_1^3 = X_1^5 Y_1^4$.

We may also state the

SECOND APPROXIMATION. *The surface* S_m *can be approximated by the "good canonical surface"* $Z_m^e + \delta^* X_m^r Y_m^s = 0$ *with* $r_0^* + s_0^* < e$ *where* r_0^* *and* s_0^* *are the residues of* r^* *and* s^* *modulo* e.

To prove the " $\Leftarrow$ " part of the algebraic criterion in the notation, assume that $\theta < 1$ and $G(X, Y) = \epsilon X^{*A} Y^{*B}$. Now we can easily see that $Z^* = X^* = 0$ and $Z^* = Y^* = 0$ are among the irreducible components of Δ passing through P. Moreover, the first one (resp. the second one) is present iff $A > 0$ (resp. $B > 0$). Also we clearly have $A + B > 0$. Thus P is pregood. If $A > 0$ (resp. $B > 0$), then by monoidally blowing up $Z^* = X^* = 0$ (resp. $Z^* = Y^* = 0$), the new surface is exactly of the same form as S except that A (resp. B) drops by 1. Therefore, P is a good point of S. (Moreover, the multiplicity drops after exactly $A + B$ such steps.)

Exercise. Prove the " $\Rightarrow$ " part of the algebraic criterion.

SURFACE RESOLUTION. *Thus we have shown how to decrease the multiplicity d if it is nondivisible by the characteristic. Hence in particular, we have completed a proof of resolution of singularities of algebraic surfaces in characteristic zero.*

The above proof is taken from my good points paper [A30] where we have also discussed the

Possibility of generalizing this procedure to higher dimensions. Suppose S is replaced by a solid $T : g(X, Y, Z, W) = 0$. Again let d be the maximum multiplicity. Now a pregood point is defined to be a d-fold point P where the d-fold locus consists of one, two, or three surfaces, each of which is required to have a simple point at P, and collectively, they are required to meet each other normally at P. Just as in the case of surfaces, this leads to a definition of good point. Similarly, we can define bad points and also worse points. Good points would fill up surfaces, bad points would fill up curves, and worse points would be like isolated d-fold points. In a similar manner we can consider varieties of dimension > 3.

Problem. Consider solids. Restricting to characteristic zero or complex numbers if necessary, obtain a "descriptive proof" of Resolution of Singularities of the above type. For doing this, you will probably need to find an algebraic criterion for bad points.

Finitistic philosophy. Now from a finitistic viewpoint we cannot conceive of a power series. But for making the SDT $Z^* = Z + \frac{\hat{f}_1(X, Y)}{d}$, it suffices to compute $\hat{f}_1(X, Y)$ only up to a certain order. Higher order terms may be neglected. We have the following methods.

(M_1) Use WPT and kill $\hat{f}_1$ completely by completing the square.

(M_2) Use approximate WPT and kill approximate $\hat{f}_1$. But there is a third method.

(M_3) Slow SDT; slowly complete the square. It is possible to slowly complete the squares without applying Weierstrass. What we are referring to is the method described in the Springer monograph [A28] titled *Weighted expansions for canonical desingularization*. An alternate title for this monograph could have been *Exercises in completing the square*. Thus, the idea is: using Slow SDTs write a given function in nice form, or in fact, the best possible form, and do it finitistically. Moreover, another dictum guiding the constructions in the above monograph was the following

Dictum. Nothing should be ad-hoc. Every step must follow naturally, it should be very deterministic.

Let us illustrate this by considering $f(X, Y, Z)$. Here $f_d(X, Y, Z)$ is the tangent cone. Now our choice of variables X, Y, Z should also be canonical. If the polynomial is written in some manner, then we should write it using a canonical choice of variables. Already this is illustrated by the plane curve case by studying the difference between nodes and cusps. (See Table 26.1.)

TABLE 26.1

Node	Cusp
$Y^2 - X^2 - X^3$.	$Y^2 + 2XY + X^2 - X^3$.
This is fine as it is.	This is $(Y + X)^2 - X^3$.
Initial form is $Y^2 - X^2$	Initial form is $(Y + X)^2$
which depends on X and Y both.	which depends only on $Y + X$.
X and Y	$X^* = X$ and $Y^* = Y + X$
are good variables.	are good variables.
Form $Y^2 - X^2 - X^3$ is good.	So good form is $Y^{*2} - X^{*3}$.

Thus we choose coordinates so that the initial form is expressed in the least number of variables. For example,

$$(X + Y + Z)^d \quad \rightarrow \quad \text{1 variable is sufficient.}$$
$$(X + Y)^d + Z^d \quad \rightarrow \quad \text{2 variables suffice.}$$
$$X^d + Y^d + Z^d \quad \rightarrow \quad \text{3 variables are needed.}$$

These "least number of variables" may be called the "essential number of variables" on which the initial form depends. Here the philosophy guiding us is: nodes are better than cusps. More to the point, we have seen that if the intrinsic section of the tangent cone is nonsingular, then the singularity gets resolved after one quadratic transformation. This indicates that more the essential number of variables required by the initial form, the better the singularity. To enlarge this number, we try to give weights to the variables so that the *weighted initial form* depends on the largest possible number of variables.

Weighted initial forms of plane curves. To explain the philosophy, let us consider a plane curve $C : F(X, Y) = 0$ having a d-fold singular point at the origin. Assuming that the Y-axis is not tangent to C at the origin, we have

$$F(X, Y) = \cdots + \tilde{F}_{-1}(X)Y^{d+1} + \tilde{F}_0(X)Y^d + \cdots + \tilde{F}_i(X)Y^{d-i} + \cdots + \tilde{F}_d(X)$$

with $\tilde{F}_0(0) \neq 0 = \tilde{F}_1(0) = \cdots = \tilde{F}_d(0)$. Now we can apply WPT to write $F(X, Y) = \hat{\epsilon}(X, Y)\hat{F}(X, Y)$ where $\hat{\epsilon}(X, Y)$ is a unit in $k[[X, Y]]$ and $\hat{F}(X, Y)$ is a distinguished monic polynomial of degree d in Y with co-efficients in $k[[X]]$ which are 0 at 0. Or better still, in place of WPT, let us invoke the approximate WPT by which we mean do nothing to the above two-sided expression. Simply "think of" its "normalized" left piece $[\cdots + \tilde{F}_{-1}(X)Y^{d+1} + \tilde{F}_0(X)Y^d]/Y^d = [\cdots + \tilde{F}_{-1}(X)Y + \tilde{F}_0(X)]$ as a substitute for the unit $\hat{\epsilon}(X, Y)$, which is to be neglected, and the right hand piece $Y^d + \tilde{F}_1(X)Y^{d-1} + \cdots + \tilde{F}_d(X)$ to be the substitute for the distinguished polynomial $\hat{F}(X, Y)$. Now by giving weight $u = \min_{1 \leq i \leq d} \frac{\text{ord } \tilde{F}_i(X)}{i}$ to X and 1 to Y, the weighted initial form of $F(X, Y)$ involves Y^d and at least one more term. Actually, $(u, 1)$ is the only such weight system. If the said weighted initial form is not a dth power, then we stop. If it is a dth power, then u must be a positive integer, and the weighted initial form looks like $\gamma^*(Y + \gamma_u X^u)^d$ with nonzero elements γ^* and γ_u in k. Now we make the transformation $Y' = Y + \gamma_u X^u$ and find the new values u' and $\gamma_{u'}$ of u and γ_u. It will turn out that $u' > u$, and so on. Thus we get a sequence of positive integers $u < u' < u'' < \cdots$, and a sequence of nonzero elements $\gamma_u, \gamma_{u'}, \gamma_{u''}, \ldots$ in k. In this manner we have found a good variable $Y^* = Y + \gamma_u X^u + \gamma_{u'} X^{u'} + \gamma_{u''} X^{u''} + \cdots$. If the sequence is infinite, then $F(X, Y) = \epsilon^* Y^{*d}$ where ϵ^* is a unit in $k[[X, Y]]$. Clearly the "correct" initial form of F depends only on the variable Y^*. If the sequence is finite, then the relevant weighted initial form *requires* the two essential variables (X, Y^*). Moreover, (d, du^*) is called the "first characteristic pair" of C at P, where u^* is the relevant weight of X.

Weighted initial forms of surfaces. To carry over the above discussion to the surface $S : f(X, Y, Z) = 0$, assuming that the d-fold point P of S is at the origin and the Z-axis is not tangent to C at P, we expand $f(X, Y, Z)$ in the form

$$\cdots + \tilde{f}_{-1}(X, Y)Z^{d+1} + \tilde{f}_0(X, Y)Z^d + \cdots + \tilde{f}_i(X, Y)Z^{d-i} + \cdots + \tilde{f}_d(X, Y)$$

with $\tilde{f}_0(0, 0) \neq 0 = \tilde{f}_1(0, 0) = \cdots = \tilde{f}_d(0, 0)$. Again, by applying the approximate WPT, we replace the unit $\hat{\delta}(X, Y, Z)$ in the above Weier-strass form of f by the "normalized" left piece $[\cdots + \tilde{f}_{-1}(X, Y)Z^{d+1} + \tilde{f}_0(X, Y)Z^d]/Z^d = [\cdots + \tilde{f}_{-1}(X, Y)Z + \tilde{f}_0(X, Y)]$ of the expression. We replace the distinguished polynomial $\hat{f}(X, Y, Z)$ by the right piece $Z^d + \tilde{f}_1(X, Y)Z^{d-1} + \cdots + \tilde{f}_d(X, Y)$ of the expression. Now we give weight 1 to Z and suitable weights to X and Y, so that the weighted initial form of f ..., and so on.

Weighted initial forms for higher dimensional varieties. For higher dimensional algebraic varieties, the process is generalized by making repeated slow SDTs. Very briefly, let $W : h_1(X_1, \ldots, X_n) = h_2(X_1, \ldots, X_n) = \cdots = 0$ be any algebraic variety. Then, at any given point Q of W, we complete the squares as much as we can. This generates a whole set of data giving different boxes of variables and sets of relative weights and so on. At any rate, we get a finite but large array of numbers. We order this data in a suitable manner. This leads to the idea of the "first characteristic data," or the "supermultiplicity," at any point Q of the variety W. By the very nature of the definitions, the maximum supermultiplicity locus is a nonsingular subvariety. By monoidally blowing it up once, the supermultiplicity drops. At least this is so in zero characteristic. For details, see the Springer monograph [A28]. Now an irreducible analytic plane curve has a finite number of characteristic pairs. The first corresponds to the "first characteristic data." For higher dimensional varieties, how do we enlarge the "first characteristic data" into the "complete system of characteristic data," which should be like the entire set of characteristic pairs? Here we have the

AMBITIOUS GOAL. *Find the complete system of characteristic data for varieties of any dimension. This system should be a sequence or a tree of characteristic data. Now what should this complete system of characteristic data do? The answer is: everything! For example, for any algebraic variety, we should be able to obtain genus formulas, do resolution of singularities, or what have you. Presently this is just a dream.*

What is the difference between characteristic 0 and characteristic p? As indicated in the binomial argument, the basic difference is that, in characteristic p, various coefficients in the binomial expansion

$$(Z + c)^d = Z^d + dcZ^{d-1} + \frac{d(d-1)}{2} c^2 Z^{d-2} + \cdots + c^d$$

are divisible by p.

Calculus. All the formulas of differential calculus can be derived from the binomial theorem. The derivative of Z^n is calculated using the binomial theorem and now from this, with the intervention of Taylor's Theorem, we can differentiate any function. So the importance of calculus is perhaps exaggerated. For a further philosophical discussion of the importance of the binomial theorem for resolution of singularities, see my Moscow lecture [A17], as well as my Bombay colloquium [A19].

Exercise. For characteristic $p \neq 0$, let closure (d) be the set of integers v, with $0 \leq v \leq d$, for which the binomial coefficient $\frac{d(d-1)\cdots(d-v+1)}{v!}$ is nondivisible by p. Describe the set closure (d), and prove that this is a closure operation, i.e.: $w \in$ closure (v) and $v \in$ closure (d) $\Rightarrow$ $w \in$ closure (d).

Arithmetic case. In this case, it is not enough to know which binomial coefficients are divisible by p, but how much. What are the highest powers of p which divide the various coefficients? Lots of other amusing things can also happen in this case. We have for example the p-adic numbers. Say $p = 3$. We want to think of integers as power series in 3. Thus, to add 7 and 8 we write

$$7 = 1 + 2 \cdot 3$$
$$8 = 2 + 2 \cdot 3.$$

We can add these as power series:

$$1 + 2 \cdot 3$$
$$+\underline{\quad 2 + 2 \cdot 3 \quad}$$
$$3 + 4 \cdot 3 = 5 \cdot 3$$
$$= 2 \cdot 3 + 1 \cdot 3^2.$$

Moreover, expressions such as

$$1 + 3 + 3^2 + 3^3 + \cdots$$

are also allowed. The theory of p-adic numbers was created by Hensel [**Hen**]. We have given a proof of Hensel's Lemma that says that if $\Theta(X, Z) = Z^n + \Theta_1(X)Z^{n-1} + \cdots + \Theta_n(X)$ is such that $\Theta(0, Z) = \overline{G}(Z)\overline{H}(Z)$ with $\mathrm{GCD}\,(\overline{G}(Z), \overline{H}(Z)) = 1$, then the factorization lifts. Originally Hensel proved this when the coefficients Θ_i's are integers and, instead of putting $X = 0$, the coefficients are reduced modulo a prime p (e.g., $p = 3$). Now we need not stop at one variable since we've seen that Hensel's Lemma holds when X is replaced by (X, Y). We can consider

$$\Theta(X, Y, Z) = Z^n + \Theta_1(X, Y)Y^{n-1} + \cdots + \Theta_n(X, Y)$$

and let one of the variables take prime values. For example $p = 3$ and $\Theta(X, 3, Z) = Z^2 + (3X + 6)Z + (7X + 8)$. Now by putting $X = 0$ and reducing modulo 3 we get $Z^2 + 2$ which can be factored as $(Z + 1)(Z + 2)$ over $\mathbb{Z}/3\mathbb{Z}$. Hence, by Hensel's Lemma the original polynomial can be factored in the form $(Z + \lambda)(Z + \mu)$, where λ and μ are power series in X with coefficients in the 3-adic field.

NOTE. Hardly anything is known about the life of Shreedharacharya, who around 500 A.D. made extensive use of completing the square arguments. So we have to be satisfied with an anecdote about Bhaskaracharya who refers to Shreedharacharya in his versified book on algebra, *Beejaganit*, composed around 1100 A.D. This *Beejaganit* is the second book in his treatise on astronomy called Siddhantashiromani. The first book of that treatise, which deals with arithmetic and pythagorean geometry, is named after his daughter, Leelavati, to whom Bhaskara taught mathematics when she became a widow at an early age. Knowing that the stars were not particularly auspicious for her wedding, Bhaskara tried to outwit them by arranging to get her married at a certain specially auspicious precise moment. To mark that precise moment, Bhaskara had started the water clock by placing in a water container a jar with a small hole. When sufficient water would come out through that hole, the jar would sink indicating the precise auspicious moment. Being of a curious nature, Leelavati peaked in that contraption. While doing so, a pearl from her earring fell into the jar, clogging the hole. This delayed the sinking of the jar beyond the auspicious moment, and so she was widowed anyway. ... More to the point, Bhaskara's *Beejaganit* certainly contains a wealth of algebraical insight. The latter books in his astronomical treatise also show his pre-Newton mastery of the principle of gravitation.

Birational and Polyrational Transformations

For resolving the singularities of a surface $S : f(X, Y, Z) = 0$ over a field k, we made transformations of the type

$$T_0 : \begin{cases} X = u(X', Y', Z') \\ Y = v(X', Y', Z') \\ Z = w(X', Y', Z') \end{cases} , \quad \begin{matrix} \text{which can be} \\ \text{reversed as} \end{matrix} \quad \tilde{T}_0 : \begin{cases} X' = \tilde{u}(X, Y, Z) \\ Y' = \tilde{v}(X, Y, Z) \\ Z' = \tilde{w}(X, Y, Z) \end{cases}$$

where u, v, w are functions of X', Y', Z' and $\tilde{u}, \tilde{v}, \tilde{w}$ are functions of X, Y, Z.

Question. What kind of functions are we talking about?

Answer. We left that loose. If we require $u, v, w, \tilde{u}, \tilde{v}, \tilde{w}$ to be polynomials, then we cannot get rid of the singular points. Our goal has been to arrange matters so that, by such a transformation, the resulting surface $S' : f'(X', Y', Z') = f(u(\ ), v(\ ), w(\ )) = 0$, is nonsingular, has simpler singularities, or is less singular. We want to apply a succession of such transformations

$$S \to S' \to S'' \to \cdots \to S^{(100)} = S^*$$

so that, say after 100 steps, the transform S^* becomes nonsingular, and hence at each step it gets a little better. At any rate, the complete data is more useful than just the final transform. If T_0 and $\tilde{T}_0$ are linear then that's not too good, because the multiplicities do not change. Moreover, if T_0 and $\tilde{T}_0$ are polynomials (if their constituent functions $u, v, w, \tilde{u}, \tilde{v}, \tilde{w}$ are polynomials) then it is more subtle but still true that the multiplicities of various points do not change.

Exercise. Interpret and solve!

Types of transformations. The discussion remains valid if we pass from surfaces in 3-space to higher dimensional varieties in n-space for any n. Note that then we would be considering transformations of the type

$$T : \begin{cases} X_1 = u_1(X'_1, \ldots, X'_n) \\ \vdots \\ X_n = u_n(X'_1, \ldots, X'_n) \end{cases} , \quad \begin{matrix} \text{which can be} \\ \text{reversed as} \end{matrix} \quad \tilde{T} : \begin{cases} X'_1 = \tilde{u}_1(X_1, \ldots, X_n) \\ \vdots \\ X'_n = \tilde{u}_n(X_1, \ldots, X_n) \end{cases}$$

where $u_1, \ldots, u_n$ are functions of $X'_1, \ldots, X'_n$, and $\tilde{u}_1, \ldots, \tilde{u}_n$ are functions of $X_1, \ldots, X_n$. So far we have discussed (1) bilinear, and (2) bipoly-

235

nomial transformations where bilinear means T and $\tilde{T}$ are both linear. Bipolynomial means T and $\tilde{T}$ are both polynomial. Roughly speaking, for getting rid of the singularities, bilinear or bipolynomial won't do, but polyrational would do. Polyrational means T polynomial and $\tilde{T}$ rational; $u_1, \ldots, u_n$ are polynomials, and $\tilde{u}_1, \ldots, \tilde{u}_n$ are rational functions. So let us add to the list (3) polyrational transformations. Now these three are special cases of (4) birational transformations, by which we mean the situation when T and $\tilde{T}$ are both rational.

Simple polyrational transformations. We cannot profitably use linear ones. The next best are quadratic transformations. If we have only two variables X, Y, then these are

$$\left\{\begin{array}{l} X = X' \\ Y = X'Y' \end{array}\right\} \quad \text{or inversely} \quad \left\{\begin{array}{l} X' = X \\ Y' = Y/X \end{array}\right\}.$$

These generalize to three variables X, Y, Z in two different ways. One is

$$\left\{\begin{array}{l} X = X' \\ Y = X'Y' \\ Z = X'Z' \end{array}\right\} \quad \text{or inversely} \quad \left\{\begin{array}{l} X' = X \\ Y' = Y/X \\ Z' = Z/X \end{array}\right\}.$$

This is called a quadratic transformation (QDT). The other is

$$\left\{\begin{array}{l} X = X' \\ Y = X'Y' \\ Z = Z' \end{array}\right\} \quad \text{or inversely} \quad \left\{\begin{array}{l} X' = X \\ Y' = Y/X \\ Z' = Z \end{array}\right\}.$$

This is called a monoidal transformation (MDT). In the general case of n variables, an MDT $(X_1, \ldots, X_n) \to (X'_1, \ldots, X'_n)$ is given by

$$\left\{\begin{array}{l} X_1 = X'_1 \\ X_2 = X'_1 X'_2 \\ \vdots \\ X_m = X'_1 X'_m \\ X_{m+1} = X'_{m+1} \\ \vdots \\ X_n = X'_n \end{array}\right\} \quad \text{or inversely by} \quad \left\{\begin{array}{l} X'_1 = X_1 \\ X'_2 = X_2/X_1 \\ \vdots \\ X'_m = X_m/X_1 \\ X'_{m+1} = X_{m+1} \\ \vdots \\ X'_n = X_n \end{array}\right\}.$$

We note that the MDT is a QDT if $m = n$.

SUPREME THEOREM. *Now the ideal situation is that by making one transformation, the singularity at some point becomes somewhat better and no singularities become worse.*

If we can prove this, then we have achieved a good deal in analyzing singularities, and questions such as parametrizability can be answered more easily. A result to this effect may be called a supreme theorem. But if we cannot prove this, then we can try to prove the following

PRECURSOR TO THE SUPREME THEOREM. *After say* 100 (*or a finite number of*) *transformations, the singularities are removed.*

Question. Is it true that any polyrational transformation is a succession of monoidal transformations?

Answer. Briefly the question is does polyrational ⇒ iterated monoidal? The answer may be given as follows because we are dealing with n-dimensional varieties.

$$n = 2 \quad \text{yes} \begin{cases} \text{some cases} & \text{Zariski [Za5]} \\ \text{remaining cases} & \text{Abhyankar [A01].} \end{cases}$$
$$n = 3 \quad \text{no} \qquad\qquad\qquad \text{Shannon [Sha]}$$

For $n = 3$, Shannon's 1972 Purdue thesis [Sha] shows that the answer is no. We may ask a

Question. For $n = 3$, given a polyrational transformation T, if T is not an iterated monoidal, then is it true that T followed by some other polyrational transformation T^* is an iterated monoidal?

Answer. Half yes. Christensen proved this in his 1978 Purdue thesis [Chr]. There are several unsolved questions in this direction. For example, can the half yes be converted into yes?

Related questions. We considered bipolynomial transformations and decided these are useless for resolution. Of course, that doesn't mean they are useless as such. So consider some simple bipolynomial transformations in two variables. For example, for any polynomial $p(X)$ in X we can consider the transformation

$$\begin{cases} X' = X \\ Y' = Y + p(X) \end{cases}, \quad \text{which can be inverted as} \quad \begin{cases} X = X' \\ Y = Y' - p(X') \end{cases}.$$

Now we can iterate this. Of course, if we follow it by another transformation of exactly the same type, we won't get anything new. But we can permute the variables, however, to make a transformation of the type

$$\begin{cases} X' = X + q(Y) \\ Y' = Y. \end{cases}, \quad \text{which can be inverted as} \quad \begin{cases} X = X' - q(Y') \\ Y = Y' \end{cases},$$

where $q(Y)$ is any polynomial in Y. Transformations of these two types may be called *elementary transformations*. A product of a finite number of elementary and linear transformations is called a *tame transformation*. Now a question is: does bipolynomial ⇒ tame? For more than two variables, this is a completely mysterious territory. But for two variables, we have the following

AUTOMORPHISM THEOREM (A_2') (Jung [Jun], van der Kulk [Kul], Nagata [Na3], Abhyankar [A27]). *Every k-automorphism of the two variable polynomial ring $k[X, Y]$ is tame.*

In general we have the

FACTORIZATION PROBLEM. *Given a class of objects, find nice building blocks or bricks!*

One of the oldest substantial theorems of this type is

NOETHER'S FACTORIZATION THEOREM (G_2). *Every plane cremona transformation is a product of cremona quadratic transformations.*

Plane cremona transformations. By a *plane cremona transformation* we mean a birational transformation of $\mathbb{P}^2$ onto itself.

Cremona quadratic transformations. Now there are ∞^5 conics. In other words, since a homogeneous equation of degree 2 in 3 variables has six coefficients determined up to proportionality, conics in the plane form 5-dimensional projective space. Consequently, conics through three fixed points form a "net." A *net* is a 2-dimensional family. Lines in the plane form a net, and we can read off points from this net. (A *pencil* is a 1-dimensional family, and a *web* is a 3-dimensional family.) But we can use different nets to read off points in a plane. (Likewise, we can use webs to read off points in 3-space.) Now a new coordinate system can be thought of as giving a transformation of the plane onto itself. In particular, the net of conics through three points is a curvilinear coordinate system. The corresponding transformation of the plane onto itself is called a *cremona quadratic transformation*. In effect, such a transformation is given by $(\mathscr{X}' : \mathscr{Y}' : \mathscr{Z}') = (f_2(\mathscr{X}, \mathscr{Y}, \mathscr{Z}) : g_2(\mathscr{X}, \mathscr{Y}, \mathscr{Z}) : h_2(\mathscr{X}, \mathscr{Y}, \mathscr{Z}))$ where f_2, g_2, h_2 are homogeneous polynomials of degree 2, and where colons indicate proportionality. The resulting correspondence is also written as $(\mathscr{X}' : \mathscr{Y}' : \mathscr{Z}') \leftrightarrow (\mathscr{X} : \mathscr{Y} : \mathscr{Z})$. Now we can make a homogeneous linear transformation to bring these three fixed points to be at the vertices of the fundamental triangle, i.e., we can arrange matters so that the three fixed points are $(1, 0, 0)$, $(0, 1, 0)$, and $(0, 0, 1)$. An example of a triple of "conics" passing through these three points is $\mathscr{X}\mathscr{Y}, \mathscr{X}\mathscr{Z}, \mathscr{Y}\mathscr{Z}$. Thus an example of a cremona quadratic transformation would be $(\mathscr{X}' : \mathscr{Y}' : \mathscr{Z}') = (\mathscr{Y}\mathscr{Z} : \mathscr{X}\mathscr{Z} : \mathscr{X}\mathscr{Y})$. Note that by putting $X = \mathscr{X}/\mathscr{Z}$, $Y = \mathscr{Y}/\mathscr{Z}$ and $X' = \mathscr{X}'/\mathscr{Z}'$, $Y' = \mathscr{Y}'/\mathscr{Z}'$, this transformation can be written as $X' = 1/X$, $Y' = 1/Y$. This is a cremona quadratic transformations of the *first kind*. We get a cremona quadratic transformations of the *second kind* when two of the three fixed points are coincidental. Two points becoming coincidental determines a direction. In this case we would be considering conics passing through the two fixed points and having a fixed tangent direction at one of the points. Finally, we get a cremona quadratic transformations of the *third kind* when all the three fixed points are coincidental. In this case, we would be considering conics passing through that point, having a fixed tangent direction at that point, and having a fixed tangent direction at the point in the first neighborhood of the given point determined by the tangent direction we first spoke of. For a discussion of these three kinds of cremona quadratic transformations, see my cubic surface paper [**A06**].

Earlier we quoted a theorem for $n = 2$ proved by Zariski [**Za5**] and Abhyankar [**A01**]. This result in a sense goes back to the Noether's Factorization Theorem (G_2). The theorem of Zariski-Abhyankar deals however with polyrational transformations. The above automorphism theorem (A_2') is also closely related to (G_2). But the exact algebraic equivalent of (G_2) is the following

THEOREM (A_2). *The fractional linear automorphisms*

$$X \to \frac{aX + bY + c}{\alpha X + \beta Y + \gamma} \quad and \quad Y \to \frac{a'X + b'Y + c'}{\alpha X + \beta Y + \gamma} \quad with \ \det \begin{pmatrix} a & b & c \\ a' & b' & c' \\ \alpha & \beta & \gamma \end{pmatrix} \neq 0,$$

together with the automorphism

$$X \to X \quad and \quad Y \to \frac{1}{Y}$$

generate all the k-automorphisms of $k(X, Y)$.

For a proof of Noether's factorization theorem or Theorem (A_2), see page 150 of Hudson's book, *Cremona Transformations* [**Hud**], or page 50 of volume 1 of Godeaux's book, *Geométrie Algebraique* [**God**], or Nagata's Paper [**Na1**]. As a preliminary precursor to Noether's Factorization Theorem we have the

FUNDAMENTAL THEOREM FOR PROJECTIVE PLANE. *Any four points in a projective plane can be mapped to any other four points by a unique projective transformation provided no three points among each of the two sets of four points are collinear.*

In general, for a projective space of dimension n, we have that any $n + 2$ points can be mapped onto any other $n + 2$ points in a unique manner provided that any $n + 1$ of these, former as well as latter, points are "independent". No three of them lie on a line, no four of them lie on a plane, and so on.

For a nice geometric discussion of n-dimensional projective space, see *Ruled surfaces* by Edge, published by the Cambridge University Press in 1931 [**Edg**]. Two other companion books were published around the same time and by the same publisher, *Determinantal varieties* by Room [**Roo**] and *Cremona transformations* by Hudson [**Hud**]. All these three authors were students of H. F. Baker who wrote a series of books on geometry during 1922–33 entitled *Principles of geometry*, vols. 1 to 6 [**Ba1**]. Baker also wrote another book, *Abel's Theorem* [**Ba2**]. This theorem of Abel's may be called the second (significant) theorem of algebraic geometry. Bezout's Theorem is the first, and Noether's Factorization Theorem is the third. This book of Baker on Abel's Theorem is about 700 pages. Therein he assumes the contents of another long book, Forsyth's *Theory of functions of a complex variable*, Cambridge (1893) [**For**]. Another book written by students of Baker which appeared a little later than the books of Edge, Room, and Hudson is Semple and Roth's *Introduction to algebraic geometry*, Oxford, (1949) [**SRo**], which we have already referred to several times.

The one-dimensional version of the above fundamental theorem for projective plane is the following

FUNDAMENTAL THEOREM OF PROJECTIVE GEOMETRY (G_1). *Any three points in a projective line can be mapped to any other three points by a unique projective transformation.*

Projective plane. In order to describe what a projective transformation is, let us note that there are two ways to define a projective plane, the synthetic approach, and the analytic approach. In the synthetic approach, a projective plane is defined as a collection of "lines" and "points" satisfying certain axioms, whereas in the analytic or algebraic definition, the projective plane $\mathbb{P}^2_\kappa$ over a skew field κ is defined to be the set of all equivalence classes of nonzero triples of elements of κ where two triples are regarded equivalent if they are proportional. Here a *skew field* is a field that is not necessarily commutative. It is a *division ring*.

In the synthetic approach, if we assume the so called Desargue's Axiom (which is a theorem, called Desargue's Theorem, in the analytic approach), then we can introduce a coordinate system so that the synthetically defined projective plane can be thought of as $\mathbb{P}^2_\kappa$ for some skew field κ. Moreover, it can be proved that this skew field κ is commutative if and only if the fundamental theorem holds for any line in the projective plane. For example, see Hodge-Pedoe [**HPe**]. Projective transformations of a line in a projective plane can be defined as follows:

Perspectivity. Let ℓ be a line and let P be any point which is not on this line. Now we can project from the point P giving a map of the line ℓ onto another line ℓ'. Such a map is called a *perspectivity*. Now we can take a point which is neither on ℓ nor on ℓ', and project from this point to obtain another line ℓ''. We can repeat this procedure transforming from one line onto another. See Figure 27.1. In this way, we map ℓ onto a line ℓ^*. If ℓ^* equals ℓ, i.e., if we map eventually onto the line we started with, then we get a one-to-one transformation of the line ℓ onto itself. Such transformations of a line onto itself are called *projective transformations*.

Now the algebraic equivalent of the fundamental theorem of projective geometry is the following

THEOREM (A_1). *The only k-automorphisms of the field $k(X)$ are the fractional linear transformations, or equivalently*

$$k(X) = k(Y) \Leftrightarrow Y = \frac{\alpha X + \beta}{\gamma X + \delta} \quad \text{with } \det\begin{pmatrix} \alpha & \beta \\ \gamma & \delta \end{pmatrix} \neq 0.$$

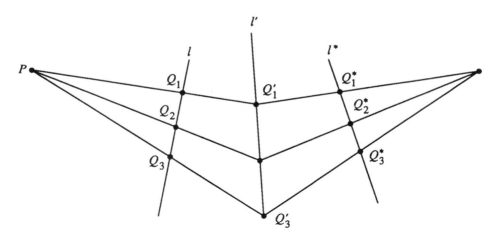

FIGURE 27.1

Note that a fractional linear transformation is determined by fixing three values of X and the corresponding values of Y. Also note that (i) a fractional linear transformation of $k(X)$, (ii) a birational transformation of the projective line $\mathbb{P}^1_k$ onto itself, (iii) a projective transformation of the projective line $\mathbb{P}^1_k$, and (iv) a homogeneous linear transformation of the underlying 2-dimensional vector space over k, are equivalent objects.

REMARK. Notice that our discussion of the factorization problem in the case of bipoly or birational transformations was more precise, than in the case of polyrational transformations because these really involve projective considerations. One way of making all this precise is by using the language of models. In that language a polyrational transformation $(X_1, \ldots, X_n) \to (X'_1, \ldots, X'_n)$ corresponds to a domination map $\widehat{W} \leftarrow \widehat{W}^*$ between two nonsingular projective models of an algebraic function field K/k. An MDT corresponds to a "simple" domination map. A version of the factorization problem corresponds to asking if a class of domination maps $\widehat{W} \leftarrow \widehat{W}^*$ are compositions $\widehat{W} = \widehat{W}_0 \leftarrow \widehat{W}_1 \leftarrow \cdots \leftarrow \widehat{W}_d = \widehat{W}^*$ of "special" domination maps $\widehat{W}_{i-1} \leftarrow \widehat{W}_i$, which may serve as "building blocks" or "bricks." We shall discuss these matters in the next lecture.

NOTE. This whole lecture was like a note, so it is not necessary to have another note!

Valuations and Birational Correspondence

Consider the parametrization of the circle $X^2 + Y^2 - 1 = 0$ given by

$$\begin{cases} X = \dfrac{1 - t^2}{1 + t^2} \\ Y = \dfrac{2t}{1 + t^2}. \end{cases}$$

Here is a discrepancy. The point $(-1, 0)$ is on the circle, but we don't obtain it by giving any value to the parameter t. We can explain this discrepancy by noting that we get the point $(-1, 0)$ as $t \to \infty$. Likewise for parametrizations of other curves.

For surfaces, more serious discrepancies arise which cannot be explained so easily. In other words, for a surface $f(X, Y, Z) = 0$ with a rational parametrization

$$\begin{cases} X = \dfrac{p(t, \tau)}{s(t, \tau)} \\ Y = \dfrac{q(t, \tau)}{s(t, \tau)} \ , \\ Z = \dfrac{r(t, \tau)}{s(t, \tau)} \end{cases}$$

it may happen that certain points of the surface are not obtainable by giving any values to the parameters t and τ. For certain values of t and τ the parametric equations may be reduced to the indeterminate form

$$\begin{cases} X = 0/0 \\ Y = 0/0 \\ Z = 0/0. \end{cases}$$

For example, the hyperboloid $X^2 + YZ = 2X$ has the rational parametrization

$$\begin{cases} X = \dfrac{2t^2}{t^2 + \tau^2 - 1} \\ Y = \dfrac{2t(\tau - 1)}{t^2 + \tau^2 - 1} \\ Z = \dfrac{2t(\tau + 1)}{t^2 + \tau^2 - 1}. \end{cases}$$

At the two points $(0, 1)$ and $(0, -1)$ in the (t, τ)-plane the above parametric equations reduce to the indeterminate form

$$\begin{cases} X = 0/0 \\ Y = 0/0 \\ Z = 0/0. \end{cases}$$

If x, y, z are the functions on the hyperboloid "induced" by X, Y, Z, then the parameters t and τ can be expressed by the equations

$$\begin{cases} t = \dfrac{2x}{z - y} \\ \tau = \dfrac{z + y}{z - y}. \end{cases}$$

Using these equations we can see that although the two lines $X = Y = 0$ and $X = Z = 0$ lie on the hyperboloid, most points on them are not obtainable from any good values of t and τ other than the values $(0, 1)$ and $(0, -1)$. By most points on the lines $X = Y = 0$ and $X = Z = 0$ we mean all except the origin $(0, 0, 0)$ in the (X, Y, Z)-space, which is their point of intersection. The origin $(0, 0, 0)$ in the (X, Y, Z)-space is actually obtained from every good point of the line $t = 0$ in the (t, τ)-plane other than the points $(0, 1)$ and $(0, -1)$. Equivalently, the entire line $t = 0$ in the (t, τ)-plane shrinks to the origin $(0, 0, 0)$ in the (X, Y, Z)-space. Thus we suspect that most points on the lines $X = Y = 0$ and $X = Z = 0$ may in some mysterious sense be obtainable from the points $(0, 1)$ and $(0, -1)$ in the (t, τ)-plane. This mysterious sense can be made precise either by using ideas of limits from analysis or by using valuations that are the algebraic counterpart of limits. Although for curves it suffices to consider discrete valuations, for surfaces and higher dimensional "algebraic varieties" we need more general valuations which can be obtained by suitably modifying the definition of discrete valuations by replacing $\mathbb{Z}$ by any ordered abelian group.

In greater detail, a *valuation* of a field K is a map $V : K \to \Gamma \cup \{\infty\}$, where Γ is an ordered abelian group, such that for all u and u' in K we have

$$V(uu') = V(u) + V(u') \quad \text{and} \quad V(u + u') \geq \min\{V(u), V(u')\},$$

and for any u in K we have

$$V(u) = \infty \Leftrightarrow u = 0.$$

(By convention, for any $a \in \Gamma$ we have $\infty + a = \infty + \infty = \infty > a$.) By the *value group* of V we mean the subgroup $V(K \setminus \{0\})$ of Γ. Given a subfield k of K, we say that V is *trivial* on k, or that V is a valuation of K/k, to mean that

$$V(k) = \{0, \infty\}.$$

We say that V is *residually rational* over k to mean that

$$u \in K \quad \text{with } V(u) \geq 0 \Rightarrow V(u - \gamma) > 0 \quad \text{for some } \gamma \in k.$$

Note that an *ordered abelian group* Γ is an additive abelian group that is also a totally ordered set such that for all a, b, c in Γ we have $a \leq b \Rightarrow a + c \leq b + c$. Recall that a *total order* on a set Ω is a binary relation $a \leq b$ which holds between some elements a, b of Ω such that: (i) (dichotomy) for all a, b in Ω we have either $a \leq b$ or $b \leq a$; (ii) (antisymmetry) for all a, b in Ω we have $a \leq b$ and $b \leq a \Leftrightarrow a = b$; and (iii) (transitivity) for all a, b, c in Ω we have $a \leq b$ and $b \leq c \Rightarrow a \leq c$.

Some examples of ordered abelian groups are $\mathbb{Z}$, $\mathbb{Q}$, $\mathbb{R}$. For any ordered abelian group Γ and any positive integer m we get the ordered abelian group Γ^m consisting of all *lexicographically ordered* m-tuples $(a_1, \ldots, a_m)$ of elements in Γ where $(a_1, \ldots, a_m) + (b_1, \ldots, b_m) = (a_1 + b_1, \ldots, a_m + b_m)$ and where $(a_1, \ldots, a_m) < (b_1, \ldots, b_m) \Leftrightarrow$ for some integer i with $1 \leq i \leq m$ we have $a_1 = b_1, \ldots, a_{i-1} = b_{i-1}$ and $a_i < b_i$.

By the *real rank* of V we mean the smallest non-negative integer q, if it exists, such that the value group of V is *order isomorphic* (i.e., isomorphic as an ordered abelian group) to a subgroup of $\mathbb{R}^q$. If no such q exists then we say that the real rank of V is ∞. By the *rational rank* of V we mean the smallest non-negative integer q, if it exists, such that the value group of V is isomorphic (only as a group, disregarding order) to a subgroup of $\mathbb{Q}^q$. If no such q exists then we say that the rational rank of V is ∞. Observe that clearly

$$\text{real rank } V \leq \text{ rational rank } V.$$

We call V *trivial* if it is trivial on K, i.e., if its value group is $\{0\}$. Obviously, this is so $\Leftrightarrow$ the real rank V is zero $\Leftrightarrow$ the rational rank of V is zero. Note that two similar algebraic structures, such as groups, rings etc., are *isomorphic*, meaning there exists a bijective correspondence between them that preserves all the relevant operations and relations. For example, two additive abelian groups Γ_0 and Γ_0' are isomorphic if there exists a bijective map $\Phi : \Gamma_0 \to \Gamma_0'$ such that $\Phi(a + b) = \Phi(a) + \Phi(b)$ for all a and b in Γ_0. Similarly, two ordered abelian groups Γ and Γ' are order isomorphic if there exists a (group) isomorphism $\Phi : \Gamma \to \Gamma'$, such that for all $a \leq b$ in Γ we have $\Phi(a) \leq \Phi(b)$. As an example of an ordered abelian group whose real rank is 1 and rational rank is 2, we have the subgroup $\{i + j\pi : i, j \in \mathbb{Z}\}$ of $\mathbb{R}$. Note that this is (group) isomorphic to $\mathbb{Z}^2$, but not order isomorphic to $\mathbb{Z}^2$.

Given any other valuation V' of K, we say that V and V' are *equivalent* if there exists an order isomorphism Φ of the value group of V onto the value group of V' such that for all $0 \neq u \in K$ we have $V'(u) = \Phi(V(u))$. It can easily be seen that V and V' are equivalent iff $R_V = R_{V'}$, where the *valuation ring* R_V of V is defined to be the subring of K given by

$$R_V = \{u \in K : V(u) \geq 0\}.$$

Now R_V is a quasilocal ring with quotient field K, and the unique maximal ideal $M(R_V)$ in R_V is given by

$$M(R_V) = \{u \in K : V(u) > 0\}.$$

Observe that V is trivial on a subfield k of K iff $k \subset R_V$. In that case we may identify k with a subfield of the *residue field* $R_V/M(R_V)$ of V. We define

$$\text{restrdeg}_k V = \textit{residual transcendence degree of } V \text{ over } k$$
$$= \text{trdeg}_k R_V/M(R_V).$$

We say that V is *residually algebraic* over k or *residually transcendental* over k according as $\text{restrdeg}_k K = 0$ or $\text{restrdeg}_k K > 0$. In the present set up we clearly have $k = R_V/M(R_V) \Leftrightarrow V$ is residually rational over k.

Recall that an m-variable (or m-dimensional) algebraic function field is a finitely generated field extension K'/k' with $\text{trdeg}_{k'} K' = m$. Moreover, an algebraic function field is simply a finitely generated field extension.

Exercise. If a field K is an algebraic function field over two different algebraically closed subfields k and k', show that $k = k'$. (Hint: Use valuations.)

REMARK. For studying m-dimensional algebraic varieties, the only valuations arising are those V for which

$$\text{real rank } V \leq \text{rational rank } V \leq m.$$

More precisely, if K is an m-variable algebraic function field over a subfield k, then for any valuation V of K/k we have

$$\text{restrdeg}_k V + \text{real rank } V \leq \text{restrdeg}_k V + \text{rational rank } V \leq m.$$

Moreover, in this case it can also be shown that

$$\text{restrdeg}_k V + \text{real rank } V = m$$
$$\Rightarrow \text{value group of } V \text{ is order isomorphic to } \mathbb{Z}^m$$

and

$$\text{restrdeg}_k V + \text{rational rank } V = m$$
$$\Rightarrow \text{value group of } V \text{ is (group) isomorphic to } \mathbb{Z}^m$$
$$\text{(but not necessarily order isomorphic to } \mathbb{Z}^m\text{)}.$$

For this and other facts about valuations of algebraic function fields see [A01] and [A04].

Example. To construct an example of a valuation of real rank 2, for any $0 \neq U(X, Y) = \frac{G(X, Y)}{H(X, Y)}$ where $G(X, Y)$ and $H(X, Y)$ are nonzero polynomials in X and Y with coefficients in a field k, we can write

$$U(X, Y) = \frac{X^a G'(X, Y)}{X^b H'(X, Y)} \quad \text{where} \quad \begin{cases} G'(X, Y) \text{ and } H'(X, Y) \text{ in } k[X, Y] \\ \text{with } G'(0, Y) \neq 0 \neq H'(0, Y). \end{cases}$$

Now we define

$$V(U(X, Y)) = (a - b, \ \operatorname{ord} G'(0, Y) - \operatorname{ord} H'(0, Y)) \in \mathbb{Z}^2.$$

For example,

$$V\left(X^2 \left[Y^3 + Y^4 + XY\right]\right) = (2, 3).$$

Thus we have a valuation V of $k(X, Y)/k$ whose value group is order isomorphic to $\mathbb{Z}^2$.

Arcs. Given a field k and a valuation V of $k(X, Y)/k$ which is residually rational over k, in case $V(X) \geq 0$ and $V(Y) \geq 0$, by the center of V on the (X, Y)-plane we mean the point (α, β) with α and β in k such that $V(X - \alpha) > 0$ and $V(Y - \beta) > 0$. If either $V(X) < 0$ or $V(Y) < 0$ then we define the center to be the appropriate point at ∞. Intuitively, a valuation V of $k(X, Y)/k$ centered at a point (say, the origin) is like an arc emanating from that point. For example,

$$\begin{cases} X = \lambda(T) = T^2 + T^3 + T^4 + \cdots \\ Y = \mu(T) = \dfrac{T^3}{3!} + \dfrac{T^4}{4!} + \dfrac{T^5}{5!} + \cdots \end{cases}$$

is a "transcendental" arc emanating from the origin in the (X, Y)-plane. By putting $V(U(X, Y)) = \operatorname{ord} U(\lambda(T), \mu(T))$ for all $U(X, Y) \in k(X, Y)$ we get a valuation V of $k(X, Y)/k$ centered at the origin. Moreover, there could be valuations of $k(X, Y)/k$ that arise from more general arcs. For example,

$$\begin{cases} X = T \\ Y = T^\pi, \end{cases}$$

$$\begin{cases} X = T \\ Y = T^{\sqrt{2}} + T^\pi + T^6 + \cdots, \end{cases}$$

$$\begin{cases} X = T \\ Y = T^{1.1} + T^{1.11} + T^{1.111} + \cdots + T^2 + T^{2.1} + T^{2.2} + \cdots + \cdots, \end{cases}$$

or

$$\begin{cases} X = T \\ Y = T^{(1, 1)} + T^{(1, 2)} + \cdots + T^{(2, 1)} + T^{(2, 2)} + \cdots \\ \quad (\text{where } (1, 1) < (1, 2) < \cdots < (2, 1) < (2, 2) < \cdots). \end{cases}$$

Note that in the last two examples, the series for Y is not of the usual type. Nevertheless, it is "well ordered."

Indeterminate forms. We can similarly define the center of a valuation on a surface $f(X, Y, Z) = 0$. Suppose we have a rational parametrization of this surface, given by

$$
\begin{cases}
X = \dfrac{p(t, \tau)}{s(t, \tau)} \\[2mm]
Y = \dfrac{q(t, \tau)}{s(t, \tau)} \\[2mm]
Z = \dfrac{r(t, \tau)}{s(t, \tau)}.
\end{cases}
$$

Now to obtain all the points on the surface $f(X, Y, Z) = 0$, instead of giving values to (t, τ), we consider (residually rational) valuations of $k(t, \tau)/k$ centered at any point in the (t, τ)-plane and then take their centers on the surface to obtain the corresponding points of the surface. Thus for example, if V is a valuation on $k(t, \tau)$ such that $V(p) \geq V(s)$, $V(q) \geq V(s)$, $V(r) \geq V(s)$, then $V(p/s) \geq 0$, $V(q/s) \geq 0$, $V(r/s) \geq 0$. Hence, we can find unique elements α, β, γ in k such that $V\left(\frac{p}{s} - \alpha\right) > 0$, $V\left(\frac{q}{s} - \beta\right) > 0$, $V\left(\frac{r}{s} - \gamma\right) > 0$. Now (α, β, γ) is the corresponding point on the surface. In this way, we can obtain all the points on the surface. Then there are no discrepancies. This is in fact a special case of a more general phenomenon, which we shall discuss in a moment. Note that rationally parametrizing the surface $f(X, Y, Z) = 0$ simply means that we have $k(x, y, z) = k(t, \tau)$, where $k(x, y, z)$ is the field of rational functions on the surface.

L'Hospital's rule. Now L'Hospital's rule says that there are really no 1–variable indeterminate forms. Moreover, valuations are essentially like arcs, that is, they are 1-variable animals. This is the basic reason why valuations enable us to evaluate indeterminate forms in several variables. Ring theoretically, the valuation ring R_V of a valuation V of field K has the characteristic property $z \in K \setminus R_V \Rightarrow 1/z \in R_V$. Conversely, it can be shown that if R is any subring of K such that for every $z \in K \setminus R$ we have $1/z \in R$, then $R = R_V$ for some valuation V of K. So by a *valuation ring* of a field K we mean a subring R of K having the property $0 \neq z \in K \Rightarrow$ either $z \in R$ or $1/z \in R$. Then valuation rings of a field K are in a one-to-one correspondence with equivalence classes of valuations of K. Given a subfield k of a field K, by a *valuation ring* of K/k we mean a valuation ring of K which contains k. We note that if k is algebraically closed and $\operatorname{trdeg}_k K = 1$, then $V \mapsto R_V$ gives a bijective map of $\mathscr{R}(K)$ onto the set of all valuation rings of K/k other than the "trivial valuation ring" K itself.

Exercise. Show that a valuation ring is a normal domain. Show that the valuation ring of a valuation is Noetherian iff its value group is isomorphic to $\mathbb{Z}$. Show that a one-dimensional local domain is normal iff it is regular iff it is a valuation ring.

Birational correspondence. To explain how valuations can be used to remove discrepancies in a more general set up, let us discuss birational correspondences between general algebraic varieties. So let $W : f_1(X_1, \ldots, X_n) = f_2(X_1, \ldots, X_n) = \cdots = 0$ be any irreducible algebraic variety in n-space with

$f_i(X_1, \ldots, X_n) \in R_n = k[X_1, \ldots, X_n]$, and let $W' : f_1'(X_1, \ldots, X_{n'}) = f_2'(X_1, \ldots, X_{n'}) = \cdots = 0$ be any other irreducible algebraic variety in n'-space with $f_j'(X_1, \ldots, X_{n'}) \in R_{n'} = k[X_1, \ldots, X_{n'}]$. Also, let $k[W] = k[x_1, \ldots, x_n]$ be the affine coordinate ring of W where x_i is the function on W induced by X_i, and let $k(W) = k(x_1, \ldots, x_n)$ be the function field of W. Likewise, let $k[W'] = k[x_1', \ldots, x_{n'}']$ be the affine coordinate ring of W' where x_j' is the function on W' induced by X_j, and let $k(W') = k(x_1', \ldots, x_{n'}')$ be the function field of W'. The (algebraic) varieties W and W' are said to be *birationally equivalent* if their function fields $k(W)$ and $k(W')$ are k-isomorphic, or equivalently if, after "identifying" $k(W)$ with $k(W')$, we can express $x_1, \ldots, x_n$ as rational function in $x_1', \ldots, x_{n'}'$ and vice versa. As exemplified by the rational parametrization of a hyperboloid, the correspondence between the points of birationally equivalent varieties W and W' may not be one-to-one. In fact it need not even be a map in the modern sense of the word. (That is what makes algebraic geometry so interesting!) However, we can always find proper subvarieties Δ and Δ' of W and W' respectively such that the resulting correspondence $W \setminus \Delta \leftrightarrow W' \setminus \Delta'$ is bijective.

The correspondence between the points of W and W' is obtained as follows. Given a point P on W, it may correspond to several points of W', which are found thus.

Consider valuations V of $k(W)/k$ whose center on W is at P. (We shall define the notion of center in a moment. It will turn out that every point of W is the center of at least one valuation of $k(W)/k$.) Then consider the center P' of V on W'. Thus we get points P' on W' corresponding to the point P of W. Moreover, these points P' form a subvariety of W'.

Valuations give a clean algebraic way to deal with the problem of making this correspondence precise. There is also the old

Analysis way. Given a point P on W, if its image P' cannot be uniquely determined, then consider sequences of points converging to P such that their images are well defined and they converge to some limit. See Figure 28.1. In this way we may obtain several limits—but that's okay. These are all the points P' on W' corresponding to P.

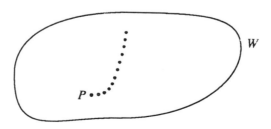

FIGURE 28.1

Now let us turn to the

Valuation way. First we define a correspondence

variety $W \leftrightarrow$ model (collection of local rings).

To every point $P = (\alpha_1, \ldots, \alpha_n)$ of W we associate the local ring $\mathscr{L}(P, W)$, which is called the local ring of P on W. It is defined by putting

$$\mathscr{L}(P, W) = k[W]_{\mathscr{I}(P, W)},$$

where $\mathscr{I}(P, W)$ is a maximal ideal in $k[W]$. This is called the prime ideal of P on W and is defined by putting

$$\mathscr{I}(P, W) = \{g(x_1, \ldots, x_n) : g(X_1, \ldots, X_n) \in R_n$$
$$\text{such that } g(\alpha_1, \ldots, \alpha_n) = 0\}.$$

Likewise, to an irreducible subvariety E of W we associate the local ring $\mathscr{L}(E, W)$, which is called the local ring of E on W and is defined by putting

$$\mathscr{L}(E, W) = k[W]_{\mathscr{I}(E, W)},$$

where $\mathscr{I}(E, W)$ is a prime ideal in $k[W]$, which is called the prime ideal of E on W and is defined by putting

$$\mathscr{I}(E, W) = \{g(x_1, \ldots, x_n) : g(X_1, \ldots, X_n) \in R_n \text{ such that}$$
$$g(\alpha_1, \ldots, \alpha_n) = 0 \text{ for all } (\alpha_1, \ldots, \alpha_n) \in E\}.$$

Note that points of W are special cases of irreducible subvarieties of W. Thus $E \mapsto \mathscr{L}(E, W)$ gives a bijection of the set of all irreducible subvarieties of W onto $\mathfrak{V}(k[W])$, where for any domain A we put

$\mathfrak{V}(A) = $ the set of all quasilocal domains A_N as N varies over $\mathrm{Spec}(A)$.

We recall that $\mathrm{Spec}(A)$ is the set of all prime ideals in A, and we remark that $\mathfrak{V}$ (German V) is meant to remind us of a "variety." Strictly speaking, $E \mapsto \mathscr{L}(E, W)$ gives the said bijection only when k is algebraically closed. However, even if k is not algebraically closed, we take $\mathfrak{V}(k[W])$ anyway! To take care of points and irreducible subvarieties of W at ∞, we enlarge $\mathfrak{V}(k[W])$ to the set of local rings $\mathfrak{W}(k; x_0, x_1, \ldots, x_n)$ given by

$$\mathfrak{W}(k; x_0, x_1, \ldots, x_n) = \bigcup_{\substack{i=0 \\ x_i \neq 0}}^{n} \mathfrak{V}\left(k\left[\frac{x_0}{x_i}, \frac{x_1}{x_i}, \ldots, \frac{x_n}{x_i}\right]\right),$$

where $x_0 = 1$. We remark that $\mathfrak{W}$ (German W) is meant to suggest a "projective model." Now the set of local rings $\mathfrak{W}(k; x_0, x_1, \ldots, x_n)$ with quotient field $k(W)$ may be called the *projective model* of $k(W)/k$ associated with the variety W. It is a complete model of $k(W)/k$ in the sense we shall make precise in a moment. As a preamble to the definition of models, let us first talk about

Domination. Given quasilocal rings R and R^*, we say that R^* *dominates* R, or R is *dominated by* R^*, if R is a subring of R^* and $M(R) \subset M(R^*)$. Given a quasilocal ring R^* and a set Λ of quasilocal rings, we say that R^* *dominates* Λ, or Λ is *dominated by* R^*, if R^* dominates some member of Λ. Given sets Λ and Λ^* of quasilocal rings, we say that Λ^* *dominates* Λ, or Λ is *dominated by* Λ^*, if every member of Λ^* dominates Λ. By a *premodel* of a field K we mean a nonempty set of quasilocal domains with quotient field K. Given a valuation V of K and a quasilocal domain R, we say that V *dominates* R, or R is *dominated by* V, if R_V dominates R. Given a valuation V of K and a set Λ of quasilocal rings, we say that V *dominates* Λ, or Λ is *dominated by* V, if R_V dominates Λ. Given a set Λ^* of valuations of K and a set Λ of quasilocal rings, we say that Λ^* *dominates* Λ, or Λ is *dominated by* Λ^*, if every member of Λ^* dominates Λ. A premodel $\widehat{W}$ of K is said to be *irredundant* if every valuation V of K dominates at most one member of $\widehat{W}$, which we then call the *center* of V on $\widehat{W}$. Given any irredundant premodel $\widehat{W}$ of K and given any quasilocal domain R^*, it is easily seen that R^* dominates at most one member of $\widehat{W}$, which is then called the *center* of R^* on $\widehat{W}$. Now we can come to

Models. Given any algebraic function field K/k' and any Noetherian domain A with quotient field k', by a *model* of K/A (read: model of K over A) we mean an irredundant premodel $\widehat{W}$ of K which can be expressed as a finite union $\widehat{W} = \mathfrak{V}(A_1) \cup \cdots \cup \mathfrak{V}(A_l)$ with $A_i = A[z_{i1}, \ldots, z_{in_i}]$ where $z_{i1}, \ldots, z_{in_i}$ are a finite number of elements in K, for $1 \leq i \leq l$. Since A is Noetherian, a model $\widehat{W}$ of K/A is a collection of local rings, and we put

$$\dim \widehat{W} = \max_{R \in \widehat{W}} \dim R$$

and

$$\mathfrak{S}(\widehat{W}) = \text{ the singular locus of } \widehat{W}$$
$$= \{R \in \widehat{W} : R \text{ is not regular}\}.$$

We say that $\widehat{W}$ is *nonsingular* if $\mathfrak{S}(\widehat{W}) = \varnothing$. A model $\widehat{W}$ of K/A is said to be *complete* if every valuation V of K, with $A \subset R_V$, dominates $\widehat{W}$. Note that a model $\widehat{W}$ of K/A is complete iff $\widehat{W}$ is dominated by the set of all valuation rings R^* of K with $A \subset R^*$. A premodel $\widehat{W}$ of K/A is said to be a *projective model* of K/A if there exists a finite number of elements $y_0, y_1, \ldots, y_{n^*}$ in an overfield of K such that $\widehat{W} = \mathfrak{W}(A; y_0, y_1, \ldots, y_{n^*})$ where,

$$\mathfrak{W}(A; y_0, y_1, \ldots, y_{n^*}) = \bigcup_{\substack{i=0 \\ y_i \neq 0}}^{n^*} \mathfrak{V}\left(A\left[\frac{y_0}{y_i}, \frac{y_1}{y_i}, \ldots, \frac{y_{n^*}}{y_i}\right]\right).$$

It can easily be seen that a projective model of K/A is a complete model of K/A. Analogous to the definition of a projective model, for any finitely generated nonzero A-submodule J of K we define

$$\mathfrak{W}(A, J) = \bigcup_{0 \neq x \in J} \mathfrak{V}(A[Jx^{-1}]),$$

where $A[Jx^{-1}]$ denotes the smallest subring of K that contains A and y/x for all $y \in J$. We note that if $z_0, z_1, \ldots, z_\nu$ is any finite set of A-generators of J then

$$\mathfrak{W}(A, J) = \mathfrak{W}(A; z_0, z_1, \ldots, z_\nu).$$

Hence $\mathfrak{W}(A, J)$ is a projective model of K'/A where K' is the (common) quotient field of $A[z_0/z_i, z_1/z_i, \ldots, z_\nu/z_i]$ for all i with $z_i \neq 0$.

Models and varieties. Going back to the consideration of an irreducible algebraic variety W over the field k, it now follows that every valuation V of $k(W)/k$ has a unique center on $\mathfrak{W}(k; x_0, x_1, \ldots, x_n)$. By the *center* of V on W we mean the corresponding point or irreducible subvariety of W, at finite distance or at ∞. Thus the study of birational correspondence between W and W' is reduced to the study of the projective models $\mathfrak{W}(k; x_0, x_1, \ldots, x_n)$ and $\mathfrak{W}(k; x_0', x_1', \ldots, x_{n'}')$ of $k(W) = k(W')$ over k. In particular, if the model $\mathfrak{W}(k; x_0, x_1, \ldots, x_n)$ dominates the model $\mathfrak{W}(k; x_0', x_1', \ldots, x_{n'}')$, then the said birational correspondence gives rise to a "regular map" (polyrational map!) from W to W'. At any rate, given any algebraic function field K/k' and any Noetherian domain A with quotient field k', we see that if $\widehat{W}$ and $\widehat{W}'$ are any two projective models of K/A such that $\widehat{W}'$ dominates $\widehat{W}$, then we get a surjective map from $\widehat{W}'$ to $\widehat{W}$ which sends every member of $\widehat{W}'$ to its center on $\widehat{W}$. We call this the *domination map* of $\widehat{W}'$ onto $\widehat{W}$. It can be shown that any such domination map can be obtained by "blowing up" an "ideal" I on $\widehat{W}$. We have $\widehat{W}' = \mathfrak{W}(\widehat{W}, I)$ where this $\mathfrak{W}$ is essentially the same as the $\mathfrak{W}$ used in defining a projective model. The concept of a projective model and the concept of blowing up, of which monoidal transformations and iterated monoidal transformations are special cases, can be brought under one umbrella. So let us formally introduce

Blowing-ups. Again consider a projective model $\widehat{W}$ of K/A where A is a Noetherian domain with quotient field k' and K is an algebraic function field over k'. By a *preideal* on $\widehat{W}$ we mean a function I which to every $R \in \widehat{W}$ associates an ideal $I(R)$ in R. A preideal I on $\widehat{W}$ is said to be an *ideal* on $\widehat{W}$ if for every subdomain A^* of K, such that A^* is an affine ring over A and $\mathfrak{V}(A^*) \in \widehat{W}$, we have

$$\left(\bigcap_{R \in \mathfrak{V}(A^*)} I(R) \right) R = I(R) \quad \text{for all } R \in \mathfrak{V}(A^*).$$

Given any ideal I on $\widehat{W}$, we define its *zeroset* $\mathfrak{z}(I)$ by putting

$$\mathfrak{z}(I) = \{R \in \widehat{W} : I(R) \neq R\}.$$

By a *subvariety* of $\widehat{W}$ we mean the zeroset of some ideal on $\widehat{W}$. A subvariety is said to be *irreducible* if it cannot be expressed as the union of two proper subvarieties. Given any nonzero ideal I on $\widehat{W}$, i.e., an ideal I on $\widehat{W}$ with $I(R) \neq \{0\}$ for all $R \subset \widehat{W}$, we put

$$\mathfrak{W}(\widehat{W}, I) = \text{ blow-up of } \widehat{W} \text{ with center } I$$
$$= \bigcup_{R \in \widehat{W}} \mathfrak{W}(R, I(R)).$$

It can be shown that then $\mathfrak{W}(\widehat{W}, I)$ is a projective model of K/A dominating $\widehat{W}$. Conversely, given any projective model $\widehat{W}'$ of K/A dominating $\widehat{W}$, there exists a nonzero ideal I on $\widehat{W}$ such that $\widehat{W}' = \mathfrak{W}(\widehat{W}, I)$. A projective model $\widehat{W}'$ of K/A is said to be a *monoidal transform* of $\widehat{W}$ if $\widehat{W}' = \mathfrak{W}(\widehat{W}, I)$ for some ideal I on $\widehat{W}$ such that the local rings R and $R/I(R)$ are regular for all $R \in \widehat{W}$ for which $I(R) \neq R$. By a *quadratic transform* of $\widehat{W}$ we mean a monoidal transform $\mathfrak{W}(\widehat{W}, I)$ of $\widehat{W}$ such that $3(I)$ is a finite set. A projective model $\widehat{W}^*$ of K/A is said to be an *iterated monoidal transform* of $\widehat{W}$ if there exists a finite sequence of projective models $\widehat{W}_0, \widehat{W}_1, \ldots, \widehat{W}_d$ such that $\widehat{W}_0 = \widehat{W}$, $\widehat{W}_i$ is a monoidal transform of $\widehat{W}_{i-1}$ for $1 \leq i \leq d$, and $\widehat{W}_d = \widehat{W}^*$. Note that a monoidal transform of a nonsingular projective model is again nonsingular, and therefore, so is an iterated monoidal transform.

Resolution and factorization. To elucidate the remark at the end of the lecture on birational and polyrational transformations and the note at the end of the lecture on hypersurfaces, let $A = k = $ a field or $A = \mathbb{Z} \subset \mathbb{Q} = k$. Also let K/k be an algebraic function field and let

$$m = \begin{cases} \operatorname{trdeg}_k K & \text{if } A = k = \text{a field} \\ 1 + \operatorname{trdeg}_k K & \text{if } A = \mathbb{Z} \subset \mathbb{Q} = k. \end{cases}$$

Note that for every projective model $\widehat{W}$ of K/A we have $\dim \widehat{W} = m$. Now here are the model-theoretic versions of the relavent problems together with their status.

RESOLUTION PROBLEM. Given a projective model $\widehat{W}$ of K/A, does there exist a nonsingular projective model $\widehat{W}^*$ of K/A such that $\widehat{W}^*$ dominates $\widehat{W}$? *Status:* Yes for (1.1) $m = 1$ (Riemann [**Rie**], Noether [**No1**], Dedekind [**DWe**]); (1.2) $m = 2$ and $A = \mathbb{C}$ (Walker [**Wal**]); (1.3) $m = 2$ and $A = $ a field of zero characteristic (Zariski [**Za4**]); (1.4) $m = 3$ and $A = $ a field of zero characteristic (Zariski [**Za5**]); (1.5) $m = 2$ and $A = $ a field of nonzero characteristic (Abhyankar [**A02**]); (1.6) $m = 2$ and $A = \mathbb{Z}$ (Abhyankar [**A10**]); (1.7) any m and $A = $ a field of zero characteristic (Hironaka [**Hir**]); (1.8) $m = 3$ and $A = $ a field of any characteristic (Abhyankar [**A12**] for characteristic $\neq 2, 3, 5$ and Abhyankar [**A18**] for characteristic $= 2, 3, 5$).

FACTORIZATION PROBLEM. Given any two nonsingular projective models $\widehat{W}$ and $\widehat{W}^*$ of K/A such that $\widehat{W}^*$ dominates $\widehat{W}$, is $\widehat{W}^*$ necessarily an iterated monoidal transform of $\widetilde{W}$? *Status:* (2.1) Yes for $m = 2$ (Zariski [**Za5**], Abhyankar [**A01**]). (2.2) No for $m = 3$ (Shannon [**Sha**]).

DOMINATION PROBLEM. Given any two projective models $\widehat{W}$ and $\widetilde{W}$ of K/A such that $\widehat{W}$ is nonsingular, does there exist an iterated monoidal transform $\widehat{W}^*$ of $\widehat{W}$ such that $\widehat{W}^*$ dominates $\widetilde{W}$? *Status:* Yes for (3.1) $m \leq 3$ and $A = $ a field of zero characteristic (Zariski [**Za5**]); (3.2) $m = 2$ and $A = \mathbb{Z}$ (Abhyankar [**A10**]); (3.3) $m \leq 3$ and $A = $ a field of any characteristic (Abhyankar [**A12**]).

SEMIFACTORIZATION PROBLEM. Given any two nonsingular projective models $\widehat{W}$ and $\widetilde{W}$ of K/A, does there exist an iterated monoidal transform $\widehat{W}^*$ of $\widehat{W}$ such that $\widehat{W}^*$ is also an iterated monoidal transform of $\widetilde{W}$? *Status:* A partial yes for $m = 3$ (Christensen [**Chr**]).

NOTE. You may like to browse in my resolution book [**A12**] where, exclusively, the language of models is used. As far as valuations are concerned, in 1991, the journal *Aequationes Mathematicae* is celebrating the one hundredth birthday of Ostrowski, the founder of valuations. See his fundamental papers [**Os1**] and [**Os2**].

Rational top Cylinders through a Variety

Today we shall discuss how to pass a rational surface through a curve in 3-space, and how to pass a rational hypercylinder through a curve in 4-space, and so on.

Recall that the circle $X^2 + Y^2 = 1$ can be rationally parametrized by taking $X = \frac{1-t^2}{1+t^2}$ and $Y = \frac{2t}{1+t^2}$. Likewise, every conic in the plane as well as every quadric in 3-space, and more generally every hyperquadric in n-space, can be rationally parametrized. In other words, given any hyperquadric $Q : g(X_1, \ldots, X_n) = 0$, where $g(X_1, \ldots, X_n)$ is an irreducible polynomial of degree two with coefficients in a field k, we can find elements $p_1, \ldots, p_n$ in $k(T_1, \ldots, T_{n-1})$ such that $g(p_1, \ldots, p_n) = 0$. To ensure that this parametrization "covers most of" Q, we require that the transcendence degree of $k(p_1, \ldots, p_n)$ over k be $n - 1$. Now the kernel of the k-homomorphism $R_n = k[X_1, \ldots, X_n] \to k(T_1, \ldots, T_{n-1})$, which sends X_i to p_i for $1 \le i \le n$, is generated by $g(X_1, \ldots, X_n)$. We may then regard $p_1, \ldots, p_n$ as the functions on Q induced by $X_1, \ldots, X_n$ respectively. Thus giving a rational parametrization of Q amounts to "embedding" the function field $k(Q)$ of Q as a subfield of the function field $k(T_1, \ldots, T_{n-1})$ of the $(n - 1)$-space of the variables $T_1, \ldots, T_{n-1}$. To concretely find $p_1, \ldots, p_n$, we can take a point $P = (\alpha_1, \ldots, \alpha_n)$ on Q and then take lines

$$L_T : \begin{cases} X_1 = \alpha_1 + T_1 t \\ \vdots \\ X_{n-1} = \alpha_{n-1} + T_{n-1} t \\ X_n = \alpha_n + t \end{cases}$$

through P. Now L_T will meet Q in exactly one point $P_T = (p_1, \ldots, p_n)$ other than P. It follows that the coordinates $p_1, \ldots, p_n$ of P_T are rational functions of $T_1, \ldots, T_{n-1}$.

Quite generally, let $S : h(X_1, \ldots, X_n) = 0$ be a hypersurface in n-space where $h(X_1, \ldots, X_n)$ is an irreducible polynomial of degree e with coefficients in the field k. By a *rational parametrization* of S we mean an n-tuple $(q_1, \ldots, q_n)$ with $q_1, \ldots, q_n$ in $k(z_1, \ldots, z_{n-1})$ such that the kernel of the k-homomorphism $R_n \to k(z_1, \ldots, z_{n-1})$ that sends X_i to q_i for $1 \le i \le n$ is generated by $h(X_1, \ldots, X_n)$. We may regard q_i to be the function on S induced by X_i for $1 \leq i \leq n$. So again, giving such a rational parametrization of S amounts to "embedding" the function field $k(S)$ of S as a subfield of the function field $k(z_1, \ldots, z_{n-1})$ of the $(n-1)$-space of the variables $z_1, \ldots, z_{n-1}$. Such a rational parametrization need not be faithful, i.e., a point of S may correspond to several points of the $(z_1, \ldots, z_{n-1})$-space. Indeed, it can be shown that the vector space dimension $[k(z_1, \ldots, z_{n-1}): k(S)]$ is a positive integer ϵ, and "most" points of S correspond to ϵ points of the $(z_1, \ldots, z_{n-1})$-space. We call this a *faithful rational parametrization* if $\epsilon = 1$, i.e., equivalently if $k(S) = k(z_1, \ldots, z_{n-1})$. Let us say that S is *unirational* if it has a rational parametrization. Let us say that S is *rational* if it has a faithful rational parametrization. As noted before, a classical Theorem of Lüroth says that for curves, that is, when $n = 2$, unirationality implies rationality. This was generalized by Castelnuovo (see [**Za3**]) for surfaces, i.e., when $n = 3$, under the assumption of k being an algebraically closed field of characteristic zero. Recently, Clemens and Griffiths [**CGr**] gave a counterexample for solids. When $n = 4$, the cubic hypersurface $X_1^3 + X_2^3 + X_3^3 + X_4^3 = 1$ in 4-space is unirational but not rational.

Monoids. The irreducible hypersurface S of degree e is said to be a *monoidal* hypersurface if it has an $(e-1)$-fold point, which is then called the *center* of the monoidal hypersurface. By a *hypermonoid* we mean a monoidal hypersurface. A hypermonoid in 3-space may be called a monoid. Now the above method of dealing with a hyperquadric can be generalized to rationally parametrize the irreducible hypersurface S of degree e assuming it to be a hypermonoid with center $P = (\alpha_1, \ldots, \alpha_n)$. The line

$$L_z: \begin{cases} X_1 = \alpha_1 + z_1 t \\ \vdots \\ X_{n-1} = \alpha_{n-1} + z_{n-1} t \\ X_n = \alpha_n + t \end{cases}$$

through P will meet S again in exactly one point P_z other than P. To find where L_z meets S, we consider

$$\phi(t) = h(\alpha_1 + z_1 t, \ldots, \alpha_{n-1} + z_{n-1} t, \alpha_n + t).$$

To "translate" the point P to the origin we put

$$h(\alpha_1 + X_1', \ldots, \alpha_n + X_n')$$
$$= h'(X_1', \ldots, X_n') = h'_{e-1}(X_1', \ldots, X_n') + h'_e(X_1', \ldots, X_n'),$$

where $h'_{e-1}(X_1', \ldots, X_n')$ and $h'_e(X_1', \ldots, X_n')$ are nonzero homogeneous polynomials of degree $e - 1$ and e respectively. Now

$$\phi(t) = h'(z_1 t, \ldots, z_{n-1} t, t)$$
$$= t^{e-1} h'_{e-1}(z_1, \ldots, z_{n-1}, 1) + t^e h'_e(z_1, \ldots, z_{n-1}, 1).$$

Hence, by setting $\phi(t) = 0$ and disregarding the root $t = 0$, which corresponds to P, we get

$$h'_{e-1}(z_1, \ldots, z_{n-1}, 1) + th'_e(z_1, \ldots, z_{n-1}, 1) = 0.$$

Hence,

$$t = -\frac{h'_{e-1}(z_1, \ldots, z_{n-1}, 1)}{h'_e(z_1, \ldots, z_{n-1}, 1)}.$$

Substituting this in the equations of L_z we conclude that $P_z = (q_1, \ldots, q_n)$ where

$$\begin{cases} q_1 = \alpha_1 - \dfrac{h'_{e-1}(z_1, \ldots, z_{n-1}, 1)z_1}{h'_e(z_1, \ldots, z_{n-1}, 1)} \\[1em] \vdots \\[1em] q_{n-1} = \alpha_{n-1} - \dfrac{h'_{e-1}(z_1, \ldots, z_{n-1}, 1)z_{n-1}}{h'_e(z_1, \ldots, z_{n-1}, 1)} \\[1em] q_n = \alpha_n - \dfrac{h'_{e-1}(z_1, \ldots, z_{n-1}, 1)}{h'_e(z_1, \ldots, z_{n-1}, 1)}. \end{cases}$$

Clearly $(q_1, \ldots, q_n)$ is a faithful rational parametrization of S.

Alternatively, by homogenizing and dehomogenizing we can send the point $P = (\alpha_1, \ldots, \alpha_n) = (1, \alpha_1, \ldots, \alpha_n)$ to the point $P_\infty = (0, \ldots, 0, 1)$, which is the point at ∞ on the X_n-axis. This can also be achieved by a fractional linear transformation. Let us say that S is a *special monoidal* hypersurface, or that S is a *special hypermonoid*, if S is a hypermonoid with center P_∞. Clearly, S is a special hypermonoid $\Leftrightarrow$

$$h(X_1, \ldots, X_n) = h_{e-1}(X_1, \ldots, X_{n-1})X_n + h_e(X_1, \ldots, X_{n-1}),$$

where h_{e-1} and h_e are polynomials of degree at most $e - 1$ and e respectively. Then the line

$$L'_z : X_1 = z_1, \ldots, X_{n-1} = z_{n-1}, X_n = t$$

through P_∞ meets S, outside P_∞, in the point $P'_z = (q'_1, \ldots, q'_n)$ where

$$q'_1 = z_1, \ldots, q'_{n-1} = z_{n-1}, q'_n = -\frac{h_{e-1}(z_1, \ldots, z_{n-1})}{h_e(z_1, \ldots, z_{n-1})}.$$

This gives the faithful rational parametrization $(q'_1, \ldots, q'_n)$ of S.

Cylinders. In the (X, Y, Z)-space, the familiar circular cylinder of radius 1 and with the Z-axis as its axis is given by $X^2 + Y^2 = 1$. See Figure 29.1 on page 258. By analogy, in n-space, the hypersurface S (without assuming it to be rational or monoidal etc.) may be called a *hypercylinder* on an $(n^* - 1)$-dimensional base if there exists a hypersurface $S^* : h^*(X_1^*, \ldots, X_{n^*}^*) = 0$ in n^*-dimensional space together with a "nonsingular" linear transformation

$$X_i^* = \sum_{j=1}^{n} a_{ij} X_j$$

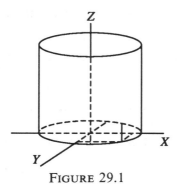

FIGURE 29.1

(where for the n by n "matrix" (a_{ij}) with entries in k we have $\det(a_{ij}) \neq 0$) such that

$$h^*(X_1^*, \ldots, X_{n^*}^*) = h(X_1, \ldots, X_n).$$

Here S^* may be called the *base* of S, and the $(n - n^*)$-dimensional linear subspace $X_1^* = \cdots = X_{n^*}^* = 0$ (of the n-space) may be called the *axis* of S. Note that clearly $\deg S^* = \deg S = e$. If S^* can be chosen to be rational (resp. monoidal, special monoidal,...) then we may call S a hypercylinder on an $(n^* - 1)$-dimensional *rational* (resp. *monoidal, special monoidal*,...) *base*.

In a moment we shall show how to find a hypercylinder on an $(m + 1)$-dimensional base passing through any given m-dimensional variety W in n-space, provided $m + 2 \leq n$. First let us discuss some generalities about

Varieties. We start by recalling that any given variety W in n-space can be uniquely written as a finite union $W_1 \cup \cdots \cup W_r$ of irreducible subvarieties $W_1, \ldots, W_r$ with $W_i \not\subset W_j$ for all $i \neq j$. The irreducible subvarieties $W_1, \ldots, W_r$ are called the *irreducible components* of W. We have

$$\dim W = \max_{1 \leq i \leq r} \dim W_i = \max_{1 \leq i \leq r} \operatorname{trdeg}_k k(W_i) = \text{(Krull)} \dim k[W].$$

We note that a *curve* is a variety of dimension 1, a *surface* is a variety of dimension 2, a *solid* is a variety of dimension 3, ... , and so on. Moreover, a hypersurface in n-space is a variety of pure codimension 1, i.e., all its irreducible components are of dimension $n - 1$. Now it can be seen that there exists a unique positive integer δ such that "most" complementary dimensional linear subspaces (i.e., "most" $(n - m)$-dimensional linear subspaces where $m = \dim W$) meet W in δ points. We call δ the *degree* of W and denote it by $\deg W$. It can easily be verified that in case W is a hypersurface, δ coincides with the degree of the defining polynomial equation of W. Moreover, in the general case, it can be shown that

$$\deg W = \sum_{\substack{i=1 \\ \dim W_i = \dim W}}^{r} \deg W_i$$

where the summation is over all maximum dimensional irreducible components of W. Now, literally speaking, this definition of the degree of a variety W works only when k is algebraically closed. However, by counting properly, it not only works without assuming k to be algebraically closed but even for "all" complementary dimensional linear subspaces instead of just "most." The quotes around all are to insure that the linear subspace doesn't meet the variety W in a positive dimensional subvariety.

Homogeneous coordinates. Going over to homogeneous coordinates $(\mathscr{X}_0, \ldots, \mathscr{X}_n)$, with $X_i = \frac{\mathscr{X}_i}{\mathscr{X}_0}$, let $\mathscr{I}^*(W, \mathbb{P}_k^n)$ be the homogeneous ideal of W in the projective n-space $\mathbb{P}_k^n$. That is, $\mathscr{I}^*(W, \mathbb{P}_k^n)$ is the set of all homogeneous polynomials $f^*(\mathscr{X}_0, \ldots, \mathscr{X}_n) \in R^* = k[\mathscr{X}_0, \ldots, \mathscr{X}_n]$ such that $f^*(1, \beta_1, \ldots, \beta_n) = 0$ for all $(\beta_1, \ldots, \beta_n) \in W$. Let $\mathscr{I}_d^*(W, \mathbb{P}_k^n)$ be the set of all hypersurfaces of degree d in $\mathbb{P}_k^n$ passing through W, i.e., $\mathscr{I}_d^*(W, \mathbb{P}_k^n) = R_d^* \cap \mathscr{I}^*(W, \mathbb{P}_k^n)$ where R_d^* is the set of all homogeneous polynomials of degree d in $\mathscr{X}_0, \ldots, \mathscr{X}_n$ (including the zero polynomial). Note that analogous to the concept of projective plane, by the *projective n-space* $\mathbb{P}_k^n$ over the field k we mean the set of all equivalence classes of $(n + 1)$-tuples where two $(n + 1)$-tuples are regarded equivalent if they are proportional. Let

$$\mathscr{H}_W(d) = \text{the vector space dimension } [R_d^*/\mathscr{I}_d^*(W, \mathbb{P}_k^n) : k].$$

Now $\mathscr{H}_W(d)$ is called the *Hilbert function* of W, and as proved by Hilbert, there exists a (unique) polynomial $\mathscr{H}_W^*(d)$ with coefficients in $\mathbb{Q}$ such that

$$\mathscr{H}_W(d) = \mathscr{H}_W^*(d) \quad \text{for all large enough } d.$$

For the polynomial $\mathscr{H}_W^*(d)$, which is called the *Hilbert polynomial* of W, Hilbert proved that

$$\mathscr{H}_W^*(d) = \frac{\delta d^m}{m!} + \bullet d^{m-1} + \cdots + \bullet d + \blacksquare,$$

where

$$m = \dim W \quad \text{and} \quad \delta = \deg W$$

and the bullets $\bullet$ are in $\mathbb{Q}$, and the gun $\blacksquare$ is in $\mathbb{Z}$. Hilbert also proved that

$$W = \text{a nonsingular curve } C \implies (-1)^m(\blacksquare - 1) = g(C) \text{ with } m = 1.$$

By analogy we define

$$p_a(W) = \text{the *arithmetic genus* of } W = (-1)^m(\blacksquare - 1).$$

Note that in the lecture on intersection multiplicity we have already met a local version of Hilbert function and Hilbert polynomial.

Transcendence bases. For a moment assume that W is irreducible and let $m = \dim W$. To explain how to find a transcendence basis of an algebraic function field, say $k(W)/k$, consider the linear transformation

$$X_i^* = \sum_{j=1}^n a_{ij} X_j$$

where (a_{ij}) is an n by n nonsingular matrix with entries in k, and let $x_1^*, \ldots, x_n^*$ be the functions induced on W by $X_1^*, \ldots, X_n^*$ respectively. Borrowing the proof of the theorem of primitive element and other related "Kroneckerian" arguments from college algebra, and assuming k to be algebraically closed, we can see that for "most" linear transformations (a_{ij}) we have the following:

(1) $(x_1^*, \ldots, x_m^*)$ is a separating transcendence basis of $k(W)/k$,

(2) $k(W) = k(x_1^*, \ldots, x_n^*) = k(x_1^*, \ldots, x_{m+1}^*)$,

(3) $k[W]$ is integral over $k[x_1^*, \ldots, x_m^*]$, and

(4) $[k(W) : k(x_1^*, \ldots, x_m^*)] = \deg W$.

Here "most" means there exists a nonzero polynomial $\phi_W(A_{11}, \ldots, A_{nn})$ in the n^2 variables $A_{11}, \ldots, A_{nn}$ with coefficients in k, such that any nonsingular linear transformation (a_{ij}) will do the job provided $\phi_W(a_{11}, \ldots, a_{nn}) \neq 0$.

In connection with (1) we note that a *separating transcendence basis* of an algebraic function field K'/k' of transcendence degree m' is a transcendence basis $x_1', \ldots, x_{m'}'$ of K'/k' such that K' is a separable algebraic field extension of $k'(x_1', \ldots, x_{m'}')$. That is, every $y \in K'$ satisfies an equation of the form $\theta(y) = 0$ where $\theta(Y)$ is a monic polynomial in Y with coefficients in $k'(x_1', \ldots, x_{m'}')$ such that $\dot{\theta}(y) \neq 0$ where $\dot{\theta}(Y)$ is the Y-derivative of $\theta(Y)$. An algebraic function field K'/k' is said to be *separably generated* if it has a separating transcendence basis. It is known that if k' is either a characteristic zero field, or an algebraically closed field, or a finite field, then every algebraic function field K' over k' is separably generated.

We also note that, in conjunction with (1), equations (2) are equivalent to saying that x_{m+1}^* is a *primitive element* of $k(W)$ over $k(x_1^*, \ldots, x_m^*)$.

Finally, in connection with (3) we note that a ring B is said to be *integral* over a subring A if every $y \in B$ is *integral* over A. That is, it satisfies an equation of the form $\theta(y) = 0$ where $\theta(Y)$ is a monic polynomial in Y with coefficients in A.

For further details concerning the above discussion, see §12 of Chapter 3 of the resolution book [A12]. Now here is some

Philosophy. A geometric problem $\mathscr{P}$ in n-space is a *good problem* if there exists $0 \neq \phi_{\mathscr{P}}(A_{11}, \ldots, A_{nn}) \in k[A_{11}, \ldots, A_{nn}]$ such that, relative to the nonsingular linear transformation $X_i^* = \sum_{j=1}^n a_{ij} X_j$, the problem $\mathscr{P}$ has a solution whenever $\phi_{\mathscr{P}}(a_{11}, \ldots, a_{nn}) \neq 0$. Since a finite product of nonzero polynomials is a nonzero polynomial, a finite union of good problems is a good problem. For example, finding a separating transcendence basis of an algebraic function field over an algebraically closed field is a good problem, and so is the problem of finding a primitive element of a finitely generated separable algebraic field extension. Likewise the problem of finding a transcendence basis over which a given finitely generated ring is integral is also a good problem. This is the so called Noether Normalization Lemma. Therefore, the simultaneous problem involving the first three items of the above display is a good problem. Note that the basic case of the theorem of primitive element says that $k(u, v)/k$ is a separable algebraic field extension with infinite $k \Rightarrow k(u, v) = k(u + cv)$ for most $c \in k$. Actually, this is also true for finite k provided we change $u + cv$ to $u + v^c$ with $0 < c \in \mathbb{Z}$. In this connection, see van der Waerden's *Modern algebra* [**Wa2**] and Zariski-Samuel's *Commutative algebra* [**ZSa**]. If you want to read a relevant original memoir, see Steinitz [**Ste**].

Irreducible varieties. To continue what we were discussing before the philosophical interlude, assume that W is irreducible and let $m = \dim W$. Also assume that k is algebraically closed and $m + 2 \le n$. Now $x_{m+2}^* \in k(x_1^*, \ldots, x_{m+1}^*)$. Hence,

$$x_{m+2}^* = \frac{u(x_1^*, \ldots, x_{m+1}^*)}{v(x_1^*, \ldots, x_{m+1}^*)}$$

with $u(X_1^*, \ldots, X_{m+1}^*)$ and $v(X_1^*, \ldots, X_{m+1}^*)$ in $k[X_1^*, \ldots, X_{m+1}^*]$ such that $v(x_1^*, \ldots, x_{m+1}^*) \ne 0$. Cross multiplying, we get the polynomial

$$s^*(X_1^*, \ldots, X_{m+2}^*) = v(X_1^*, \ldots, X_{m+1}^*)X_{m+2}^* - u(X_1^*, \ldots, X_{m+1}^*)$$

in $X_1^*, \ldots, X_{m+2}^*$ with coefficients in k. Now in the n-space of the variables $X_1, \ldots X_n$, the hypersurface $S : s^*(X_1^*, \ldots, X_{m+2}^*) = 0$ is a hypercylinder on a special monoidal $(m + 1)$-dimensional base. We clearly have $W \subset S$. We have shown that

Given any m-dimensional irreducible variety W in n-space over an algebraically closed "ground" field k, with $m + 2 \le n$, we can always pass a rational hypersurface S through W. Moreover S can be chosen to be a hypercylinder on a special monoidal $(m + 1)$-dimensional base.

Reducible varieties. Continuing with the hypothesis that k is algebraically closed and $\dim W = m$ with $m + 2 \le n$, let us show that we can pass rational (monoidal) hypercylinders through W without assuming W to be irreducible. We start by choosing distinct irreducible m-dimensional varieties $W_1', \ldots, W_r'$ such that $W_i \subset W_i'$ for $1 \le i \le r$. Applying the philosophy of good problems to $W_1', \ldots, W_r'$ we can find $0 \ne \phi'(A_{11}, \ldots, A_{nn}) \in k[A_{11}, \ldots, A_{nn}]$ such that for every (a_{ij}) with $\phi'(a_{11}, \ldots, a_{nn}) \ne 0$, upon letting $x_1^{(l)}, \ldots, x_n^{(l)}$ to be the functions induced on W_l' by $X_1^*, \ldots X_n^*$, we simultaneously have:

(1_l) $(x_1^{(l)}, \ldots, x_m^{(l)})$ is a separating transcendence basis of $k(W_l')/k$,

(2_l) $k(W_l') = k(x_1^{(l)}, \ldots, x_n^{(l)}) = k(x_1^{(l)}, \ldots, x_{m+1}^{(l)})$, and

(3_l) $k[W_l']$ is integral over $k[x_1^{(l)}, \ldots, x_m^{(l)}]$,

for $1 \le l \le r$. Now, for $1 \le l \le r$, clearly there exists a unique monic irreducible polynomial $f_l(X_1^*, \ldots X_{m+1}^*)$ in X_{m+1}^* with coefficients in $k[X_1^*, \ldots, X_m^*]$ such that $f_l(x_1^{(l)}, \ldots, x_{m+1}^{(l)}) = 0$. By another application of the philosophy of good problems, we can find a nonzero multiple $\phi(A_{11}, \ldots, A_{nn})$ of $\phi'(A_{11}, \ldots, A_{nn})$ in $k[A_{11}, \ldots, A_{nn}]$ so that for all (a_{ij}) with $\phi(a_{11}, \ldots, a_{nn}) \neq 0$; in addition to

$$(1_1), (2_1), (3_1), \ldots, (1_r), (2_r), (3_r),$$

we have that the polynomials

$$f_1(X_1^*, \ldots, X_{m+1}^*), \ldots, f_r(X_1^*, \ldots, X_{m+1}^*)$$

are pairwise distinct. In view of (2_l), for $1 \leq l \leq r$, we have

$$x_{m+2}^{(l)} \in k(x_1^{(l)}, \ldots, x_{m+1}^{(l)}).$$

Hence,

$$x_{m+2}^{(l)} = \frac{u_l^*(x_1^{(l)}, \ldots, x_{m+1}^{(l)})}{v_l^*(x_1^{(l)}, \ldots, x_{m+1}^{(l)})},$$

where $u_l^*(X_1^*, \ldots, X_{m+1}^*)$ and $v_l^*(X_1^*, \ldots, X_{m+1}^*)$ are each polynomials in $X_1^*, \ldots, X_{m+1}^*$ with coefficients in k such that $v_l^*(x_1^{(l)}, \ldots, x_{m+1}^{(l)}) \neq 0$. Now for all $u_l'(X_1^*, \ldots, X_{m+1}^*)$ and $v_l'(X_1^*, \ldots, X_{m+1}^*)$ in $k[X_1^*, \ldots, X_{m+1}^*]$ we have

$$\begin{aligned}
x_{m+2}^{(l)} &= \frac{u_l^*(x_1^{(l)}, \ldots, x_{m+1}^{(l)})}{v_l^*(x_1^{(l)}, \ldots, x_{m+1}^{(l)})} \\
&= \frac{u_l^*(x_1^{(l)}, \ldots, x_{m+1}^{(l)}) + u_l'(x_1^{(l)}, \ldots, x_{m+1}^{(l)})f_l(x_1^{(l)}, \ldots, x_{m+1}^{(l)})}{v_l^*(x_1^{(l)}, \ldots, x_{m+1}^{(l)}) + v_l'(x_1^{(l)}, \ldots, x_{m+1}^{(l)})f_l(x_1^{(l)}, \ldots, x_{m+1}^{(l)})}.
\end{aligned}$$

Clearly, we can choose $v_1'(X_1^*, \ldots, X_{m+1}^*), \ldots, v_r'(X_1^*, \ldots, X_{m+1}^*)$ in $k[X_1^*, \ldots, X_{m+1}^*]$, such that, for $1 \leq l \leq r$, upon letting

$$\begin{aligned}
&v_l(X_1^*, \ldots, X_{m+1}^*) \\
&= v_l^*(X_1^*, \ldots, X_{m+1}^*) + v_l'(X_1^*, \ldots, X_{m+1}^*)f_l(X_1^*, \ldots, X_{m+1}^*),
\end{aligned}$$

we have

$$v_l(X_1^*, \ldots, X_{m+1}^*) \notin f_{l'}(X_1^*, \ldots, X_{m+1}^*)k[X_1^*, \ldots X_{m+1}^*] \quad \text{for } 1 \leq l' \leq r.$$

Let

$$v'(X_1^*, \ldots, X_{m+1}^*) = \prod_{l=1}^{r} v_l(X_1^*, \ldots, X_{m+1}^*)$$

and for $1 \leq l \leq r$ let

$$u_l(X_1^*, \ldots, X_{m+1}^*) = u_l^*(X_1^*, \ldots, X_{m+1}^*) \prod_{\substack{l'=1 \\ l' \neq l}}^{r} v_{l'}(X_1^*, \ldots, X_{m+1}^*).$$

Now $u_1(X_1^*, \ldots, X_{m+1}^*), \ldots, u_r(X_1^*, \ldots, X_{m+1}^*)$ and $v'(X_1^*, \ldots, X_{m+1}^*)$ are polynomials in $X_1^*, \ldots, X_{m+1}^*$ with coefficients in k such that for $1 \leq l \leq r$ we have $v'(x_1^{(l)}, \ldots, x_{m+1}^{(l)}) \neq 0$ and

$$x_{m+2}^{(l)} = \frac{u_l(x_1^{(l)}, \ldots, x_{m+1}^{(l)})}{v'(x_1^{(l)}, \ldots, x_{m+1}^{(l)})}.$$

Since $f_1, \ldots, f_r$ are pairwise distinct monic irreducible polynomials in X_{m+1}^* with coefficients in $k[X_1^*, \ldots, X_m^*]$, by the Chinese Remainder Theorem (the grade school method of finding GCD by long division) we can find

$$u_1'(X_1^*, \ldots, X_{m+1}^*), \ldots, u_r'(X_1^*, \ldots, X_{m+1}^*) \text{ and}$$
$$u(X_1^*, \ldots, X_{m+1}^*) \quad \text{in } k[X_1^*, \ldots, X_{m+1}^*],$$

and

$$0 \neq v''(X_1^*, \ldots, X_m^*) \in k[X_1^*, \ldots, X_m^*],$$

such that for $1 \leq l \leq r$ we have

$$u(X_1^*, \ldots, X_{m+1}^*) = u_l(X_1^*, \ldots, X_{m+1}^*)v''(X_1^*, \ldots, X_m^*)$$
$$+ u_l'(X_1^*, \ldots, X_{m+1}^*)f_l(X_1^*, \ldots, X_{m+1}^*).$$

Now upon letting

$$v(X_1^*, \ldots, X_{m+1}^*)$$
$$= v'(X_1^*, \ldots, X_{m+1}^*)v''(X_1^*, \ldots, X_m^*) \in k[X_1^*, \ldots, X_{m+1}^*]$$

we obviously have

$$x_{m+2}^{(l)} = \frac{u(x_1^{(l)}, \ldots, x_{m+1}^{(l)})}{v(x_1^{(l)}, \ldots, x_{m+1}^{(l)})} \quad \text{for } 1 \leq l \leq r.$$

By cross multiplying we get the polynomial

$$s^*(X_1^*, \ldots, X_{m+2}^*) = v(X_1^*, \ldots, X_{m+1}^*)X_{m+2}^* - u(X_1^*, \ldots, X_{m+1}^*)$$

in $X_1^*, \ldots, X_{m+2}^*$ with coefficients in k. Now in the n-space of the variables $X_1, \ldots X_n$, the hypersurface $S : s^*(X_1^*, \ldots, X_{m+2}^*) = 0$ is a hypercylinder on a special monoidal $(m + 1)$-dimensional base. We clearly have $W \subset S$. We have shown that

Given any m-dimensional variety W in n-space over an algebraically closed "ground" field k, with $m + 2 \leq n$, but without assuming W to be irreducible, we can always pass a rational hypersurface S through W. Moreover S can be chosen to be a hypercylinder on a special monoidal $(m + 1)$-dimensional base.

Space curves. Taking $m = 1$ and $n = 3$ we see that given any curve in 3-space over an algebraically closed field, we can always pass a rational surface S through C. Moreover, S can be chosen to be a special monoid.

Exercise. Show that

$$[R_d^* : k] = \text{ the binomial coefficient } \binom{d+n}{n}$$

$$= \frac{d^n}{n!} + \bullet d^{n-1} + \cdots + \bullet d + 1.$$

Deduce that $\dim \mathbb{P}_k^n = n$, $\deg \mathbb{P}_k^n = 1$, and $p_a(\mathbb{P}_k^n) = 0$. Moreover, for a hypersurface $S : h(X_1, \ldots, X_n) = 0$, with $\deg h = e$, show that

$$\mathscr{H}_S(d) = \begin{cases} \binom{d+n}{n} & \text{if } d < e \\ \binom{d+n}{n} - \binom{d-e+n}{n} & \text{if } d \geq e. \end{cases}$$

Hence,

$$\mathscr{H}_S^*(d) = \binom{d+n}{n} - \binom{d-e+n}{n}$$

$$= \frac{e d^{n-1}}{(n-1)!} + \bullet d^{n-2} + \cdots + \bullet d + \blacksquare$$

with

$$\blacksquare = \left[\text{what we get by putting } d = 0 \text{ in } \binom{d+n}{n} - \binom{d-e+n}{n}\right]$$

$$= \left[\frac{\text{what we get by putting } d = 0 \text{ in}}{(d+1)(d+2)\cdots(d+n) - (d-e+1)(d-e+2)\cdots(d-e+n)}{n!}\right]$$

$$= \frac{n! - (-1)^n (e-1)(e-2)\cdots(e-n)}{n!}$$

$$= 1 + (-1)^{n-1}\frac{(e-1)(e-2)\cdots(e-n)}{n!}.$$

Therefore ,

$$\dim S = n - 1 \quad \text{and} \quad \deg h = e = \delta = \deg S$$

and

$$p_a(S) = \frac{(e-1)(e-2)\cdots(e-n)}{n!}.$$

So $p_a(S) = 0$ for all $e \leq n$. In particular, assuming S to be a nonsingular plane curve C of degree e we get

$$p_a(C) = \frac{(e-1)(e-2)}{2} = g(C).$$

Geometric genus. The idea of genus can be generalized from a curve to a nonsingular irreducible algebraic variety W of any dimension m in several ways. One of them is the *arithmetic genus* $p_a(W)$ defined as the "normalized" constant term of the Hilbert polynomial of W. Another is the *geometric genus* $p_g(W)$ defined as the number of linearly independent m-fold differentials of $k(W)$ having no poles on W. This is a direct generalization of Jacobi's definition of the genus of a curve. Actually, by considering the number $g_i(W)$ of linearly independent i-fold differentials of $k(W)$ having no poles on W, for $1 \leq i \leq m$, we get m incarnations of Jacobi's definition, where linear independence is over the field k which is assumed to be an algebraically closed field of characteristic zero. Note that now $g_m(W) = p_g(W)$. Concerning these various genera (genuses), Kodaira-Spencer [KSp] proved the equation $p_a(W) = g_m(W) - g_{m-1}(W) + \cdots + (-1)^{m-1}g_1(W)$, which was conjectured by Severi [Se1]. Also see [Ser]. Actually, there is yet another generalization of p_g called the jth plurigenus $P_j(W)$ for every positive integer j so that $P_1(W) = p_g(W)$. In terms of these genera, we can state the Castelnuovo-Enriques Criterion [CEn] of rationality of a surface, which says that for $m = 2$ we have that W is rational $\Leftrightarrow p_a(W) = P_2(W) = 0$. For a sketch of these matters, you may see my Castelnuovo Centenary Paper [A14] and Zariski's book [Za3].

So what are i-fold differentials? Heuristically, by an i-fold differential we mean the integrand of an i-fold integral. To get an idea of an i-fold integral, remember that Stokes' Theorem equates a triple integral (3-fold integral) on a solid to a double integral (2-fold integral) on its surface. Likewise, Green's Theorem equates a double integral on a planer region to a line integral (1-fold integral) on its boundary. Going down to the very bottom, the fundamental theorem of calculus equates an integral on a segment (1-fold integral) to something (0-fold integral) on the two boundary points. These three theorems generalize into De Rham's Theorem, which says that the integral of the derivative of a $(j - 1)$-fold differential form on a j-dimensional region equals the integral of the differential form on the boundary of the region. For details see the books of De Rham [Rha], Hirzebruch [Hiz], and Hodge [Hod].

NOTE. The interpretation of the geometric concepts of the dimension of a variety, its degree, its (arithmetic) genus , ... , as the transcendence degree of a finitely generated field extension, degree of a finite algebraic field extension, modified constant term of the Hilbert polynomial , ... , is what makes modern commutative algebra meaningful!

Resultants

The etymology of the term "resultant" is "result of elimination." In general, we have several polynomial equations in several variables

$$f_i(X_1, \ldots, X_r, Y_1, \ldots, Y_s) = 0, \quad i = 1, \ldots, q.$$

We wish to eliminate some of the variables, say $Y_1, \ldots, Y_s$, from these equations to arrive at a finite number of polynomial expressions

$$R_1(X_1, \ldots, X_r), \ldots, R_l(X_1, \ldots, X_r),$$

in the remaining variables $X_1, \ldots, X_r$, so that the vanishing of $R_1, \ldots, R_l$ gives a necessary and sufficient condition for given equations $f_1 = 0, \ldots, f_q = 0$ to have a common solution. In other words, the "vertical projection" of the $(X_1, \ldots, X_r, Y_1, \ldots, Y_s)$-space onto the $(X_1, \ldots, X_r)$-space should map the zeros of the polynomials $f_1, \ldots, f_q$ onto the zeros of the polynomials $R_1, \ldots, R_l$.

However, the problem stated in this generality may not have a solution unless we pay proper attention to things at ∞, which can for instance be done by requiring the equations to be homogeneous, and so on.

For example, the projection of the hyperbola $XY - 1 = 0$ onto the X-axis is not describable by polynomial equations, since it consists of the X-axis minus the origin. See Figure 30.1.

FIGURE 30.1

Thus, in addition to equations, we may be forced to consider inequations to describe a locus such as the above.

Recall that the Y-resultant of two polynomials

$$F(Y) = a_0 Y^n + a_1 Y^{n-1} + \cdots + a_n$$
$$G(Y) = b_0 Y^m + b_1 Y^{m-1} + \cdots + b_m$$

is given by

$$\operatorname{Res}_Y(F, G) = \det \begin{pmatrix} a_0 & a_1 & \cdot & \cdot & \cdot & \cdot & a_n & 0 & \cdot & \cdot & \cdot & \cdot & 0 \\ 0 & a_0 & a_1 & \cdot & \cdot & \cdot & \cdot & a_n & 0 & \cdot & \cdot & \cdot & 0 \\ \cdot & \cdot & \cdot & \cdot & \cdot & \cdot & \cdot & \cdot & \cdot & \cdot & \cdot & \cdot & \cdot \\ 0 & 0 & \cdot & \cdot & \cdot & a_0 & a_1 & \cdot & \cdot & \cdot & \cdot & \cdot & a_n \\ b_0 & b_1 & \cdot & \cdot & \cdot & \cdot & b_m & 0 & \cdot & \cdot & \cdot & \cdot & 0 \\ 0 & b_0 & b_1 & \cdot & \cdot & \cdot & \cdot & b_m & 0 & \cdot & \cdot & \cdot & 0 \\ \cdot & \cdot & \cdot & \cdot & \cdot & \cdot & \cdot & \cdot & \cdot & \cdot & \cdot & \cdot & \cdot \\ \cdot & \cdot & \cdot & \cdot & \cdot & \cdot & \cdot & \cdot & \cdot & \cdot & \cdot & \cdot & \cdot \\ 0 & 0 & \cdot & \cdot & \cdot & b_0 & b_1 & \cdot & \cdot & \cdot & \cdot & \cdot & b_m \end{pmatrix}.$$

This is sometimes called the Sylvester resultant of F and G because it was introduced by Sylvester in his 1840 paper [Syl] where he showed that $\operatorname{Res}_Y(F, G) = 0$ if and only if either $a_0 = 0 = b_0$ or F and G have a common root. Thus we are not assuming that the degrees of F and G are exactly n and m respectively. So strictly speaking, for $\operatorname{Res}_Y(F, G)$ we should write something like $\operatorname{Res}_Y^{n, m}(F, G)$. For an elaborate discussion of resultants involving such a strict viewpoint, see pages 374–404 of the Kyoto paper [A27], where the definition of resultant is made to come out in a very natural manner out of the idea of multiplication of two polynomials. If that be any comfort, the treatment given in the Kyoto paper applies to resultants over rings with zerodivisors. Sylvester's own interest in resultants must have been related to his, and his friend Cayley's, preoccupation with invariant theory. For the role of resultants in invariant theory, you may consult the book of Alfred Young entitled *Algebra of invariants* [GYo], which he coauthored with his friend Grace at the turn of the century. This book of Young is surely the best way to learn invariant theory.

Now the coefficients of F and G could be polynomials in X. We could have

$$F(X, Y) = a_0(X)Y^n + a_1(X)Y^{n-1} + \cdots + a_n(X)$$
$$G(X, Y) = b_0(X)Y^m + b_1(X)Y^{m-1} + \cdots + b_m(X).$$

Then for $R(X) = \operatorname{Res}_Y(F, G)$ and for any value α of X we would have $R(\alpha) = 0 \Leftrightarrow$ either $a_0(\alpha) = 0 = b_0(\alpha)$ or $F(\alpha, \beta) = 0 = G(\alpha, \beta)$ for some β. Thus the roots of $R(X)$ are the projections of the points of intersection of F and G. In fact, the resultant gives more precise information. Namely, if the order of the zero α of $R(X)$ is e, i.e., if $R(X) = (X - \alpha)^e D(X)$ with $D(\alpha) \neq 0$, then, counting properly, there are exactly e points of intersection of F and G lying above $X = \alpha$. See Figure 30.2. Here, assuming the

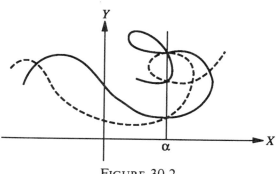

FIGURE 30.2

ground field to be algebraically closed, the "counting properly" amounts to saying that

$$e = \text{ord}_\alpha R(X) = \sum_P I(F, G; P),$$

where the summation is taken over all points P lying above $X = \alpha$.

For a proof of this formula see the Montreal notes [**A21**] where we establish the following precise

PROPOSITION. *If the coefficients* $a_0, \ldots, a_n, b_0, \ldots, b_m$ *belong to the valuation ring* R_V *of a discrete valuation* V *of a field* K, *and if either* $V(a_0) = 0$ *or* $V(b_0) = 0$, *then*

$$V(\text{Res}_Y(F, G)) = \sum_N I(F, G; N)J(N),$$

where

(i) *the summation is extended over all height two prime ideals* N *in* $R_V[Y]$ *such that* $N \cap R_V = M(R_V)$;

(ii) $I(F, G; N) = \text{length}_A(F, G)A$ *and* $J(N) = [A/M(A) : R_V/M(R_V)]$ *where* $A = R_V[Y]_N$;

(iii) *the convention is made that the equality holds if both sides are infinite, and that* $I(F, G, N)$ *is infinite if* $(F, G)A$ *is contained in but not primary for* $M(A)$ *where* $A = R_V[Y]_N$.

As an application of the proposition, we can obtain a proof of Bezout's Theorem for plane curves (viz., $C_n \cdot C_m = mn$ points), or for two surfaces in 3-space, or more generally for two hypersurfaces in d-space for any d.

In the classical literature (Salmon [**Sa1**], Coolidge [**Coo**], Semple, Roth [**SRo**], etc.), a proof of Bezout's Theorem reads as follows:

The curves C_n *and* C_m *can be degenerated into* n *lines and* m *lines respectively. These lines clearly meet in* mn *points, and hence so do the original curves.*

The above mentioned general form of Bezout's Theorem for two hypersurfaces $S_n : f(Z_1, \ldots, Z_d) = 0$ and $T_m : g(Z_1, \ldots, Z_d) = 0$ of respective degrees n and m in d-space, for any d, says that the two hypersurfaces, assumed devoid of common components, intersect along a finite number of $(d-2)$-dimensional irreducible subvarieties $U_1, \ldots, U_\kappa$ (some of which may be at infinity) of degrees $\delta_1, \ldots, \delta_\kappa$, and upon letting e_j be the intersection multiplicity $I(S, T; U_j)$ of S and T at U_j, we have

$$\sum_{j=1}^{\kappa} e_j \delta_j = mn.$$

Here the *intersection multiplicity* $I(S, T; U_j)$ is defined by "reducing to the plane curve case," that is, by considering the local ring A_j of U_j in the d-dimensional ambient space, by noting that A_j is a 2-dimensional regular local ring, and by putting

$$I(S, T; U_j) = \text{length}_{A_j}(f, g)A_j.$$

Now, using resultants, we can answer the following
Question. Given any two polynomials u and v in X, Y, that is,

$$\begin{cases} u = F(X, Y) \\ v = G(X, Y), \end{cases}$$

when can X and Y be expressed as rational functions of u and v? Moreover, when can X and Y be expressed as polynomials in u and v?
Answer. Let

$$\phi(u, v, X) = \text{Res}_Y(F(X, Y) - u, \ G(X, Y) - v)$$
$$\psi(u, v, Y) = \text{Res}_X(F(X, Y) - u, \ G(X, Y) - v).$$

Assume that the coefficients of the highest degree terms in F and G when regarded as polynomials in Y do not have a nonconstant polynomial in X as a common factor. Also assume that the coefficients of the highest degree terms in F and G when regarded as polynomials in X do not have a nonconstant polynomial in Y as a common factor. (Note that these are not serious assumptions since we can always arrange them to be satisfied by a rotation in the (X, Y)-plane.) Then, as a consequence of the proposition, it can be shown that

(1) X and Y are rational functions of u and v $\Leftrightarrow$ $\deg_X \phi = 1 = \deg_Y \psi$. Moreover, $X = \frac{\rho(u, v)}{\sigma(u, v)}$ and $Y = \frac{\rho'(u, v)}{\sigma'(u, v)}$ where $\rho(u, v)$, $\rho'(u, v)$, $\sigma(u, v)$, $\sigma'(u, v)$ are polynomials with $\sigma(u, v) \neq 0 \neq \sigma'(u, v)$ $\Leftrightarrow$ $\phi(u, v, X) = c\sigma(u, v)X - c\rho(u, v)$ and $\psi(u, v, Y) = c'\sigma'(u, v)Y - c'\rho'(u, v)$ where c and c' are nonzero constants.

(2) X and Y are polynomials in u and v $\Leftrightarrow$ $\deg_X \phi = 1 = \deg_Y \psi$ and the coefficient of X in $\phi(u, v, X)$ is a constant and the coefficient of Y in $\psi(u, v, Y)$ is a constant. Moreover, $X = \rho(u, v)$ and $Y = \rho'(u, v)$ where $\rho(u, v)$ and $\rho'(u, v)$ are polynomials $\Leftrightarrow$ $\phi(u, v, X) = cX - c\rho(u, v)$ and $\psi(u, v, Y) = c'Y - c'\rho'(u, v)$ where c and c' are nonzero constants.

So this not only gives a criterion for X and Y to be rational function or polynomial in u and v but, when that is so, it actually produces the relevant explicit expressions.

We can also consider another

Question. Given a polynomially parametrized irreducible plane curve

$$C : \begin{cases} X = p(t) \\ Y = q(t), \end{cases}$$

how do we describe it implicitly? In other words, how do we find an implicit equation $H(X, Y) = 0$ of C. That is, how do we find an irreducible polynomial $H(X, Y)$ such that $H(p(t), q(t)) = 0$? Moreover, how do we decide whether the given parametrization is faithful, and if not faithful, then how do we find the degree of "unfaithfulness," that is, the field degree $\delta = [k(t) : k(p(t), q(t))]$ where k is the coefficient field.

Answer. Here we can consider the resultant $\mathrm{Res}_t(X - p(t), Y - q(t))$, which is a polynomial $\theta(X, Y)$ in X and Y. As a consequence of the proposition, it can be shown that $\theta(X, Y) = cH(X, Y)^\delta$, where c is a nonzero constant.

So the ordinary resultant enables us to find an implicit equation $H(X, Y) = 0$ of the irreducible plane curve C whose polynomial parametrization we started with. It also enables us to decide whether t can be expressed as a rational function of $p(t)$ and $q(t)$ (exactly when $\delta = 1$). We have already explained how, by using the Taylor resultant, we can decide whether t can be expressed as a polynomial in $p(t)$ and $q(t)$.

So much for resultants of two polynomials. Now let us discuss multivariate resultants. Suppose we have three polynomials in three variables, say $f(X, Y, Z)$, $g(X, Y, Z)$, $h(X, Y, Z)$, and we would like to define their resultant with respect to two of the variables, say Y and Z, to obtain a polynomial function $E(X) = \mathrm{Res}_{Y, Z}(f, g, h)$ in the remaining variable X. This function should have the following properties.

(1) Now the surfaces f, g, h would meet, in general, in some points. The projection on the X-axis of their points of intersection should be given by the roots of $E(X)$. Moreover, the order of a root of $E(X)$ should coincide with the sum of the intersection multiplicities of f, g, h at the points lying above that root.

(2) As a more concrete test for this resultant function, the higher dimensional analogues of the above two answers should hold. Thus we should have the following.

(i) If we consider

$$\xi(u, v, w, X)$$
$$= \text{Res}_{Y,Z}(f(X, Y, Z) - u, \ g(X, Y, Z) - v, \ h(X, Y, Z) - w)$$
$$\eta(u, v, w, Y)$$
$$= \text{Res}_{X,Z}(f(X, Y, Z) - u, \ g(X, Y, Z) - v, \ h(X, Y, Z) - w)$$
$$\zeta(u, v, w, Z)$$
$$= \text{Res}_{X,Y}(f(X, Y, Z) - u, \ g(X, Y, Z) - v, \ h(X, Y, Z) - w)$$

then, with suitable assumptions on the relevant highest degree terms of f, g, h, we should have that X, Y, Z are rational functions of f, g, h $\Leftrightarrow \ \deg_X \xi = \deg_Y \eta = \deg_Z \zeta = 1$; and that X, Y, Z are polynomials in $f, g, h \ \Leftrightarrow \ \deg_X \xi = \deg_Y \eta = \deg_Z \zeta = 1$ and the coefficients of X, Y, Z in ξ, η, ζ are nonzero constants.

(ii) If

$$\begin{cases} X = \lambda(t, \tau) \\ Y = \mu(t, \tau) \\ Z = \nu(t, \tau) \end{cases}$$

is a polynomial parametrization of an irreducible surface $S : \omega(X, Y, Z) = 0$ with degree of unfaithfulness $\delta = [k(t, \tau) : k(\lambda(t, \tau), \mu(t, \tau), \nu(t, \tau))]$ then we should have:

$$\text{Res}_{t,\tau}(X - \lambda(t, \tau), \ Y - \mu(t, \tau), \ Z - \nu(t, \tau)) = c \, \omega(X, Y, Z)^{\delta} \quad \text{with } c \neq 0.$$

Problem. Carry out the above project concerning multivariate resultants. A successful project should enable you to solve the problem of determining biregular maps of 3-space, or more generally of d-space, onto itself. In other words, study the k-automorphisms of $k[X, Y, Z]$, or more generally of $k[X_1, \ldots, X_d]$. These correspond to biregular maps of affine spaces. Likewise, study the k-automorphisms of $k(X, Y, Z)$, or more generally of $k(X_1, \ldots, X_d)$. These correspond to birational maps of projective spaces which are also called Cremona transformations. At least in the case of a plane, their study was initiated by Cremona around 1860. If Max Noether was the father of algebraic geometry, then Cremona at least deserves to be called the uncle of algebraic geometry. For automorphisms of polynomial rings, and the related Jacobian problem, refer to the Kyoto paper [A27] or the TIFR Lecture notes [A26].

REMARK. In the classical theory, the resultant is often considered for homogeneous polynomials. We noted that if $F(X, Y)$ and $G(X, Y)$ are two polynomials given by

$$F(X, Y) = a_0(X)Y^n + a_1(X)Y^{n-1} + \cdots + a_n(X)$$
$$G(X, Y) = b_0(X)Y^m + b_1(X)Y^{m-1} + \cdots + b_m(X),$$

and if $R(X) = \text{Res}_Y(F, G)$ then for any value α of X, $R(\alpha) = 0 \Leftrightarrow$ either $a_0(\alpha) = 0 = b_0(\alpha)$ or $F(\alpha, \beta) = 0 = G(\alpha, \beta)$ for some β. If

$a_0(\alpha) = 0 = b_0(\alpha)$, then a point of intersection has gone to ∞. So we can go to the projective space by homogenizing. Unlike the affine spaces, projective spaces have the following nice property. *If W is an algebraic variety in a projective space $\mathbb{P}^{r+s}$ and we project it onto a projective space $\mathbb{P}^r$, then the projection of W is again an algebraic variety.*

NOTE. Elimination theory (or topics such as the study of resultants) is in the same circle of ideas which includes the following cascade of more and more general results.

Fundamental theorem of algebra. The field $\mathbb{C}$ is algebraically closed.

Liouville's Theorem. Analytic function on a compact Riemann surface is constant.

Chow's Theorem. ([**Cho**]). Analytic variety in a projective space is algebraic.

Remmert's Theorem. ([**Rem**]). The image of an analytic set, under a proper map, is analytic. (A map is *proper* if the inverse image of any compact set is compact).

Grauert's Theorem or university algebra analogue of Remmert's Theorem. ([**Gra**]). Direct images of coherent analytic sheaves are coherent.

As a general reference for this kind of thing, you may see *Local analytic geometry* [**A07**].

Bibliography

[**Ab1**] N. H. Abel, *Mémoire sur une propriété générale d'une classe très-étendue de fonctions transcendantes, 1826*, Oeuvres Complètes, New Edition **I** (1881), 145–211.

[**Ab2**] N. H. Abel, *Démonstration d'une propriété générale d'une certaine classe de fonctions transcendantes*, Crelle Journal **4** (1829), 200–201.

[**A01**] S. S. Abhyankar, *On the valuations centered in a local domain*, Amer. J. Math. **78** (1956), 321–48.

[**A02**] S. S. Abhyankar, *Local uniformization on algebraic surfaces over ground fields of characteristic $p \neq 0$*, Ann. of Math. **63** (1956), 491–526.

[**A03**] S. S. Abhyankar, *Coverings of algebraic curves*, Amer. J. Math. **79** (1957), 825–56.

[**A04**] S. S. Abhyankar, *Ramification theoretic methods in algebraic geometry*, Princeton University Press, Princeton, 1959.

[**A05**] S. S. Abhyankar, *Tame coverings and fundamental groups of algebraic varieties, Parts I to VI*, Amer. J. Math. **81, 82** (1959–60).

[**A06**] S. S. Abhyankar, *Cubic surfaces with a double line*, Memoirs of the College of Science, University of Kyoto, Series A Mathematics **32** (1960), 455–511.

[**A07**] S. S. Abhyankar, *Local analytic geometry*, Academic Press, New York, 1964.

[**A08**] S. S. Abhyankar, *Uniformization in p-cyclic extensions of algebraic surfaces over ground fields of characteristic p*, Mathematische Annalen **153** (1964), 81–96.

[**A09**] S. S. Abhyankar, *Reduction to multiplicity less than p in a p-cyclic extension of a two dimensional regular local ring*, Mathematische Annalen **154** (1964), 28–55.

[**A10**] S. S. Abhyankar, *Resolution of singularities of arithmetical surfaces*, Arithmetical Algebraic Geometry, Harper and Row (1965), 111–152.

[**A11**] S. S. Abhyankar, *Uniformization in a p-cyclic extension of a two dimensional regular local domain of residue field characteristic p*, Festschrift zur Gedächtnisfeier für Karl Weierstrass 1815–1895, Wissenschaftliche Abhandlungen des Landes Nordrhein-Westf. **33** (1966), 243–317.

[**A12**] S. S. Abhyankar, *Resolution of singularities of embedded algebraic surfaces*, Academic Press, New York, 1966.

[A13] S. S. Abhyankar, *An algorithm on polynomials in one indeterminate with coefficients in a two dimensional regular local domain*, Annali di Matematica pura ed applicata—Series 4, **71** (1966), 25–60.

[A14] S. S. Abhyankar, *On the birational invariance of arithmetic genus*, Rendiconti di Matematica **25** (1966), 77–86.

[A15] S. S. Abhyankar, *Nonsplitting of valuations in extensions of two dimensional regular local rings*, Math. Ann. **170** (1967), 87–144.

[A16] S. S. Abhyankar, *Inversion and invariance of characteristic pairs*, Amer. J. Math. **89** (1967), 363–72.

[A17] S. S. Abhyankar, *On the problem of resolution of singularities*, Proceedings of 1966 Moscow International Congress of Mathematicians, Moscow (1968), 469–81.

[A18] S. S. Abhyankar, *Three-dimensional embedded uniformization in characteristic p*, Lectures at Purdue University, Notes by M. F. Huang (1968).

[A19] S. S. Abhyankar, *Resolution of singularities of algebraic surfaces*, Proceedings of 1968 Bombay International Colloquium at the Tata Institute of Fundamental Research, Oxford University Press (1969), 1–11.

[A20] S. S. Abhyankar, *Singularities of algebraic curves*, Analytic Methods in Mathematical Physics Conference Proceedings, Gordon and Breech (1970), 3–14.

[A21] S. S. Abhyankar, *Algebraic space curves*, Les Presses de l'Université de Montreal, Montreal, Canada, 1971.

[A22] S. S. Abhyankar, *Some remarks on the Jacobian question*, Lectures at Purdue University, Notes by M. van der Put and W. Heinzer (1971–72).

[A23] S. S. Abhyankar, *On Macaulay's examples*, Conference on Commutative Algebra, Lecture Notes in Mathematics No. 311, Springer-Verlag, New York (1973), 1–16.

[A24] S. S. Abhyankar, *Lectures on algebraic geometry*, Lectures given at University of Minnesota and Purdue University, Notes by C. Christensen (1974).

[A25] S. S. Abhyankar, *Historical ramblings in algebraic geometry and related algebra*, American Mathematical Monthly **83** (1976), 409–48.

[A26] S. S. Abhyankar, *Lectures on expansion techniques in algebraic geometry*, Tata Institute of Fundamental Research, Bombay, 1977.

[A27] S. S. Abhyankar, *On the semigroup of a meromorphic curve, Part I*, Proceedings of the International Symposium on Algebraic Geometry Kyoto (1977), 240–414.

[A28] S. S. Abhyankar, *Weighted expansions for canonical desingularization*, Lecture Notes in Mathematics, No. 910, Springer-Verlag, New York, 1982.

[A29] S. S. Abhyankar, *Desingularization of plane curves*, American Mathematical Society Proceedings of Symposia in Pure Mathematics **40 Part 1** (1983), 1–45.

[A30] S. S. Abhyankar, *Good points of a hypersurface*, Adv. in Math. **68** (1988), 87–256.

[A31] S. S. Abhyankar, *What is the difference between a parabola and a hyperbola?*, Math. Intelligencer **10** (1988), 37–43.

[A32] S. S. Abhyankar, *Irreducibility criterion for germs of analytic functions of two complex variables*, Adv. in Math. **74** (1989), 190–257.

[A33] S. S. Abhyankar, *Noether's Fundamental Theorem*, 1973 Calcutta Conference of the Indian Mathematical Society, forthcoming.

[ASa] S. S. Abhyankar and A. M. Sathaye, *Geometric Theory of Algebraic Space Curves*, Lecture Notes in Mathematics, No. 423, Springer-Verlag, New York, 1974.

[Ba1] H. F. Baker, *Principles of geometry*, Vols. 1 to 6, Cambridge University Press, London, 1922–33.

[Ba2] H. F. Baker, *Abel's theorem and allied topics*, Cambridge University Press, London, 1904.

[Bee] E. T. Bell, *Men of mathematics*, Simon and Schuster, New York, 1937.

[Bel] R. J. T. Bell, *An elementary treatise on coordinate geometry of three dimensions*, Cambridge University Press, London, 1920.

[Bez] É. Bezout, *Théorie générale des équations algébriques*, Paris, 1770.

[Bha] Bhaskaracharya, *Beejaganit*, Ujjain, 1150.

[Boc] M. Bôcher, *Introduction to higher algebra*, Macmillan, New York, 1907.

[Bra] K. Brauner, *Klassifikation der singularitäten algebroider kurven*, Abhandlungen aus dem Mathematischen Seminar der Hamburgischen Universität **6** (1928).

[Bur] W. Burau, *Kennzeichnung der schlauchknoten*, Abhandlungen aus dem Mathematischen Seminar der Hamburgischen Universität **6** (1928).

[BCW] H. Bass, E. H. Connell, and D. Wright, *The Jacobian conjecture: reduction of degree and formal expansion of the inverse*, Bull. Amer. Math. Soc. **7** (1982), 287–330.

[BMa] G. Birkhoff and S. Mac Lane, *A survey of modern algebra*, Macmillan, New York, 1941.

[BNo] A. Brill and M. Noether, *Über die algebraischen functionen und ihre anwendungen in der geometrie*, Math. Ann. **7** (1873), 269–310.

[BPa] W. S. Burnside and A. W. Panton, *Theory of equations*, Vols. I and II, Dublin, Hodges, Figgis and Co., London, 1904.

[Cay] A. Cayley, *On the intersection of curves*, Math. Ann. **30** (1887), 85–90.

[Ch1] C. Chevalley, *On the theory of local rings*, Ann. of Math. **44** (1943), 690–708.

[Ch2] C. Chevalley, *Ideals in power series rings*, Trans. Amer. Math. Soc. (1944).

[Ch3] C. Chevalley, *Intersections of algebraic and algebroid varieties*,

Trans. Amer. Math. Soc. **57** (1945), 1–85.

[**Ch4**] C. Chevalley, *Introduction to the theory of algebraic functions of one variable*, American Mathematical Society, Mathematical Surveys, No. 6, New York, 1951.

[**Cho**] W. L. Chow, *On compact analytic varieties*, Amer. J. of Math. **71** (1949), 893–914.

[**Chr**] C. Christensen, *Strong domination/weak factorization of three-dimensional regular local rings*, J. Indian Math. Soc. (N.S.) **45** (1984), 21–47.

[**Chy**] G. Chrystal, *Algebra*, Vols. 1 and 2, A. and C. Black, Edinburgh, 1889.

[**Coh**] I. S. Cohen, *On the structure and ideal theory of complete local rings*, Tran. Amer. Math. Soc. **59** (1946), 54–106.

[**Coo**] J. L. Coolidge, *A treatise on algebraic plane curves*, Clarendon Press, Oxford, 1931.

[**CEn**] G. Castelnuovo and F. Enriques, *Sopra alcune questioni fondamenta nella teoria delle superficie algebriche*, Ann. Mat. Pura Appl. III s. **6** (1901).

[**CGr**] H. Clemens and P. Griffiths, *The intermediate Jacobian of the cubic threefold*, Ann. of Math. **95** (1972), 281–356.

[**CNo**] R. C. Cowsik and M. V. Nori, *Affine curves in characteristic p are set-theoretic complete intersections*, Invent. Math. **45** (1978), 111–14.

[**Ded**] R. Dedekind, *Gesammelte mathematische werke*, Erster Band, Fridr. Vieweg und Sohn, Braunschweig, 1930.

[**DWe**] R. Dedekind and H. Weber, *Theorie der algebraischen functionen einer veränderlichen*, Crelle Journal **92** (1882), 181–290.

[**Edg**] W. L. Edge, *The theory of ruled surfaces*, Cambridge University Press, London, 1931.

[**Ed1**] J. Edwards, *An elementary treatise on the differential calculus*, Macmillan, London, 1886.

[**Ed2**] J. Edwards, *A treatise on the integral calculus*, Vols. 1 and 2, Macmillan, London, 1921–22.

[**Eul**] L. Euler, *Introductio in analysin infinitorum*, Berlin Academy, 1748.

[**For**] A. R. Forsyth, *Theory of functions of a complex variable*, Cambridge University Press, London, 1893.

[**God**] L. Godeaux, *Géométrie algébrique*, Vols. 1 and 2, Sciences et Lettres, Liege, 1948–49.

[**Gra**] H. Grauert, *Ein theorem der analytischen garbentheorie und die modulräume komplexer strukturen*, Pub. Math. IHES **5** (1960), 1–64.

[**GYo**] J. H. Grace and A. Young, *The algebra of invariants*, Cambridge University Press, London, 1903.

[**Hal**] G. Halphen, *Étude sur les points singuliers des courbes algébriques planes*, Appendix to the French Edition of Salmon [**Sal**] (1884), 535–648.

[**Hen**] K. Hensel, *Theorie der algebraischen zahlen*, Teubner, Leipzig, 1908.

[**Hil**] D. Hilbert, *Über die theorie der algebraischen formen*, Mathematische Annalen **36** (1890), 473–534.

[Hi2] D. Hilbert, *Über die vollen invariantensysteme*, Math. Ann. **42** (1893), 313–73.

[Hir] H. Hironaka, *Resolution of singularities of an algebraic variety over a field of characteristic* 0, Ann. of Math. **79** (1964), 109–326.

[Hiz] F. Hirzebruch, *Neue topologische methoden in der algebraischen geometrie*, Springer, Berlin, 1962.

[Hod] W. V. D. Hodge, *The theory and applications of harmonic integrals*, Cambridge University Press, London, 1941.

[Hud] H. Hudson, *Cremona transformations*, Cambridge University Press, London, 1927.

[HPe] W. V. D. Hodge and D. Pedoe, *Methods of algebraic geometry*, Vols. I to III, Cambridge University Press, London, 1947, 1952, 1954.

[Jac] C. G. J. Jacobi, *Considerationes generales de transcentibus abelianis*, Crelle Journal **9** (1832), 394–403.

[Jun] H. W. E. Jung, *Über ganze birationale transformationen der Ebene*, Crelle Journal **184** (1942), 161–74.

[Kah] E. Kähler, *Über die verzweigung einer algebraischen funktion zweier veränderlichen in der umgebung einer singulären stelle*, Math. Z. **30** (1929).

[Kne] M. Kneser, *über die Darstellung algebaischer Raumkurven als Durchschnitte von Flächen*, Arch. Math. (Basel) **11** (1960), 157–58.

[Kro] L. Kronecker, *Grundzuge einer arithmetischen theorie der algebraischen Größen*, Crelle Journal **92** (1882), 1–122.

[Kr1] W. Krull, *Dimensionstheorie in stellenringen*, Crelle Journal **179** (1938), 204–26.

[Kr2] W. Krull, *Theorie der polynomideale und eliminationstheorie*, Encyklopädie der mathematischen wissenschaften **Band I1, 12** (1939), 1–53.

[Kr3] W. Krull, *Der allgemeine diskriminantensatz*, Math. Z. **45** (1939), 1–19.

[Kul] W. van der Kulk, *On polynomial rings in two variables*, Nieuw Arch. Wisk. **3** (1953), 33–41.

[KSp] K. Kodaira and D. C. Spencer, *On arithmetic genera of algebraic varieties*, Proc. Nat. Acad. Sci. U.S.A. **39** (1953), 641–649.

[Le1] S. Lefschetz, *L' analysis situs et la géométrie algébrique*, Gauthier-Villars, Paris, 1924.

[Le2] S. Lefschetz, *A page of mathematical autobiography*, Bull. Amer. Math. Soc. **74** (1968), 854–79.

[Mac] F. S. Macaulay, *The algebraic theory of modular systems*, Cambridge University Press, London, 1916.

[Mum] D. Mumford, *The topology of normal singularities of an algebraic surface and a criterion for simplicity*, Publ. Math. IHES **9** (1961), 229–46.

[Na1] M. Nagata, *Rational surfaces I*, Memoirs of University of Kyoto, Series A, Mathematics **32** (1959–60), 351–70.

[Na2] M. Nagata, *Local rings*, Interscience, New York.

[Na3] M. Nagata, *On automorphism group of* $k[x, y]$, Kinokuniya Bookstore Co., Tokyo, 1972.

[New] I. Newton, *Geometria analitica*, 1660.

[Noe] E. Noether, *Eliminationstheorie und allgemeine idealtheorie*, Math. Ann. **90** (1923), 229–61.

[No1] M. Noether, *Zur theorie des eindeutigen entsprechens algebraischer gebilde von beliebig vielen dimensionen*, Math. Ann. **2** (1870), 293–316.

[No2] M. Noether, *Über einen satz aus der theorie der algebraischen funcktionen*, Math. Ann. **6** (1873), 351–59.

[Os1] A. Ostrowski, *Über einige lösungen der funktionalgleichung $\phi(x)\phi(y) = \phi(xy)$*, Acta Math. **41** (1918), 271–84.

[Os2] A. Ostrowski, *Untersuchungen zur arithmetischen theorie der Körper*, Math. Z. **39** (1934), 296–404.

[Per] O. Perron, *Über das Vahlensche Beispiel zu einem Satz von Kronecker*, Math. Z. **47** (1942), 318–24.

[Pui] V. A. Puiseux, *Recherches sur les fonctions algebriques*, J. Math. Pure Appl. (9) **15** (1850), 365–480.

[Ree] J. E. Reeve, *A summary of results in the topological classification of plane algebroid singularities*, Rendiconti Seminario Matematico Universita e Politecnico Torino **14** (1955), 159–87.

[Rem] R. Remmert, *Projektionen analytischer mengen*, Math. Ann. **130** (1956), 410–41.

[Rha] G. De Rham, *Varietes differentiables*, Hermann, Paris, 1960.

[Rie] B. Riemann, *Theorie der abelschen funktionen*, J. Reine Angew. Math. **64** (1865), 115–55.

[Roo] T. G. Room, *The geometry of determinantal loci*, Cambridge University Press, London, 1938.

[Sa1] G. Salmon, *Higher plane curves*, Dublin, 1852.

[Sa2] G. Salmon, *Conic sections*, Dublin, 1859.

[Sco] C. A. Scott, *A proof of Noether's fundamental theorem*, Mathematische Annalen **52** (1899), 593–97.

[Ser] J. P. Serre, *Faisceaux algébriques cohérents*, Ann. of Math. **64** (1955), 197–278.

[Se1] F. Severi, *Fondamenti per la geometria sulle varietà algebriche*, Rendiconti Circolo Matematico Palermo **28**, 33–87.

[Se2] F. Severi, *Vorlesungen über algebraische geometrie*, Teubner, Leipzig, 1921.

[Sha] D. Shannon, *Monoidal transforms*, Amer. J. Math. **95** (1973), 294–320.

[Sie] C. L. Siegel, *Topics in complex function theory, Vols. I, II and III*, Wiley-Interscience, New York, 1969–72.

[Smi] C. Smith, *An elementary treatise on solid geometry*, Macmillan, London, 1884.

[Smt] H. J. S. Smith, *On the higher singularities of plane curves*, Proc. London Math. Soc. **6** (1873), 153–82.

[Sta] H. Stahl, *Theorie der Abel'schen funktionen*, Teubner, Leipzig, 1896.

[Ste] E. Steinitz, *Algebraische theorie der Körper*, Crelle Journal **137** (1910), 163–308, reprinted by Chelsea Publishing Company in New York in 1950.

[Syl] J. J. Sylvester, *On a general method of determining by mere inspection the derivations from two equations of any degree*, Philosophical Magazine **16** (1840), 132–135.

[SRo] J. G. Semple and L. Roth, *Introduction to algebraic geometry*, Clarendon Press, Oxford, 1949.

[STh] H. Seifert and W. Threlfall, *Lehrbuch der Topologie — Translated into English as A Textbook of Topology*, Teubner — Academic Press, Leipzig – New York, 1934-80.

[Vah] K. Th. Vahlen, *Bemerkung zur vollständigen Darstellung algebraischer Raumkurven*, Crelle Journal **108**, 346–47.

[Wa1] B. L. van der Waerden, *Einfürung in die algebraische geometrie*, Teubner, Leipzig, 1939.

[Wa2] B. L. van der Waerden, *Modern algebra*, Vols. 1 and 2, Frederick Ungar Publishing Co., New York, 1950.

[Wal] R. J. Walker, *Reduction of singularities of an algebraic surface*, Ann. of Math. **36** (1935), 336–65.

[Wei] K. Weierstrass, *Vorbereitungssatz, Berlin University Lecture of 1860, contained in: Einige auf die theorie der analytischen funktionen mehrerer veränderlichen sich beziehende sätze*, Mathematische Werke **II** (1895), 135–188.

[Wey] H. Weyl, *Die idee der Riemannschen Fläche*, Teubner, Leipzig, 1923.

[Za1] O. Zariski, *On the problem of existence of algebraic functions of two variables possessing a given branch curve*, Amer. J. Math. **51** (1929).

[Za2] O. Zariski, *On the topology of algebroid singularities*, Amer. J. Math. **54** (1932).

[Za3] O. Zariski, *Algebraic surfaces*, Ergeb. Math. Grenzgeb. **5** (1934).

[Za4] O. Zariski, *The reduction of singularities of an algebraic surface*, Ann. of Math. **40** (1939), 639–89.

[Za5] O. Zariski, *Reduction of singularities of algebraic three-dimensional varieties*, Ann. of Math. **45** (1944), 472–542.

[ZSa] O. Zariski and P. Samuel, *Commutative algebra*, Vols. I and II, Van Nostrand Publishing Company, Princeton, 1960.

Index

Merry Quizmas!

H. Becker

Scholastic Canada Ltd.
Toronto New York London Auckland Sydney
Mexico City New Delhi Hong Kong Buenos Aires

Scholastic Canada Ltd.
604 King Street West, Toronto, Ontario M5V 1E1, Canada
Scholastic Inc.
557 Broadway, New York, NY 10012, USA
Scholastic Australia Pty Limited
PO Box 579, Gosford, NSW 2250, Australia
Scholastic New Zealand Limited
Private Bag 94407, Botany, Manukau 2163, New Zealand
Scholastic Children's Books
Euston House, 24 Eversholt Street, London NW1 1DB, UK

www.scholastic.ca

Library and Archives Canada Cataloguing in Publication
Becker, Helaine, 1961-, author
Merry quizmas! / H. Becker.

ISBN 978-1-4431-4286-1 (softcover)

1. Christmas--Miscellanea--Juvenile literature. I. Title.

GT4985.5.B43 2018   j394.2663   C2018-900688-9

For photo credits please see page 96.

6 5 4 3 2 1   Printed in Canada 121   18 19 20 21 22

MIX
Paper from
responsible sources
FSC® C004071

Table of Contents

Christmas in Canada

Christmas in Canada is like nothing else on earth! How well do you know the traditions and history of our country?

1. The first Christmas tree in Canada was displayed in Sorel, Quebec, in 1781. What was it decorated with?
   a. Paper chains and cranberry garlands.
   b. Fruit and candles.
   c. Tinsel and spherical ornaments in bright colours.
2. What was special about the Queen's Christmas message in 2006?
   a. She sang "We Wish You a Merry Christmas."
   b. It was in English and French.
   c. It was available as a podcast.
3. In Newfoundland, people traditionally threw a piece of flaming yule log over their house to . . .
   a. ensure the house would be protected from fire in the following year.
   b. guide Santa to the rooftop.
   c. light up the sky with attractive fireworks.

4. Which province exports the greatest number of Christmas trees to other countries around the world?
   a. British Columbia.
   b. Alberta.
   c. Quebec.
5. Canadians drink a LOT of eggnog! How much?
   a. 6,400 litres annually.
   b. 7.8 million litres annually.
   c. 640,000 litres annually.
6. Which is NOT a real place?
   a. Christmas Island, Nova Scotia.
   b. Hohoho, Saskatchewan.
   c. Reindeer Station, Northwest Territories.

7. During the Christmas season, people in Newfoundland and Labrador put on masks, ring bells and run door to door asking for treats. What is this tradition called?
   a. Trick-or-treating.
   b. Mummering.
   c. Carolling.

8. What is the Christmas capital of Canada?
   a. Winnipeg.
   b. Ottawa.
   c. Labrador City.

9. Which Canadian prime minister was born on Christmas Day?
   a. John A. Macdonald.
   b. Wilfrid Laurier.
   c. Justin Trudeau.

10. Santa and Mrs. Claus live in Canada. What is their address?
    a. North Pole H0H 0H0, Canada.
    b. 101 Reindeer Way, Santa's Corner, NT W0W 0M1.
    c. c/o The Workshop, 12 Days Lane, Elftown, NU J0I 0H0.

SCORING

Give yourself one point for each correct answer.

1. b. The tradition originated in Germany and was brought to Canada by Baroness Frederika von Riedesel, whose husband was a military officer.
2. c
3. a
4. c. Quebec exports more than a million trees a year!
5. b. Yet 34 percent of people say they don't like the stuff!
6. b. But you can visit Sled Lake, Saskatchewan.
7. b. People dress up in outlandish outfits that include wearing their underwear on the outside of their clothes!
8. a
9. c. His brother Alexandre was also born on Christmas Day!
10. a. You don't need a stamp on a letter mailed within Canada to this address!

HOW YOU RATE . . .

0-3 Trivia trying. There's so much still to discover about Christmas in Canada — and it's so much fun finding out.

4-7 Trivia pro. You're sharp enough to hear those hooves on the roof.

8-10 Trivia tops. You're ready to apply for a job with Santa next year!

What's in Santa's Sack?

There's something for everyone in your family in Santa's sack.
Match each gift to the wrapped package you'd find it in.

1.

2.

3.

4.

5.

6.

7.

8.

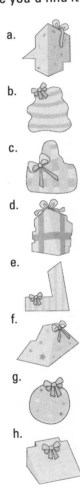

a.

b.

c.

d.

e.

f.

g.

h.

SCORING

Give yourself one point for each correct answer.

1. f
2. a
3. d
4. g
5. b
6. h
7. e
8. c

HOW YOU RATE . . .

0-1 Getting into shape. No worries! Even Santa has to work out to get into tip-top shape for Christmas fun!

2-5 In good shape. So gather round for some holiday cheer!

6+ Ship shape! Congratulations! Your holiday is shaping up to be one of the all-time greats!

Carol Calamity

You're in charge of the Carolling Crew this year. But uh-oh — the sheet music has gone missing! No one remembers the words to the songs! Can you save the day? Fill in the missing words from each popular carol.

1. Deck the halls with boughs of _____ ,
   Fa la la la la, la la la la.
2. Silent night, holy _____.
   All is ____, all is bright.
3. Good King Wenceslas looked out,
   On the Feast of _____,
   When the snow lay round about,
   Deep and crisp and even.
4. _____! The herald angels sing.
   "Glory to the newborn King!"
5. It came upon the midnight clear,
   That glorious song of ____,
6. O come, all ye faithful,
   _____ and triumphant.
7. Here we come a-wassailing
   Among the _____ so green.
   Here we come a-wand'ring
   So fair to be seen.
8. Dashing through the snow,
   In a ___ horse ____ sleigh.
   O'er the fields we go,
   Laughing all the way.
   Bells on bobtail ring,
   Making spirits bright.
   What fun it is to ride and sing
   A _____ song tonight.

9. I saw _____ ships come _____ in
   On Christmas Day, on Christmas Day.
10. On the first day of Christmas,
    my true love sent to me:
    A _____ in a pear tree.

SCORING

Give yourself one point for each correct answer.
1. holly
2. night, calm
3. Stephen
4. Hark
5. old
6. Joyful
7. leaves
8. one, open, sleighing
9. three, sailing
10. partridge

HOW YOU RATE . . .

0-5 Christmas elf. You're a valued member of the choir!

6-10 Christmas candle. You're a bright spot in the carolling crowd.

11-14 Christmas star. You're the star soloist!

What's Your Secret Elf Name?

Using the chart below, find the initial of your first name and the initials of your last name to get your secret elf name. For example, if your name is **D**ylan **V**anderKamp, then your secret elf name is Freezie Swoosh.

First Name Initial		Last Name Initial	
A	Sparkle	A	Slidelle
B	Crimson	B	Dondertail
C	Jingle	C	Sugartoes
D	Freezie	D	McBlitzen
E	Ring-a-Ding	E	Sleighbell
F	Gingerbread	F	Prancer-Dancer
G	Tinsel	G	Toeloop
H	Flaky	H	Roofrunner
I	Pom-Pom	I	Candycane
J	Cocoa	J	Angelpie
K	Peppermint	K	Snowdrop
L	Jolly	L	Kringle
M	Nog	M	Hollyberry
N	Pixie	N	Pinecone
O	Cookie	O	Highboot
P	Ringo	P	Blackbutton
Q	Twinkle	Q	Popsicle
R	Pickle	R	Tingle
S	Frisky	S	Frostwillow
T	Nimble	T	Fruitcake
U	Snip-Snap	U	Gingersnap
V	Nutsy	V	Swoosh
W	Snowsprite	W	Crackle-Log
X	Noel	X	Sprucetip
Y	Yulie	Y	Frostowitz
Z	Elvira	Z	Stripysock

How Much Christmas Spirit Do You Have?

Are you nuts about Noel? Decide which statements are true about you, then add up your "trues" to find out how high you score on Santa's Christmas Spirit Chart.

1. I start singing Christmas carols in May.
2. My favourite Halloween costume is a Santa suit.
3. I am SUPERB at wrapping presents.
4. I love to see the happy look on people's faces when they open my gifts — I always choose just the right present for each person.
5. I have a large collection of very ugly Christmas sweaters. I adore them.
6. Mmm . . . I adore the scents of pine, bayberry and clove.
7. My all-time favourite movie is a Christmas movie.
8. Fruitcake! Yum!
9. My favourite colours are red and green.
10. I wish I had a pet reindeer.
11. My favourite season is winter.
12. I secretly wish all snowmen would come to life, just like Frosty.
13. I have kept a snowball in the freezer until summer.
14. I know all the names of Santa's reindeer.
15. I know what all the gifts are for each of the twelve days of Christmas.
16. I own a reindeer-antler headband.
17. I have organized a Secret Santa gift exchange with my friends.
18. I never forget to put out cookies and milk for Santa.
19. I leave a treat for the reindeer too.
20. I wish every day of the year were Christmas!

SCORING

Give yourself one point for each true statement.

HOW YOU RATE . . .

0 Zero percent Christmas spirit. Is that you, Scrooge?

1-5 Twenty-five percent Christmas spirit. You're more of a Halloween person.

6-10 Fifty percent Christmas spirit. You enjoy the holidays, but also enjoy Easter bunnies, Canada Day fireworks and your own birthday.

11-15 Seventy-five percent Christmas spirit. You really know how to get into the spirit of the holidays! You're fun to be around and the go-to elf for all things Noel.

16-20 One hundred percent Christmas spirit. Are your parents elves? Your blood runs red and green!

What Kind of Holiday Treat Are You?

Are you nutty like a fruitcake? Or are you traditional, like shortbread?
Take this quiz to discover the secret recipe for being you.

1. What is your favourite colour?
    a. Red. > Go to question 2.
    b. White. > Go to question 3.
    c. Green. > Go to question 4.
2. Which is your favourite sport?
    a. Dogsledding. > Go to question 5.
    b. Ice-skating. > Go to question 7.
    c. Skiing. > Go to question 8.

3. Which shape?
    a. Star. > Go to question 9.
    b. Diamond. > You are SHORTBREAD.
4. Which pattern?
    a. Herringbone. > Go to question 5.
    b. Plaid. > Go to question 6.
5. People say you are . . .
    a. sweet. > You are FRUITCAKE.
    b. strange. > You are MINCEMEAT PIE.
    c. genius! > You are FIGGY PUDDING.
6. Choose one:
    a. Christmas EVE. > You are GINGERBREAD.
    b. Christmas DAY. > You are BUCHE DE NOEL
7. You prefer to . . .
    a. go to the ballet to see *The Nutcracker*. >
    You are BUCHE DE NOEL.
    b. listen to someone reading "The Night Before Christmas." > You
    are CANDY CANE.
    c. watch an animated TV special about Christmas. > You are
    EGGNOG.

8. You prefer . . .
   a. warm and toasty. > You are EGG NOG.
   b. cool and refreshing. > You are CANDY CANE.
9. You prefer . . .
   a. to try new things. > You are SPICED APPLES.
   b. the tried and true. > You are MAPLE FUDGE.

YOU ARE . . .

Buche de Noel. You are a bit old-fashioned. You are easygoing and tend to roll with the punches. You keep a journal that you call your daily log.

Candy cane. You can be both sweet and spicy. You like stripes. You have nice minty breath.

Eggnog. You love people and animals, especially chickens. You get along with everybody. You like terrible puns, brand-new notebooks and crazy hats.

Figgy pudding. You have the best sense of humour! No wonder everyone laughs when you come into the room. You have tremendous stick-to-it-iveness.

Fruitcake. You can be something of a nut. Not everyone understands you, but your close friends adore you. You enjoy mixing it up.

Gingerbread. You have a warm personality. You are very bright. You frequently leave crumbs on the table.

Maple fudge. You are loyal and sentimental. You keep scrapbooks and remember all your friends' birthdays. You can be something of a sap.

Mincemeat pie. You are something of an oddball — some might even call you flaky. You go your own way and don't let naysayers get you down. You don't mince your words. You enjoy meeting new people.

Shortbread. You prefer the simple life. You can be short tempered, but be assured — everyone likes you.

Spiced apples. You are a loving and generous person with a heart of gold. You can be unpredictable, and have a taste for the unusual. You dream of exotic travel and are not afraid to experiment.

Christmas Crackers

1. Where do snowmen keep their money?
   a. In a snow bank.
   b. In a snow squallet.
   c. Under their hats.
2. Santa has a pet cat. What is her name?
   a. Santa Paws.
   b. Santa Claws.
   c. Meowstletoe.

3. What do sheep say at Christmas?
   a. Merry Christmas to ewe.
   b. Season's bleatings.
   c. Fleece Navidad.
4. Why is it so cold at Christmas?
   a. Because it's a n*ice* time of year.
   b. Because it's Decem*brrrrrr*.
   c. No one really snows.
5. Which of Santa's reindeer has the worst manners?
   a. Rudolph, because he's rude.
   b. Cupid, because he's always flirting with the other reindeer.
   c. Dancer, because he keeps turning up the party music.
6. Do Christmas trees knit?
   a. Yes — they knit Christmas stockings.
   b. Yes — since they have plenty of needles!
   c. No — they do needlepoint.
7. What do you get when you cross reindeer with cows?
   a. Ho, ho. Uh-oh.
   b. A *Moo*rry Christmas.
   c. Sleigh bulls.

8. What is Santa's favourite snack?
   a. Peanut butter and jolly.
   b. *Brrrr*itos.
   c. Bear claws.
9. What's a dog's favourite Christmas carol?
   a. "Bark! The Herald Angels Sing."
   b. "Deck the Halls with Bow-wow-wows of Holly."
   c. "Do You Hear What I Hear?"
10. What do grizzlies use to decorate their Christmas trees?
    a. *Grrrr*lands.
    b. Holly *bear*ies.
    c. *Bear*ly anything.

SCORING

Give yourself one point for each correct answer.

1. a	3. b	5. a	7. c	9. a
2. b	4. b	6. c	8. a	10. a

HOW YOU RATE . . .

0-3 Festive funster. Ho, ho, ho! You're so funny!

4-7 Holiday hilarity. You're a real card!

8-10 Christmas cracker. You crack everyone up!

18

How Well Do You Know Santa?

Santa knows *everything* about you. But how much do you know about him? Take this jolly quiz to find out!

1. In what country was Saint Nicholas born?
    a. Turkey.
    b. Canada.
    c. France.
2. In Finland, Santa rides on . . .
    a. a sleigh of gold.
    b. a goat named Ukko.
    c. a winged horse.
3. The first appearance of the name Santa Claus in English dates from . . .
    a. 1773.
    b. 1567.
    c. 1921.
4. Santa has been around forever. But he's been wearing red and white only since . . .
    a. Mrs. Claus made him a new outfit in 1635.
    b. he moved to Canada — they're Canada's colours, after all.
    c. 1931, when he was featured in advertisements for a cola company that used red and white as its brand's colours.
5. When did Santa marry Mrs. Claus?
    a. 1894.
    b. 1980.
    c. 1849.
6. What colour are Santa's boots?
    a. Red.
    b. Black.
    c. White.

7. In the famous poem "The Night Before Christmas," how is Santa's belly described?
   a. Like a bowl full of jelly.
   b. Round as a turkey's tum.
   c. Bouncy, flouncy.
8. What is Santa called in England?
   a. Saint Peter.
   b. Father Christmas.
   c. Kris Kringle.
9. Santa keeps a list of . . .
   a. the names of all his elves, so he doesn't mix them up!
   b. all the presents children have asked for.
   c. who has been naughty and who has been nice.
10. Scientists have calculated how fast Santa's sleigh must travel to deliver all the gifts on time. How fast does it go?
    a. 2,060 kilometres per hour.
    b. At the speed of sound.
    c. At the speed of light.

SCORING

Give yourself one point for each correct answer.

1. a
2. b
3. a
4. c
5. c. Mrs. Claus was first mentioned in a short story called "A Christmas Legend" by James Rees.
6. b
7. a
8. b
9. c
10. a

HOW YOU RATE . . .

0-3 Santa's reindeer. You know Santa better than the average reindeer!

4-5 Santa's elf. You are on a first-name basis with Santa.

6-10 Santa's BFF. You know everything about Santa. Are you Mrs. Claus?

Reindeer, Elf or Frosty?

What's your Christmas personality? Are you more like a reindeer, an elf or a magical snowman?

1. It's a summer day. Where are you?
   a. Working on your back-to-school to-do list.
   b. Standing in front of the open freezer — you hate hot weather.
   c. Snacking.

2. What's your favourite article of clothing?
   a. Hat, scarf and mittens.
   b. Pointy shoes.
   c. Birthday suit.
3. Which is your favourite food?
   a. Carrots.
   b. Ice cream cone.
   c. Gingerbread.
4. Which gift would you most enjoy finding under the tree?
   a. A toboggan.
   b. A corncob pipe.
   c. A woodworking tool kit.
5. Which book would you prefer to read?
   a. *The Hobbit.*
   b. *Bambi.*
   c. *The Adventures of Pinocchio.*
6. Which vacation spot appeals to you most?
   a. Iceland.
   b. Lapland.
   c. Disneyland.

7. What is your favourite colour?
   a. Red.
   b. Green.
   c. White.

8. You prefer to . . .
   a. work alone.
   b. work in a small group.
   c. be part of a great big team.
9. Which sport do you prefer?
   a. Cross-country skiing.
   b. Softball.
   c. Hockey.
10. Your favourite school subject is . . .
   a. Phys. ed.
   b. Science.
   c. Math.

SCORING

1. a1 b3 c5	4. a5 b3 c1	7. a1 b5 c3	9. a5 b3 c1
2. a3 b1 c5	5. a1 b5 c3	8. a3 b5 c1	10. a5 b1 c3
3. a5 b3 c1	6. a3 b5 c1		

HOW YOU RATE . . .

10-25 Elf. You are cheerful and industrious. You get along well with others and enjoy working with your hands. You favour brightly coloured clothing and anything flavoured with pumpkin spice. You tend to whistle while you work. You cry easily but laugh often.

26-38 Frosty. You enjoy your own company and are happiest when you can just hang out and chill. You love warm jammies, fuzzy slippers and a hot mug of cocoa — especially if it's stirred with a peppermint stick! You can be very sentimental; cute puppies and kittens make your heart melt!

39-50 Reindeer. You're the wild one in the bunch — no party can start without you! You love the great outdoors and engaging in active team sports. You have a tremendous appetite but are thinking of becoming a vegetarian. You are a real dear — always gentle and thoughtful.

Which Christmas Song Best Describes You?

Are you a joy to the world? Or more like a noisy little drummer boy?

1. Which do you prefer?
   a. New and shiny. > Go to question 2.
   b. Worn in and comfy. > Go to question 3.
   c. Mysterious and exotic. > Go to question 4.

2. Which travel destination appeals to you most?
   a. Hong Kong. > Go to question 5.
   b. Mount Everest. > Go to question 6.
   c. Hawaii. > Go to question 7.

3. Which do you prefer?
   a. Hanging with one or two friends. > Go to question 9.
   b. Your own company. > Go to question 10.
   c. Wherever there's action! > You are "ROCKIN' AROUND THE CHRISTMAS TREE."

4. You prefer . . .
   a. hip hop > You are "WINTER WONDERLAND."
   b. K-pop. > You are "WHAT CHILD IS THIS?"

5. You are a . . .
   a. night owl. > Go to question 8.
   b. early bird. > You are "DECK THE HALLS."
   c. sun lover. > You are "JOY TO THE WORLD."

6. Which would be your favourite way to be awakened in the morning?
   a. Your dog licking your face. > You are "AWAY IN A MANGER."
   b. Tunes. > You are "THE LITTLE DRUMMER BOY."

7. You have a spare hour at school. How do you fill it?
   a. Doing a fun logic puzzle. > You are "THE TWELVE DAYS OF CHRISTMAS."
   b. Drawing and colouring funny pictures of your friends. > You are "DECK THE HALLS."
   c. Absorbed in the exciting graphic novel you took out of the library. > You are "SILENT NIGHT."
8. Which best describes you?
   a. Leader of the pack. > You are "WINTER WONDERLAND."
   b. Lead singer. > You are "ROCKIN' AROUND THE CHRISTMAS TREE."
9. Which do you prefer?
   a. Chocolate. > You are "SANTA CLAUS IS COMING TO TOWN."
   b. Vanilla. > You are "JOY TO THE WORLD."
   c. Pistachio marshmallow crunch. > You are "SILVER BELLS."
10. Which do you prefer?
    a. Penguins. > You are "SILVER AND GOLD."
    b. Puffins. > You are "SILENT NIGHT."

YOU ARE . . .

"Away in a Manger." You love dogs. And cats. And bunnies. And horses! And bugs. And birds. And . . . Your friends joke you love animals more than people. *Psst* — they're right!

"Deck the Halls." You've got style, baby. You know how to dress and how to decorate. You're the one making the artsy picture frame out of old pop bottles and turning an ugly Christmas sweater into an up-cycled fashion item. The Hall of Fame awaits you.

"Joy to the World." You've got a cheerful, optimistic personality that is utterly contagious. You always look on the bright side and find the silver lining in every situation. Your hero and inspiration? Tigger.

"The Little Drummer Boy." You snap your fingers, tap your toes, whistle while you work. Music turns your crank. You love to play it, listen to it, write it, dance to it. You march to your own drummer.

"Rockin' Around the Christmas Tree." You're the bright bulb in the string, the glitter in the globe, the icing on the gingerbread cookie. No party can start without you!

"Santa Claus Is Coming to Town." And you've made the bed, set the table, and put out his favourite foods (the reindeer's too!). You love looking after people and making them feel welcome. You are the first to stand up to bullies and invite the new kid to sit at your table. No wonder people like you!

"Silent Night." You prefer the quiet life. A cozy couch, a new pack of coloured markers, a pair of noise-cancelling headphones. You're a deep thinker with plenty of original ideas.

"Silver and Gold." You have exquisite taste. You love the finer things in life. You collect old toys, buttons and pirate treasure. Yet you don't take yourself too seriously. You have a very loud voice.

"Silver Bells." You're very versatile. You like to try new things, but you are also happy with the tried and true — you've read your favourite book at least 100 times. You can be a bit of a ding-a-ling. You have a tinkly laugh.

"The Twelve Days of Christmas." You have a logical mind. You enjoy math and science, and like nothing better than to engineer a new gadget — a roller coaster for your teddy bear or a bed-making machine. People count on you.

"What Child Is This?" You are so mysterious — a spy with a legend, a movie star with a past, a kid with — heh-heh — a secret. You like to keep 'em guessing, sort of like Waldo.

"Winter Wonderland." You are so cool! Everyone wants to be like you. Underneath your chill exterior, though, you are very warm-hearted.

Christmas Mix-Up!

Oh no! The cat knocked over the Christmas tree! Now everything is scrambled! Can you sort out the names of these popular Christmas decorations?

1. GNKITOSC
2. RNMNTOAE
3. GANLE
4. LARGADN
5. HTAERW
6. LUEY OLG
7. NESLIT
8. LIMK NDA OCOKESI

SCORING

Give yourself one point for each correct answer.
1. STOCKING
2. ORNAMENT
3. ANGEL
4. GARLAND
5. WREATH
6. YULE LOG
7. TINSEL
8. MILK AND COOKIES

HOW YOU RATE . . .

0-3 ICNE ORWK. You've got Christmas covered.

4-7 EWLL ONDE. You're a sparkly Christmas card.

8 EPRCFET. You're a Christmas ORP!

How Well Do You Know Holiday Movies?

Get out the buttered popcorn and dig in to this celebration of the all-time best holiday movies.

1. In *The Nightmare Before Christmas*, he is the King of Halloween Town. What's his name?
   a. Jack O'Lantern.
   b. Jack Skellington.
   c. Jack Holiday.
2. Which angel is trying to earn his wings in *It's a Wonderful Life*?
   a. Clarence Odbody, Angel, Second Class.
   b. Nick Elodeon.
   c. Tiny Tim.
3. Which child is left *Home Alone*?
   a. Gavin McGillicuddy.
   b. Marvin Martian.
   c. Kevin McCallister.
4. Who is Buddy?
   a. The human elf in *Elf*.
   b. The tree in *A Charlie Brown Christmas*.
   c. The dog in *Home Alone*.
5. This movie is named after a train that travels to the North Pole. What's it called?
   a. *Thomas the Tank Engine*.
   b. *The Polar Express*.
   c. *The Candy Cane Train*.
6. What is the name of the Grinch's dog?
   a. Grunt.
   b. Cindy Lou.
   c. Max.

7. In *Miracle on 34th Street*, one little girl believes she has met the real Santa. What is her name?
   a. Maud.
   b. Anne.
   c. Susan.

8. Which muppet plays Charles Dickens in *The Muppet Christmas Carol*?
   a. Gonzo the Great.
   b. Kermit.
   c. Elmo.

9. In the film *Arthur Christmas*, what is the name of Santa's son?
   a. Arthur.
   b. Nick.
   c. Kris.

10. In what movie does Scott Calvin take Santa's place after an accident?
   a. *Jingle All the Way*.
   b. *A Christmas Story*.
   c. *The Santa Clause*.

SCORING

Give yourself five points for each correct answer.

1. b	4. a	7. c	9. a
2. a	5. b	8. a	10. c
3. c	6. c		

HOW YOU RATE . . .

0-20 Film buff. You enjoy settling in with favourite movies, new and old.

25-35 Movie maven. You put the pop in popcorn.

40-50 Cinema superstar. Roll out the red carpet — you're a celebrity film critic!

Christmas by the Numbers

You can count on Christmas for holly, jolly fun. Solve this number puzzle for even more holiday joy.

1. Start with the number of Santa's reindeer (not including Rudolph).
2. ADD the number of days of Christmas.
3. ADD the date that Boxing Day falls on.
4. DIVIDE by the number of turtledoves.
5. MULTIPLY by the number of Magi.
6. SUBTRACT the number of points on a snowflake.
7. ADD the number of drummers drumming.
8. DIVIDE by the number of letters in the word ANGEL.

SCORING

Give yourself one point for each correct answer. Give yourself ten
bonus points if you got all the answers, 1 to 8, correct *without using a
calculator.*

1. **8**
2. $8 + 12 = 20$
3. $20 + 26 = 46$
4. $46 \div 2 = 23$
5. $23 \times 3 = 69$
6. $69 - 6 = 63$
7. $63 + 12 = 75$
8. $75 \div 5 = 15$

HOW YOU RATE . . .

0-2 Math magus. You are beginning to unlock
the magic of numbers!

3-6 Noel know-it-all. You can count on having plenty of holiday fun!

7-8 Santa's solver. You get a thumbs-up from the boss.

18 Shining star. You're so gifted at math!

What a Wonderful World!

Do you think you know all about Christmas? See if you can spot the fakes among these Christmas traditions from around the world. Answer true or false for each statement.

1. In Austrian legend, a scary character called a Krampus captures naughty children and takes them away in a sack.
2. In Japan, Kentucky Fried Chicken is a traditional Christmas dish.
3. In the Philippines, people say goodbye to the old year by wearing their clothes backwards.
4. Icelandic traditional lore includes mischievous Yule Lads with names like Sausage-Swiper, Meat-Hook and Bowl-Licker.
5. On Christmas Eve, Norwegians hide their brooms so evil spirits can't ride away on them.
6. Brazilians celebrate with a kid-friendly kind of firework called "The Burning Schoolhouse."
7. In Venezuela, people who live in the city of Caracas go to midnight mass on roller skates.
8. In the region of Spain called Catalonia, people "feed" a log, then beat it with sticks until it "poos."
9. In Ukraine, people drape their Christmas trees in spider webs to bring good luck.
10. In Australia, people take snow and ice out of their freezers and use it to make Christmas angels on the hot sand.

SCORING

Give yourself one point for each correct answer.

1. True	4. True	7. True	9. True
2. True	5. True	8. True	10. False
3. False	6. False		

HOW YOU RATE . . .

0-3 Doubting Thomas. You don't make up your mind until you see three references.

4-7 Santa's helper. You know the world is full of surprises.

8-10 Three wise men. With triple the wisdom, no one can fool you!

The Christmas Goose

The Christmas goose flies around the world to discover some interesting Christmas traditions (and to escape becoming Christmas dinner!). Take this quiz to find out if you know as much as the goose!

1. What does the word "mistletoe" mean?
    a. Dangling feet.
    b. Dung on a twig.
    c. Kiss me!
2. What is frankincense?
    a. An aromatic resin gathered from a tree.
    b. A hot dog with spicy chili on top.
    c. A gold box holding treasure.
3. Seven out of ten British dog owners do what on Christmas?
    a. Dress Fido in furry slippers and a Santa hat.
    b. Buy their dog a gift.
    c. Take the dog to church for a special blessing over the animals.
4. In Hungary, a traditional meal for Christmas Eve includes . . .
    a. fish soup.
    b. pancakes.
    c. meatballs.
5. On the island of Bali, in Indonesia, Christmas trees may be made out of . . .
    a. chicken feathers.
    b. lambswool.
    c. straw.
6. In Jamaica, Christmas Eve is marked by . . .
    a. solemn prayers and fasting.
    b. fireworks.
    c. a festive Grand Market.

7. Where did the three Magi come from?
   a. The north.
   b. The east.
   c. The south.

8. In China, Santa Claus is called Sheng Dan Lao Ren. What does that actually mean?
   a. Funny, kind man.
   b. Round, laughing man.
   c. Old Christmas man.

9. What is a Christmas pudding usually made of?
   a. Apples.
   b. Chocolate.
   c. Plums.

10. On the stroke of midnight on Christmas Eve, it's said that . . .
    a. animals temporarily get the gift of speech.
    b. whatever you dream of will come true.
    c. the sun and the moon bow to the Star of Bethlehem.

SCORING

Give yourself three points for each correct answer.

1. b	4. a	7. b	9. c
2. a	5. a	8. c	10. a
3. c	6. c		

HOW YOU RATE ...

0 Goose egg. Better luck next year!

3-12 Goose fat. Thanks for playing!

15-21 Goosy gander. You put the ho, ho, ho — and the honk, honk, honk! — in holiday!

24-30 Canada goose. You scour the world in search of Christmas knowledge!

Kooky Christmas Doodle

To take this quiz, you will need a blank piece of paper, a pen or pencil and a friend. Have your friend read you the instructions while you do your best to follow them. Add up your score to find out what job in Santa's workshop best suits you! Don't forget to let your friend take a turn.

1. Draw Santa's sleigh.
2. Draw Santa.
3. Add presents to your drawing.
4. It's snowing! Draw snowflakes.
5. Add a Christmas tree.
6. Put ornaments on the tree.
7. And some tinsel.
8. Almost done! Add a chimney for Santa to slide down.

SCORING

Your sleigh is . . .
- • in the centre of the paper. 5 points
- • to the left of centre. 3 points
- • to the right of centre. 2 points

Santa is . . .
- • in the sleigh. 10 points
- • not in the sleigh. 0 points

Does Santa have a beard?
- • Yes. 5 points
- • No. 0 points

Does Santa have a hat?
- • Yes. 5 points
- • No. 0 points

Is Santa smiling?
- Yes. 5 points
- No. 0 points

The presents are . . .
- ALL in the sleigh. 15 points
- SOME in the sleigh. 10 points
- NOT in the sleigh. 5 points

How many presents are there?
- Give yourself 5 points for each present.
- Add 5 more points for each present that has a ribbon or bow.

How many snowflakes did you draw?
- One or two. Give yourself 5 points for each one.
- Three. Give yourself 20 points.
- Four or more. Give yourself 25 points.
- Give yourself 3 more points for each six-pointed snowflake.

Where are the snowflakes?
- In the sky, above or beside the sleigh. 10 points
- On the ground, below the sleigh. 5 points
- In or on the sleigh. 0 points
- A combination of any of the above. 10 points

Where is your Christmas tree?
- To the left of the sleigh. 5 points
- To the right of the sleigh. 3 points
- Above the sleigh. 1 point
- Below the sleigh. 10 points
- In or on the sleigh. 0 points

Did you draw a star or angel at the top of your tree?
- Yes. 15 points
- No. 0 points

How many ornaments did you draw?
- Three or fewer. Give yourself 1 point for each.
- Four or more. Give yourself 2 points for each.

What shape are the ornaments?
- Round. 1 point each
- Any other shape. 5 points each

Are the ornaments actually ON the tree?
- Yes. 5 points
- No. 0 points
- No. They're on Santa. 10 points

Is the tinsel on the tree?
- Yes. 10 points
- No. 0 points

Where is the chimney?
- To the left of the sleigh. 5 points
- To the right of the sleigh. 3 points
- Above the sleigh. 1 point
- Below the sleigh. 10 points
- In or on the sleigh. 0 points

Is there smoke coming out of the chimney?
- Yes. Take away 5 points.
- No. Add 1 point.

HOW YOU RATE . . .

0-50 Santa's chef. You make delicious scrambled eggs!

51-100 Santa's jester. You're fun to have around.

101-150 Santa's sports coach. You're great at giving advice.

151-200 Santa's engineer. You are great at coming up with new inventions.

201-250 Santa's workshop manager. You're the genius behind the scenes.

251+ Santa's reindeer. You're flying high! That's because you took the quiz already, isn't it?

Reindeer Games

Rudolph's not the only one who enjoys playing reindeer games. Take this quiz to discover how well you know Santa's flight crew.

Match the clue to the reindeer.

1. The brightest of them all!
2. His name originated from a German word that means "thunder."
3. He is *sooo* romantic.
4. He's super at salsa and fandango.
5. He's named for a heavenly body.
6. His name is also the name for a female fox.
7. He is Dancer's bestie.
8. His favourite sporting event is the 100-metre race.
9. He adores playing defence in football.

a. Donner
b. Blitzen
c. Comet
d. Dancer
e. Prancer
f. Vixen
g. Dasher
h. Cupid
i. Rudolph

SCORING

Give yourself one point for each correct answer.

1. i	4. d	7. e
2. a	5. c	8. g
3. h	6. f	9 . b

HOW YOU RATE ...

0-3 Reindeer rookie. Don't eat that snow. Those speckles are reindeer droppings.

4-6 Reindeer pen pal. You correspond with Comet instead of Santa, don't you?

7-9 Rudolph. You know so much about reindeer, you might actually be one!

Christmas Critters

Attention all animal lovers! How much do you know about these popular Christmasy critters?

1. Reindeer are closely related to what Canadian animal?
   a. Unicorns.
   b. Moose.
   c. Caribou.
2. What animals did the three Magi ride on their way to see the baby Jesus?
   a. Horses.
   b. Donkeys.
   c. Camels.
3. Which two animals almost always appear in nativity scenes?
   a. Donkey and ox.
   b. Dog and cat.
   c. Horse and cow.
4. Turtledoves symbolize . . .
   a. peace.
   b. love.
   c. joy.
5. True or false: reindeer can have red noses.
   a. True.
   b. False.

6. How likely would you be to find a partridge in a pear tree?
   a. Very likely. It's their favourite place to hang out.
   b. Not likely. Partridges prefer to roost on the ground.
   c. Extremely likely! Partridges spend their entire lives in pear trees.

7. Partridges get their name from . . .

    a. a character in Greek mythology named Perdix who was thrown off a tall tower by a jealous teacher, but saved from death when he was turned into a bird by the gods.

    b. a character in Celtic mythology who was sent to the underworld half the year in punishment for eating a pear seed.

    c. a mountain in Wales.

8. In the song "The Twelve Days of Christmas," "five golden rings" describes the neck of what kind of animal?

    a. A bird.

    b. A tiger.

    c. A monkey.

9. "Calling birds" refers to what kind of bird?

    a. Bluebird.

    b. Blackbird.

    c. Parrot.

10. True or false: Santa's reindeer are probably all female.

    a. True.

    b. False.

SCORING

Give yourself seven points for each correct answer.

1. c
2. c
3. a
4. b
5. a. They have a dense network of blood vessels in their noses that can make them appear red.
6. b
7. a
8. a
9. b
10. True. Male reindeer typically lose their antlers by the middle of November, while female reindeer tend to keep theirs through the Christmas season.

HOW YOU RATE . . .

0-21 Animal abecedarian. You've got the ABCs of animal lore under your Santa belt.

28-42 Gifted with animals. Your knowledge of animals is anything but fuzzy!

49-70 Chief veterinarian, North Pole. Santa relies on you to keep his reindeer fit!

Reindeer-in-Training

1. Which ballet is frequently performed at Christmastime?
   a. *Swan Lake.*
   b. *The Nutcracker.*
   c. *Waltz of the Reindeer.*
2. Which is NOT a traditional Christmas dessert?
   a. Panettone.
   b. Fruitcake.
   c. Blueberry pie.
3. Which famous monarch always gives a speech on Christmas Day?
   a. Monarch butterfly.
   b. The British king or queen.
   c. The Swedish king or queen.

4. Finish this sentence: "Yes, Virginia, there is . . ."
   a. a Santa Claus.
   b. a Father Christmas.
   c. a ghost of Christmas past.
5. After Jesus was born, Mary and Joseph took him to what country?
   a. Syria.
   b. Egypt.
   c. Lebanon.
6. In Charles Dickens's *A Christmas Carol*, who said, "God bless us, every one!"
   a. Scrooge.
   b. Marley.
   c. Tiny Tim.

7. What is the name of the town featured in the classic film *It's a Wonderful Life*?
   a. Bedford Falls.
   b. Bedford Hills.
   c. Bedford Corners.
8. What is Frosty's nose made of?
   a. Coal.
   b. Corncob.
   c. Button.
9. Finish this sentence: "All I want for Christmas is . . ."
   a. my two front teeth.
   b. a kiss from you.
   c. a visit from Santa.
10. What brought Frosty the snowman to life?
    a. A kiss from Polly.
    b. A magic silk hat.
    c. A wish upon a star.

SCORING

Give yourself one point for each correct answer.

1. b	4. a	7. a	9. a
2. c	5. b	8. c	10. b
3. b	6. c		

HOW YOU RATE . . .

0-3 Cupid. You're so loveable!

4-6 Dasher. You're so quick!

7-10 Rudolph. You're so bright!

Where Should You Celebrate Christmas Next Year?

Do you love walking in a winter wonderland? Or would you prefer your holiday somewhere hot, hot, hot? Do you enjoy your holiday time calm and quiet or exciting and adventurous? Take this quiz to find out your ideal destination for celebrating Christmas.

1. Your favourite sport is . . .
   a. soccer. > Go to question 2.
   b. ice-skating. > Go to question 3.
   c. skateboarding. > Go to question 4.
2. Your favourite fruit is . . .
   a. the orange. > Go to question 5.
   b. the apple. > Go to question 6.
   c. the mango. > Go to question 7.
3. You prefer . . .
   a. coloured pencils. > Go to question 8.
   b. black permanent markers. > Go to question 9.
4. You prefer . . .
   a. painting. > Go to question 10.
   b. paintball. > Go to question 11.
5. Your favourite card game is . . .
   a. solitaire. > Next year you'll be on A DESERT ISLAND.
   b. go fish. > Next year you'll be sport fishing in the BAHAMAS.
   c. rummy. > Next year you'll be hanging with the pirates of THE CARIBBEAN.
6. Your favourite animal is . . .
   a. the horse. > Next year you'll be on A WORKING DUDE RANCH.
   b. the dog. > Next year you'll be mushing in NUNAVUT.
   c. the cat. > Next year you'll be in PARIS, FRANCE, dahling!

7. You love to . . .

a. try new things. > Next year you'll be exploring beautiful BALI.

b. solve puzzles. > Next year you'll be at NASA headquarters in HOUSTON, TEXAS.

c. surprise people. > Next year you'll be a member of Cirque du Soleil in LAS VEGAS, NEVADA.

8. Beets or carrots?

a. Beets. > Next year you'll be in WHITEHORSE to see the northern lights.

b. Carrots. > Next year you'll be in QUEBEC CITY to feast on tourtière.

9. Classic movies, or new releases?

a. Classic movies. > Next year you'll be in OLD MONTREAL for that retro vibe.

b. New releases. > Next year you'll be in VANCOUVER for a taste of big city trends.

10. Red or green?

a. Red. > Next year you'll ooh and ahh over the sights of ROME, ITALY.

b. Green. > Next year you'll find four-leaf clovers when you spend Christmas in DUBLIN, IRELAND.

11. Roller coaster or funhouse?

a. Roller coaster. > Next year you'll be in HOLLYWOOD, CALIFORNIA, to hang with the stars.

b. Funhouse. > Next year you'll be in DISNEY WORLD, the place to make your dreams come true.

I'm Dreaming of a Green Christmas?

Are you dreaming of a white Christmas or a green one? Take this *plant*astic quiz to find out how well you know these beautiful and fragrant Christmas plants.

1. What colour are the berries on mistletoe?
   a. White.
   b. Purplish black.
   c. Red.
2. Poinsettias are native to what country?
   a. Canada.
   b. United States.
   c. Mexico.
3. The Christmas cactus got its name because . . .
   a. its spines look like stars.
   b. it is native to Bethlehem, where Jesus was born.
   c. it blooms around Christmastime.
4. Which common Christmas plant is actually a parasite?
   a. Mistletoe.
   b. Holly.
   c. Poinsettia.
5. In parts of England, holly was placed around children's beds on Christmas Eve to . . .
   a. bring good luck.
   b. help Santa.
   c. keep away mischievous imps.
6. Which Christmas greenery represents eternal life?
   a. The tree.
   b. The wreath.
   c. The poinsettia.

7. The colourful part of poinsettia plants is . . .
   a. the flowers.
   b. specialized leaves called bracts.
   c. their fruit.
8. Which is the most popular Christmas tree in Canada?
   a. Scotch pine.
   b. Balsam fir.
   c. Blue spruce.
9. Which popular Christmas plant is poisonous?
   a. Mistletoe.
   b. Poinsettia.
   c. Holly.
   d. Both a and b.
   e. a, b and c.
10. In Mexico, this Christmas plant is called "holy night flower" (*flor de la noche buena*). What is it?
    a. Poinsettia, because of a legend about a girl who had no gift to bring to church on Christmas. The green weeds she picked were transformed by a miracle into a stunning flower.
    b. Christmas rose, because of its incredible beauty.
    c. Amaryllis, because it always blooms on December 25.

SCORING

Give yourself five points for each correct answer.

1. a
2. c
3. c
4. a
5. c
6. b. Because circles have no beginning or end.
7. b
8. a
9. e
10. a

HOW YOU RATE . . .

0-15 Green thumb. You're going to have a holly, jolly Christmas!

20-30 Red and green. Your Christmas spirit keeps growing!

35-50 Evergreen wreath. There's no end to your Christmas knowledge!

Which Christmas Tree Ornament Are You?

1. You are . . .
   a. great at wrapping gifts.
   b. great at making gifts.
   c. great at choosing gifts.
2. Which is your favourite Christmas treat?
   a. Hot chocolate.
   b. Candy canes.
   c. Gingerbread.
3. You tend to be . . .
   a. affectionate. You love to give lots of hugs.
   b. entertaining. You love to make others laugh.
   c. clever. You are quick to solve problems and
   come up with new ideas.
4. You prefer to read . . .
   a. a joke book you can share with friends.
   b. an absorbing novel or funny chapter book you can escape into.
   c. a book full of fascinating facts you can dip into in small doses.
5. Which do you prefer?
   a. Sparkles and glitter.
   b. Red and purple.
   c. Black and white.
6. Which job would you prefer at Santa's
   workshop?
   a. Designing the toys.
   b. Putting the toys together.
   c. Wrapping the toys attractively.
7. Which article of clothing would you hate to live without?
   a. Sunglasses. You are a star and can't be seen without them.
   b. Your Santa hat. It tells the world how much fun you are.
   c. Your undies. Undies are the best.

8. Which school supply do you like the best?
   a. Your eraser in a funny shape. It wipes out boo-boos.
   b. Your coloured markers. They let you draw fantabulous drawings of your friends.
   c. Your ruler. It keeps you on the straight and narrow!
9. Which Christmas character appeals to you most?
   a. Frosty the snowman.
   b. Rudolph the red-nosed reindeer.
   c. The Grinch.
10. You're in the Christmas play! Which character do you want to be?
    a. One of the three Magi.
    b. The donkey.
    c. The Christ child.

SCORING

1. a1 b3 c5	3. a5 b3 c1	5. a5 b3 c1	7. a5 b3 c1	9. a1 b5 c3
2. a5 b1 c3	4. a5 b1 c3	6. a3 b1 c5	8. a1 b5 c3	10. a3 b5 c1

HOW YOU RATE . . .

10-19 Icicle. You are totally cool. There you are, chillin' with your besties, making your own fun. Sometimes you can be a drip, but mostly you are incredibly sharp.

20-29 Candy cane. Once people get to know you, they're hooked! You are very well read.

30-35 Globe. You're steady and reliable. You're fun to be around — when you're in the house, everybody has a ball!

36-42 Bell. You are definitely no ding-a-ling — in truth, you are very bright! You're a little on the loud side. Sometimes your ears ring.

43-50 Star. Your personality really shines, but sometimes you miss the point. You are competitive and like to come out on top.

What's in Your Stocking?

Have you been good or . . . *ummm* . . . not quite as good as you should have been this past year? Will Santa be leaving you coal in your stocking, or a lovely little treat?

1. How often did you lose your temper?
   a. Never.
   b. Not that often.
   c. More than you'd like.
   d. Every day! You're hot tempered.
2. How often did you do your homework?
   a. Always.
   b. Most of the time.
   c. Ooops! You were more forgetful than you'd like.
   d. Homework? Never!
3. How often did you do your chores without being asked?
   a. Always.
   b. Most of the time.
   c. You frequently needed a reminder or two, but you did them then.
   d. You only did your chores with lots of reminding.
4. How often did you brush your teeth?
   a. Three times a day, without fail.
   b. At least once a day, without any reminders.
   c. Whenever someone reminded you.
   d. As little as possible. Brushing your teeth is so boring!

5. Uh-oh! You broke a friend's pen. What did you do?
   a. Told him right away and apologized.
   b. Told him you did it when he noticed.
   c. Kept it a secret.
   d. Blamed it on someone else.
6. What about "please," "thank you" and "I'm sorry"? Did you use them regularly?
   a. Of course!
   b. Most of the time.
   c. You try to be polite, but sometimes you forget.
   d. Oh. Right. You should say those.
7. How would you describe your typical behaviour?
   a. Cheerful.
   b. Calm.
   c. In your own world.
   d. Grumpy.

8. When you played games or sports with your friends, you were typically . . .
   a. easygoing. It's all for fun, right?
   b competitive but fair.
   c. determined to win at all costs.
   d. a sore loser or spoilsport.
9. How quick were you to offer help to a friend?
   a. You'd offer before they even asked, if you saw they needed help.
   b. Very quick.
   c. You'd help if you could and it wasn't too hard.
   d. Your friends don't ask you for help.

10. How strict were you with the truth?

    a. You try not to lie, unless it's a white lie so someone's feelings aren't hurt.

    b. You never lie.

    c. You lie now and then to get out of something.

    d. It's fun to make up stories! They're not really lies, are they?

SCORING

1. a5 b3 c2 d1
2. a5 b3 c2 d1
3. a5 b3 c2 d1
4. a5 b3 c2 d1
5. a5 b3 c2 d1
6. a5 b3 c2 d1
7. a5 b3 c2 d1
8. a5 b3 c2 d1
9. a5 b3 c2 d1
10. a5 b3 c2 d1

HOW YOU RATE . . .

10 Coal. You must be related to the Grinch.

11-16 One old, dusty mint. You must be related to Scrooge.

17-21 A slightly chewed dog toy. You're going to try a little harder next year, right?

22-30 An orange. You are so sweet!

31-40 An adorable, small stuffed animal. You are warm and loving.

41-50 Five golden rings. You're as good as gold.

Will You Survive Christmas Dinner?

The turkey is roasted, and the scents of gingerbread and pumpkin pie fill the air. You take your seat at Grandma Gert's side and prepare to dig in to Christmas dinner. Will you survive the meal, satisfied and happy, or will you *explode*?

1. First course! It's a lovingly prepared fish soup. You . . .
   a. tuck your napkin under your chin and grab your soup spoon. Fish soup is your fave! > Go to question 2.
   b. take a sniff. Take another sniff. *Yuck.* You fish out the one floating carrot and feed it to your dog Doozie, who licks it and flounces away. > Go to question 3.
2. Uncle Wiggly proudly brings out his tray of homemade dinner rolls. Soft and white and slathered in butter. You . . .
   a. take one. > Go to question 3.
   b. take two. > Go to question 4.
   c. take two with your left hand and one with your right. > Go to question 5.

3. The main event! Out comes the turkey, the stuffing, the cranberry sauce, the sweet potatoes, the mashed potatoes, the lasagna, the Brussels sprouts, the creamed onions and the luscious gravy. You . . .
   a. take a small serving of everything. > Go to question 6.
   b. take a large portion of everything — it's Christmas, after all! > Go to question 7.
   c. take heaping spoonfuls of mashed potatoes, sweet potatoes, stuffing and cranberries. Then slather it all with gravy. > Go to question 8.
   d. take one slice of white turkey. > Go to question 9.

4. The main event! Out comes the glazed ham, the Christmas goose stuffed with chestnuts, the sausage-and-onion dressing, the mashed turnip, the fried carp and the perogies. You . . .

a. feel slightly queasy. Must be the dozen mini-quiches, piles of smoked almonds and 76 salsa-topped chips you ate while waiting for Aunt Minnie to arrive. You ask to be excused. > Go to question 8.

b. open your eyes wide. It's all your absolute favourites! You help yourself to everything, then ask for seconds. > Go to question 7.

c. eat a few bites of goose, a perogy and one small slice of ham. You've seen what's coming for dessert and want to leave room. > Go to question 6.

5. Time for a toast! Everyone raises their glasses and sings out, "Merry Christmas!" You . . .

a. down your glass of apple cider in one swallow. All those tasty dinner rolls made you so thirsty! > Go to question 7.

b. take a small sip of your apple cider. You don't much like the stuff. > Go to question 6.

c. snag another dinner roll while no one is looking. > Go to question 8.

6. Time for dessert! Out comes the buche de Noel — the yule log — a lovely rolled chocolate cake slathered in whipped cream. You . . .

a. eat your slice happily. yule logs are one of your all-time favourite desserts. > You SURVIVE and enjoy the magic of Christmas.

b. wonder aloud where the flaming plum pudding is. Your dad promised you a flaming plum pudding! > You stomp away from the table in a snit. You eventually EXPLODE from a fit of overexcitement and bad temper.

c. eat your slice. It's very good. When no one is looking, you slip

an untouched slice from Aunt Minnie's plate and eat that too. > You lick your lips in satisfaction, but then feel a strange sensation in your tummy. It groans and knots. Uh-oh! Maybe you shouldn't have eaten that extra slice of buche. You hurl yourself away from the table and EXPLODE all over Uncle Wiggly's antique rug. He never forgives you.

7. Time for dessert! It's mincemeat pie and fruitcake! You . . .

a. accept a piece of each. YUM! > You SURVIVE, because clearly you have a cast-iron stomach and a hollow leg.

b. take the cake. You are a nut, after all. > You SURVIVE.

c. try the pie. > It is delicious. Filled with gooey brown stuff, bits of weird dried fruits, raisins. You especially like how the double scoop of vanilla ice cream melts over the slightly warm cru— *KABOOM!* You EXPLODE.

8. Everyone's got Christmas crackers. *Pop! Pop! Pop!* What's inside yours?

a. A tiny puzzle. > It's utterly charming. You become so absorbed in trying to solve the puzzle you absently nosh on whatever hits your plate. Without realizing it, you have consumed raw oysters, chestnuts in port wine gravy, crispy pig ears and a dozen mincemeat tarts. You EXPLODE.

b. A silly joke. > It's so bad it makes you lose your appetite. You wind up shoving your food around on your plate and eating barely anything. You SURVIVE, but wake up in the middle of the night extremely hungry. You sneak down to the kitchen and on your way spot ALL THE COOKIES. You devour twelve of the snowmen with the little red sprinkles on them. You EXPLODE.

c. A paper crown. > You put it on your head and make everyone laugh by doing silly faces and voices. Then you stand on your chair and start singing "Deck the Halls" a cappella. Your Aunt Minnie shouts at you, and your dad sends you to your room. You SURVIVE because you didn't get a chance to eat the mincemeat pie, which was riddled with salmonella. Luckily, you were able to call 9-1-1 and save your entire family.

9. Time for dessert! Out comes a flaming plum pudding! You . . .

a. fill up on sugar cookies in the shapes of reindeer. Do people actually expect you to eat dessert that's on fire? > The reindeer cookies are delicious, but they leave you feeling a bit jumpy. Now, with nothing but reindeer in your belly, you feel a little ill. You doze off in front of the fire for a moment. Your mom touches your head and *tsks* — it seems you have a fever. No wonder your appetite was off. She bundles you off to bed, where you have dreams of sugar plums dancing on your tummy. You DON'T EXPLODE, but wish you would.

b. look at it sideways. You're something of a picky eater, and this weird dessert is certainly outside of your comfort zone. You pass and happily eat the orange that was in your stocking. > You SURVIVE, but everyone in your family thinks you're peculiar. (Like they should talk!)

c. try a bite. > The flaming plum pudding is incredibly delicious but incredibly hot. It touches off a firestorm in your belly, igniting the single piece of white-meat turkey you ate. Who knew how flammable fowl could be? You EXPLODE, and are sent hurtling through the roof into the sky, where you land in Santa's sleigh and are delivered to a four-year-old in Minsk who'd asked Santa for a new best friend.

You're Santa!

You're Santa Claus! Will you succeed as the big guy or ruin *the whole world's* Christmas?

1. It's the day before Christmas! Hooray! You are so looking forward to your evening of fun, frolic and air travel. You . . .

    a. yawn, stretch, and jump into your red suit. > Go to question 2.

    b. sing Christmas carols as you brush your teeth. (Your favourite is "Santa Claus Is Coming to Town.") > Go to question 3.

    c. check your schedule. You'd hate to get off to a late start! > Go to question 4.

2. You head to the kitchen to make your breakfast (gingerbread cereal with hot cocoa). Mrs. Claus is nowhere to be seen. You can't possibly do everything by yourself! Christmas would be ruined without her! You . . .

    a. don't worry. You realize she must be overseeing final preparations in the workshop. > Go to question 3.

    b. worry a little, but you're hungry. You dig in to your breakfast. She'll turn up soon enough. > Go to question 22.

    c. worry a lot. You go look for her pronto! > Go to question 5.

3. You eat a hearty breakfast (gingerbread cereal with hot cocoa) AND a hearty lunch (tuna on rye with a pickle). It's time to fly! You . . .

    a. harness the reindeer and off you go! > Go to question 16.

    b. give a speech of thanks to your noble, hard-working elves. > Go to question 17.

4. According to your chief elf, Nogg, you are running 38 minutes behind in bicycle and sugar plum fairy doll manufacture. If you don't put on extra staff, you won't be finished in time for takeoff. You . . .

a. roll up your red sleeves and get to work making sugar plum fairy dolls. > Go to question 18.

b. call Polar Bear Point Winter Workshop and ask them to send over three dozen bears and a few narwhals to help complete the work. > Go to question 19.

5. You head to the workshop. No Mrs. Claus! You . . .

a. ask your chief elf if she's seen her. > Go to question 6.

b. run out into the snow, shouting her name. > Go to question 7.

6. Your chief elf says, "Mrs. Claus said you needed some glue. She took the sleigh and said she'd be back before sundown." You . . .

a. grow suspicious. Mrs. Claus has never done anything like that before, not on the day before Christmas! And you have plenty of glue. > Go to question 8.

b. breathe a sigh of relief. You know you can always count on Mrs. C. to help out in a pinch. > Go to question 9.

7. You forgot to put your boots on this morning, and the snow is deep. You begin to get cold and tired. Where is Mrs. Claus? Where are your elves? Where are your reindeer? You . . .

a. stumble and fall on your face. > You start to freeze. But then you hear a loud rumble. It is a huge, white, shaggy abominable snowman coming to eat you! You must not allow that to happen though — billions of people are depending on you! So you burrow down into the snow and kick with all your might. You are able to dig a tunnel under the snow right back to the workshop. You emerge into the workshop with the abominable snowman chasing you! But once you arrive indoors, the snow on the snowman starts to melt. It's Mrs. Claus! She'd gone out to bring the reindeer some candy canes and got lost in a sudden blizzard! You hug each other happily. CHRISTMAS IS SAVED.

b. get sleepy. > You have a magical dream in which a nutcracker comes to life and fights with a rat king. Sugar plum fairies come and dance on your face. Their feet tickle. You try to brush them away and feel something warm and solid. It is Rudolph, blowing on your face with his sweet reindeer breath! With frozen fingers, you grab on to his harness and he whisks you back to your workshop. Mrs. Claus serves you hot mulled cider and doughnuts to bring you back from the brink of becoming an ice pop. (She had only gone to the outhouse!) CHRISTMAS IS SAVED!

8. In your workshop's shed, you have a flying snowmobile, just a little runabout for quick trips to that place in Winnipeg that makes those yummy Buffalo-style wings. You rev her up and go out to look for Mrs. Claus. You clear the clouds and there she is, flying your sleigh. You . . .

a. silently glide in behind her. > Go to question 10.

b. call out, "Honeybunchkins!" > Go to question 11.

9. You ask your chief elf how the present-wrapping is going. She says, "We are running low on red bows and gold ribbon." You say . . .

a. "Impossible! I bought three billion tonnes of each on sale last year on Boxing Day." > Go to question 12.

b. "Okay. Then use the blue bows and silver ribbon." > Go to question 13.

10. You discover Mrs. Claus has twelve elves tied up in the belly of the sleigh. She is up to no good! What has happened to your loving and faithful spouse? You . . .

a. pull your snowmobile up beside her and clear your throat. > She looks at you with a strange look. You realize it is not your wife flying that sleigh after all; it is the Grinch, dressed in Mrs. Claus's clothes! And one of those elves in the back is your own dear, darling wife! You sprinkle Christmas magic dust over the Grinch and his face softens. He laughs. You gently take the reins from his hands and steer the sleigh back to the North Pole. You untie the elves and your wife, who are very grateful that you saved them. CHRISTMAS IS SAVED.

b. whistle for the reindeer. > At the familiar sound, they screech to a halt and turn toward you. Mrs. Claus crosses her arms. "Why do you get to have all the fun?" she says. "Don't I do half the work?" You apologize — it really hasn't been fair all these years. So you agree to let her deliver the presents this year. You kiss and make up. CHRISTMAS IS SAVED.

11. Mrs. Claus hears her familiar nickname, and her back stiffens. She turns and you see her eyes are spinning like whirligigs. She's been enchanted by a melting zombie snowman. And the snowman is sitting behind her in the sleigh, about to eat her brain! You . . .

a. reach into your right pocket and pull out a toy fishing rod you had stashed there last night — it needed another coat of paint. > You cast out a line and snag the zombie snowman's scarf. You yank the rod to pull the monster out of the sleigh and cut the fishing line. The zombie snowman falls over Florida, causing a freak blizzard. MRS. CLAUS AND CHRISTMAS ARE BOTH SAVED.

b. reach into your left pocket and pull out a blow-dryer meant for a teenage boy in Saskatoon. > You aim it at the zombie snowman and melt him. Mrs. Claus's eyes pop back to normal. She says, "Oh, Nick! You came just in the nick of time!" You both laugh at her silly joke, and CHRISTMAS IS SAVED.

12. Your eyes narrow. Is there a thief in Santa's workshop? You ...
a. take off your red suit and get into a teeny green elf outfit. You skulk around looking for shifty behaviour. Finally you see it! Your chief elf, Nogg, has pockets stuffed with gold ribbon! > Go to question 14.

b. shake your head. What a ridiculous thought. You turn on the workshop p.a. system and announce, "If you've seen the missing bales of red bows and gold ribbon, please report to *Santa 1* immediately." > Go to question 15.

13. The parcels decorated with blue and silver ribbons and bows look splendid next to the red-and-gold ones in the sleigh. It's now time for you to take off. You ...
a. ask your chief elf, Nogg, to check your sleigh to make sure it's flight ready. > Go to question 20.

b. click your tongue to signal the reindeer it's time to hoof it. > Go to question 21.

14. You tap Nogg on the shoulder. She turns and looks at you with an expression of utter horror, then bursts into tears! Finally you get her story. Another elf she'd never seen before came to her with a written order to put aside all the red bows and gold ribbon. She thrusts the paper into your hand — it is supposedly signed

by you, but it is a forged signature. You laugh, "Ho, ho, ho!" Someone's playing a practical joke! It must be that old trickster elf, Plink. > You rescue the bows and ribbon and SAVE CHRISTMAS.

15. Your chief elf comes running up to you. "I'm sorry, Santa. I forgot to log in the delivery of red bows and gold ribbon. They're all right here!" You clap Nogg on the shoulder and instruct the error-prone elf to bring the supplies to the assembly line. > All the presents are wrapped in the Saint Nick of time, and CHRISTMAS IS SAVED!

16. It is a beautiful night for a sleigh flight. You swing down low to admire the quiet woodlands. A big white blob rises up from the ground. It is a friendly yeti! Its hairy paw grabs a hold of your sleigh, and the sleigh wobbles and tips! You . . .

a. shout, "Gee! Haw!" to your reindeer and they rise higher into the air. > It works! You break free of the yeti's grip. Only one gift is shaken loose — it's the one for the yeti! It's a hairbrush. CHRISTMAS IS SAVED!

b. shout, "Run, run, Rudolph!" > But Rudolph can't run. His little legs are scrabbling uselessly as the yeti tugs the sleigh. The sleigh begins to tip and all the gifts tumble from your sack! They land in a heap in a clearing. You sigh and suggest to the reindeer they take a break. You land in the clearing. All the woodland animals come out to help you sort the gifts. CHRISTMAS IS SAVED.

17. Your elves shout, "Huzzah!" and give you a round of applause. You climb into your sleigh and click your tongue. The reindeer take off

and you ride up, up, up into the sky! But oh no! You accidentally left your magical sack of gifts in the workshop! You will be late if you go back to get it! You . . .

a. call Mrs. Claus on your Santa satellite phone. > She says, "You forgot your magical sack of gifts again, didn't you? Don't worry, I'm right behind you on the spare snowmobile." You look over your shoulder and there she is, swinging the magical sack. CHRISTMAS IS SAVED!

b. say, "Rudolph, it's up to you!" > Rudolph nods and says, "I'm on it." He slips from his harness and scampers back to the workshop, where the elves are craning their necks to the sky and praying for your return. Through the gloom they see Rudolph's red nose. "Oh no," Nogg says. "It's Rudy. We're doomed." But Rudolph does not care what the mean old elves say about him. He is brave! He is strong! He is a klutz! He crash-lands into the sack of presents. The tie at the top gets tangled in his antlers. While all the elves laugh, he snorts, stamps his foot, and emits a gentle toot from his hindquarters. The magical toot lifts him gently into the air and wafts him toward your sleigh, which is circling in a holding pattern. You untangle the sack from Rudolph's antlers and give him TWO packages of pretzels, knowing your trust in this odd deer was the right choice. CHRISTMAS IS SAVED!

18. Your arms pump. You grab each half-finished fairy doll off the assembly line and add wings, rhinestones and a lick of magical

plum paint faster than the speed of light. You do the work of
46 ½ elves, and the sugar plum fairy dolls are finished before
snacktime. Then you . . .

a. get to work on the bicycles. > Alas, your arms
are so tired! You make mistake after mistake and
you are now 42 minutes behind! Luckily
you are not alone. Mrs. Claus is there to help.
She works at your side, as she has done all these
years, to make Christmas a success. The bikes
are finished in the Saint Nick of time, and
CHRISTMAS IS SAVED!

b. take a break for a sip of peppermint tea and a
bite of fruitcake. > You get sleepy and doze off.
You wake up in a panic — it's already dark and there are still
three million bicycles to make! But then you hear the sweet,
soft sounds of reindeer singing "Over the River and Through the
Woods." While you slept, all the
polar animals came to help build
the bikes. Christmas
magic worked its charms, and
the bikes are finished and
loaded into the sleigh. "Merry
Christmas, Santa," the polar
bears growl. "Merry Christmas,"
sing the beluga whales. It is the
BEST CHRISTMAS EVER.

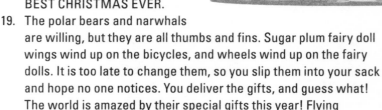

19. The polar bears and narwhals
    are willing, but they are all thumbs and fins. Sugar plum fairy doll
    wings wind up on the bicycles, and wheels wind up on the fairy
    dolls. It is too late to change them, so you slip them into your sack
    and hope no one notices. You deliver the gifts, and guess what!
    The world is amazed by their special gifts this year! Flying

bicycles! Rolling fairy dolls. CHRISTMAS IS SAVED!

20. Nogg checks your sleigh. She says, "You're good to go, Santa Baby!" You . . .

a. narrow your eyes. There is something weird about this elf. > She is green. You jump out of the sleigh and yank off her hat. She is not an elf — she is the Grinch! *Ahhh!* Your elves come running and tie the Grinch up with green ribbons. You check your sleigh and discover one of the runners was loose! That evil Grinch! Yet there was no harm done. CHRISTMAS IS SAVED!

b. thank her but decide to check the sleigh yourself, just to be sure. > You discover a tiny icicle lodged in the left runner. It could have been a disaster! Thanks to your safety awareness and sharp eyes, CHRISTMAS IS SAVED!

21. The reindeer take off! You soar into the air, higher and higher. Suddenly, the sleigh slows down. Something's gone wrong! You . . .

a. climb over the edge to check the runners. > There's some tinsel wrapped around the left one! You pluck it off. The sleigh lurches forward, free again! You clamber back into the sleigh and continue on your journey. CHRISTMAS IS SAVED!

b. keep flying until you arrive at the first house on your list. > You land on the roof. You shimmy down the chimney, instructing Rudolph to use his red flashlight nose to check the sleigh. When you return, Rudolph says you were out of moose gas, and he offers to supply you with some reindeer gas. You laugh, "Ho, ho, ho!" and accept his offer. CHRISTMAS IS SAVED, even if you have to hold your nose!

22. Mrs. Claus comes in from gathering wood and sees you eating breakfast. She starts to cry. You ask her what is wrong and she

says, "You never make me breakfast! You are so selfish!" You . . .

a. feel terrible. She is right. You apologize, make her North Pole toast, and happily eat breakfast together before heading off to the workshop. > Go to question 9.

b. say, "How can I, the great giver of gifts, be selfish?" She says, "I dare you to do Christmas deliveries without my help!" and stomps out. > Go to question 23.

23. You feel very hurt and sad, but also annoyed. Of course you can manage Christmas without Mrs. Claus! You're Santa, after all! But nothing seems to go right. You trip in the snow on your way to the workshop and give yourself a black eye. When you get there, you discover the elves have knocked off work and there are still thousands of presents to finish! You . . .

a. call Mrs. Claus on the Santa satellite phone. > Go to question 24.

b. decide to finish the presents by yourself. > Go to question 25.

24. Mrs. Claus accepts your apology. She returns to the workshop and as soon as the elves see her, they hop-to! They know who's the real boss of Santa's workshop! CHRISTMAS IS SAVED!

25. Your arms pump. You must make 3,000 tricycles and 76,000 puppets before noon if you are to set off on time. But you're beginning to get sleepy. And your fingers are blistering. Your chief elf, Nogg, returns and begs you to make up with Mrs. Claus — she will help and everything will turn out okay! You refuse. Your stubbornness makes Nogg shake her head. "What has happened to the kind and generous Santa I know?" You shake your head. What? You're not kind and generous? Then how can you be Santa? Maybe you're a pretender? Maybe you're just an ordinary kid dreaming that you're Santa? You rub your eyes and suddenly you find yourself in your own bed. It is dawn on Christmas morning. You hear hooves on the roof. Thank goodness! The real Santa never had an argument with Mrs. Claus! CHRISTMAS IS SAVED!

Tree-Trimmer in Training

1. How does Santa get back up the chimney?
   a. He places his finger on the side of his nose, then gives a smile and a nod.
   b. He calls for his pipe, he calls for his bowl, and he calls for his fiddlers three.
   c. He sprinkles reindeer droppings on the hearth and *woosh!* up he goes.
2. What is Scrooge's first name?
   a. Screech.
   b. Ebenezer.
   c. Grinch.
3. In which direction should you stir mincemeat for good luck?
   a. What? You NEVER stir mincemeat!
   b. Clockwise.
   c. Counter-clockwise.
4. The Keller machine was used to make what familiar Christmas items?
   a. Candy canes.
   b. Tinsel.
   c. Candles.
5. What is the all-time bestselling Christmas single?
   a. "Jingle Bells."
   b. "Holly Jolly Christmas."
   c. "White Christmas."
6. What is wassail?
   a. A kind of ale.
   b. A kind of cake.
   c. A kind of spiced cider.

7. Which traditional dish was once popular for Christmas in England?

   a. Pigs' heads in mustard.

   b. Roast swan.

   c. Both a and b.

8. Are Christmas trees edible?

   a. Yes, completely. If you are a beaver.

   b. Some parts. Spruce tips, for example, are very nutritious.

   c. No.

9. Which Christmas song was broadcast from space in 1965?

   a. "White Christmas."

   b. "Jingle Bells."

   c. "Let It Snow."

10. True or false: December 25 is mentioned in the Bible as the date of Jesus's birth.

    a. True.

    b. False.

SCORING

Give yourself twelve points for each correct answer.

1. a	4. a	7. c	9. b
2. b	5. c	8. b	10. b
3. b	6. c		

HOW YOU RATE . . .

0-36 Christmas C. Your beautiful singing voice is an asset for carolling.

48-60 Christmas B. Your hard work is welcome for decking all those halls!

72-120 Christmas tree. Your knowledge of Christmas trivia isn't *tree*vial.

Silly Santa Riddles

1. What comes at the end of Christmas Day?
   a. The letter Y!
   b. Christmas Eve.
   c. Reindeer poo on the roof.
2. Where do Santa's elves go to become famous?
   a. Hollywood!
   b. Bethlehem. You see stars there.
   c. The top of the tree.
3. What did the dog breeder get when she crossed an Irish setter with a pointer at Christmastime?
   a. A headache.
   b. A Labradoodle.
   c. A point*setter*!

4. What do wild animals sing at Christmastime?
   a. "Noisy Night, Howly Night."
   b. "Jungle Bells."
   c. "Arooo in the Manger."
5. What's the best thing to put into Christmas dinner?
   a. Your teeth!
   b. *Great*vy.
   c. Pie do you ask?
6. What do you call an elf who steals gift wrap from the rich and gives it to the poor?
   a. Elf Capone.
   b. Ribbon Hood.
   c. Boxing Day Bob.
7. What's red, white and blue at Christmastime?
   a. A Habs candy cane.
   b. An American candy cane.
   c. A sad candy cane.

8. Who is Santa's favourite singer?
   a. Michael Bauble.
   b. Elfish Presley.
   c. Carole King.
9. What did Mrs. Claus say to Santa Claus the night before Christmas?
   a. Looks like rain, dear!
   b. Merry Christmas!
   c. Don't forget your sleigh keys!
10. What sport is Santa great at?
    a. Bobsledding, of course!
    b. Karate, because he has a black belt.
    c. All of them, because he's gifted in everything.

SCORING

Give yourself two points for each correct answer.

1. a	4. b	7. c	9. a
2. a	5. a	8. b	10. b
3. c	6. b		

HOW YOU RATE . . .

0-6 Ring-a-ding riddler. You're a funny one!

8-14 Yule time yuk yuk. You can make even the Grinch chuckle.

16-20 Holly, jolly jokester. You've got a gift for holiday humour.

Are You a Christmas Trivia Star?

1. What famous scientist was born on Christmas Day?
2. What country is the birthplace of eggnog?
3. In the song "Jingle Bells," how many horses pull the sleigh?
4. Who wrote *A Christmas Carol*?
5. True or false: celebrating Christmas was once banned in England.
6. Harry Potter receives a Christmas present while he is at Hogwarts. It has a note pinned to it. What is the present?
7. Elves wear this ornament on the tips of their shoes. What is it?
8. How many points are there on a snowflake?
9. In *A Charlie Brown Christmas*, who kissed Lucy?
10. Who is Tiny Tim's father?
11. What famous wizard wishes people would give him socks for Christmas?
12. What date does Twelfth Night fall on?
13. Who was the first ghost to visit Ebenezer Scrooge?
14. Which mushroom is sometimes called the Christmas mushroom?
15. "The Adventure of the Blue Carbuncle" is a Christmas story featuring what famous detective?
16. What three ingredients are the essential components of shortbread?
17. What is the French name for Santa?
18. Which saint introduced the idea of singing carols in church?
19. Which two enemy armies declared a temporary truce for Christmas in 1914?
20. In the song "Grandma Got Run Over by a Reindeer," what did Grandma go home to get?

SCORING

Give yourself one point for each correct answer.

1. Isaac Newton.
2. England.
3. One.
4. Charles Dickens.
5. True, in 1647.
6. An invisibility cloak.
7. Bell.
8. Six.
9. Snoopy.
10. Bob Cratchit.
11. Dumbledore.
12. January 5.
13. Jacob Marley.
14. The enoki mushroom.
15. Sherlock Holmes.
16. Butter, sugar, flour.
17. Père Noël.
18. Francis of Assisi.
19. British and German.
20. Her medication.

HOW YOU RATE . . .

0-8 Stocking stuffer. Your head is stuffed with Christmas knowledge!

9-16 Christmas light. You are incredibly bright!

17-20 Tree topper. You are tops at trivia!

Nightmare Before Christmas

There's no snow. When your family went to pick out your tree, the only one left was small and scraggly! And the power has gone out! Will you survive your Christmas disaster, or will it turn out to be the Worst Christmas Ever?

1. You've been so busy with last-minute stuff at school — the holiday concert, your science test, giving out Secret Santa gifts — you were a bit of a scatterbrain. As a result you . . .

    a. accidentally sent your Christmas wish list to the South Pole! > Go to question 2.
    b. tripped bringing up the antique tree ornaments from the basement. > Go to question 3.
    c. discovered your family accidentally left you home alone for Christmas! > Go to question 4.

2. You realize it's too late to get your list to Santa in time, so you . . .
    a. resolve to go see him yourself. You decide to travel by bicycle. > Go to question 5.
    b. resolve to go see him yourself. You decide to travel by dogsled. > Go to question 14.
    c. collapse in a weepy heap. > Go to question 23.

3. There are broken ornaments all over the floor! You . . .
    a. quickly sweep up the broken ornaments before anyone sees the mess. > Go to question 15.
    b. shout, "Oh no!" > Go to question 16.

4. You feel . . .
   a. afraid. > Go to question 18.
   b. sad. > Go to question 19.
   c. excited! You're going to have some fun! > Go to question 22.

5. You set off on your bicycle trip with three pairs of mittens, two scarves and a tall pair of boots. You stay warm and toasty until you arrive in North Bay, Ontario. Then your teeth start to chatter. You . . .
   a. go to a store called Mel's Cheap Boutique to buy warmer clothes. > Go to question 6.
   b. pedal faster to stay warm and get to the North Pole sooner. > Go to question 7.

6. Mel turns out to be a kind, wee lady with a broom. She suggests you may have trouble cycling through the boreal forest — the North Pole is awfully far away. She offers to take you on her broomstick. You . . .
   a. agree. > Go to question 8.
   b. say, "No thanks," and pedal off. > Go to question 9.

7. You pedal and pedal and before long you have pedalled all the way to Churchill, Manitoba. You . . .
   a. take a break to drink some water. > Go to question 12.
   b. take a break to look out over the frozen expanse of Hudson Bay. > Go to question 13.

8. You hop on the broomstick behind your new bestie, Esmerelda, and up you go into the air. It's just like Santa's sleigh! You pass over beautiful snowy forests. Then you see a clearing ahead. There's a house there. But wait. Is it made out of gingerbread? Could your new pal actually be a witch? Alas, she is a witch, and your days are numbered. So sorry. It's the WORST CHRISTMAS EVER.

9. Your brief stop has helped warm you up and you now have extra energy. You keep pedalling. Before long you come to a mountainous area. You have to walk your bike up the hill. It is cold. It is steep. You are thirsty and tired. Suddenly a snowman appears in your path. You . . .

   a. ask its name. > Go to question 10.

   b. tremble in fright. > Go to question 11.

10. The snowman answers, "Frosty. I've just magically come to life. You need some help?" You say yes, and Frosty gives you his magic scarf. It wraps around you and lifts you into the air, whisking you to Santa's village and workshop. Once you are there, you explain the mishap with the letter, and Santa says, "No worries, my friend Pingu the penguin sent it to me by Freezair. I have it right here! And you've been such a good kid, you are getting everything on it! Since you're here, would you like to help me deliver presents? And then I will bring you home." You wind up having the BEST CHRISTMAS EVER.

11. Perhaps you should have run instead of standing there quaking in your boots, because the snowman turns out to be an abominable one who wants to smother you in snow and take you home to be a house elf. You live out the rest of your days mopping up puddles. So sorry. It's the WORST CHRISTMAS EVER.

12. The water in your bottle has frozen. You start to cry. Your tears freeze on your face. The snow freezes on your back. Before long you are covered in shaggy white icicles. A police officer comes by and mistakes you for a polar bear. You are hoisted up and tossed into the Churchill polar bear jail, where you remain for the rest of your frosty life. So sorry. It's the WORST CHRISTMAS EVER.

13. You see a blurry white shape on the ice. It's a polar bear! And it's magnificent! You ride your bike out onto the ice to greet the beast. The beast is entranced by your bike. You let her try it. Together you pedal north on the ice until you reach the North Pole and Santa. Your new best friend, Georgie, gets a lovely pair of earmuffs from Santa. You get all the items on your list and an express trip home on Santa's sleigh. BEST CHRISTMAS EVER!

14. Unfortunately, your dog, Spot, has no interest in pulling your sleigh. You get no farther than the broken flagstone in your front walk. So sorry. It's the WORST CHRISTMAS EVER.

15. You resolve to repair and replace all the broken ornaments. You get out tubes of glue and painstakingly stick antlers on deer and carrots on snowmen. Everything is good as new! You hang them on the tree. Your cat, Sneaky, immediately knocks the tree over. You . . .
    a. yell, "Oh no!" > Go to question 16.
    b. lay your head in your arms and start to weep. > Go to question 17.

16. Uncle Cornelius happens to be visiting. He comes running at your shout. He says, "Don't worry. We can stand the tree back up and fix all the ornaments." You have so much fun putting the tree back together with Uncle Cornelius! And when you are finished, it looks more beautiful than it ever did before. Everyone *oohs* and *aahs* and gives you plenty of thanks and hugs when Uncle C. tells them about all your hard work. It turns out to be the BEST CHRISTMAS EVER!

17. A Christmas angel hears your crying. She appears in front of you in a shimmering golden glow. "Never fear, for I am here." She does a twirl and magically the Christmas tree rights itself and all the ornaments fly back to their spots, whole and perfect — as if they'd never broken. A pile of gleaming gifts for everyone in your family also magically appears under the tree. It turns out to be the BEST CHRISTMAS EVER!

18. You barricade yourself in your house and booby trap it in case any burglars decide to pay you a Christmas visit. They do! Your booby traps work, and you catch the bumblers no problem. Just as you put the finishing touches on their ribbon handcuffs, your parents appear at the front door. They realized they forgot you before they got on the plane. Off you go with your very apologetic folks to a whirlwind holiday that includes tropical beaches, amusement parks and scuba diving. It turns out to be the BEST CHRISTMAS EVER!

19. Of course you do. You get sadder and sadder. You start to feel yourself turning mean and green. Are you turning into — *gasp!* — the Grinch? You . . .

    a. decide to do something very nice right away to stop grinching ASAP. > Go to question 20.
    b. check in a mirror. > Go to question 21.

20. You hurriedly make your way to a nearby food bank and volunteer to help hand out Christmas baskets to needy people. While you work, the people at the food bank contact your parents. They arrive just as the last baskets are given away. They are so proud of you and so sorry to have left you by accident! They take you to a fancy hotel where you spend the BEST CHRISTMAS EVER!

21. You are not the Grinch — you've turned into Kermit the Frog. For the rest of your life you speak in a funny voice and suffer the indignity of being green. So sorry. It's the WORST CHRISTMAS EVER.

22. You do all the things you're not supposed to do: Slide down the bannister. Turn the TV up really loud. Eat dessert first. But after a while you get bored and lonely. You . . .

a. call your BFF and explain what has happened. > They say, "You can't be alone on Christmas! We'll be right over to get you!" Your friend's folks come and get you and set you a place at their dinner table. The food is fabulous. You eat and eat all your favourites — turkey and stuffing, mashed potatoes and sweet potatoes. But you kind of wish your parents were there too. And then there's a knock at the door. It's your folks! They realized they

forgot you before they left town! They came back for you! Your friend makes room for your parents at the table and you all toast Merry Christmas. It turns out to be the BEST CHRISTMAS EVER!

b. decide to build a snowman. > You go outside and roll one giant snowball, one medium snowball and one small snowball. You put them together, then add twig arms, a button nose and eyes and a holly berry smile. You think your snowman looks cold, so you give it your hat. Suddenly it comes to life! Your snowman and you have lots of adventures together. It turns out to be the BEST CHRISTMAS EVER!

23. Lucky for you, a very kind penguin named Snafu opens your letter. Snafu has always wanted to see a real live puffin — are they truly cuter than penguins? So Snafu decides to take your letter to the North Pole. Penguins are slow swimmers, though, so Snafu does not deliver your letter to Santa until the year you turn 30. That year you are extremely surprised on Christmas morning to receive ALL THE GIFTS you asked for back in the day — the video game, the markers, the drone. It turns out to be the BEST CHRISTMAS EVER, just a bit delayed.

Holiday Headscratchers

Can you solve these brain busters before Santa slides down the chimney?

1. There are 365 days in a year. What number is Christmas Day? (8 points)
2. Altogether, how many hooves do Santa's reindeer have? (Don't forget Rudolph!) (5 points)
3. If you received all the gifts in "The Twelve Days of Christmas," how many gifts would you receive? (8 points)
4. Can you figure out what number each letter stands for to make this equation true? (20 points)

   ```
     S A I N T
   + _N I C K
   = G I F T S
   ```

5. In chess a knight can move in any direction, either one space horizontally and two vertically or two horizontally and one vertically, making an L pattern. Beginning on an M square, guide your knight through the board below so that the spots it lands on spell out "Merry Christmas." (10 points)

M	T	S	X	C	C	G	H
E	L	I	W	V	R	B	I
S	E	M	Y	A	H	S	M
A	S	A	R	Y	T	R	T
M	T	A	E	E	I	C	T
S	H	C	S	E	A	S	R
M	T	E	M	R	Y	R	Y
C	H	A	Y	H	D	E	A

6. How many pairs of adjacent numbers that add up to 17 can you find? (5 points)

7. Decode this puzzle to come up with a Christmas wish: (5 points)
ppppp
EARTH

SCORING

Check your answers and add up your score.

1. 359
2. 36
3. 364
4.   79345
     + 4312
     = 83657
   (There may be other solutions. Give yourself an additional 30 points if you can find a second!)
5.

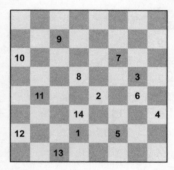

6. 14.
7. Peace on Earth

HOW YOU RATE . . .

0-30 Handbell ringer. You deserve a hand!

31-60 Caroller. Everyone's singing your praises!

61-91 Nutcracker. You've cracked the quiz!

Awesome Gift Giver or Gift-Giving Flop?

Do your friends and family open gifts from you with special excitement, because they know that whatever is under that shiny paper will be just what they wanted? Or do they give you the side eye and wonder what you were thinking? Take this quiz to find out what your peeps really think about their prezzies from you — if you dare.

1. When do you start making your list of people you intend to give Christmas gifts to?

   a. December 26. Right after Christmas ends, you start on next year's plans.

   b. Sometime in early fall.

   c. Ummm . . . the day before Christmas?

2. You always . . .

   a. know exactly what each person on your list would love!

   b. fret about what to give each person.

   c. get people gifts that you would like to receive.

3. You don't have lots of cash so you . . .

   a. don't give many gifts. But the ones you give are awesome!

   b. give a token, inexpensive gift to a few people.

   c. make personalized gifts for everyone you love.

4. "It's better to give than receive." Do you agree or disagree?

   a. Agree a million percent.

   b. You should probably say you agree . . .

   c. Who are you kidding? Getting presents is way better than giving them.

5. Do you make giving something to a charity part of your Christmas plan?
   a. Yes. Always.
   b. Sometimes.
   c. You haven't thought about it.

6. You worked hard making a gift for a friend, but they didn't give you a gift back! How do you feel?
   a. Fine. You gave them the gift because you wanted to, not because you expected one in return.
   b. Disappointed. You expect to always get something back when you give something.
   c. Worried. Maybe they didn't like your prezzie.

7. Which gift would you give?
   a. A fake spider to your friend who's afraid of spiders. It would be so funny!
   b. A green scarf to your pal who only wears black. They need a little colour in their wardrobe.
   c. An ABC board book to your newborn niece. It was your fave when you were a baby.

8. A kid you don't know that well gave you a Christmas gift, but you have nothing for them! What do you do?
   a. Just feel weird.
   b. Say thank you and apologize for not having a gift for them.
   c. Ask them what flavour of gum they like best, and get them a pack as a way of saying thank you.

9. Which is the WORST gift you could give someone?
   a. A cold shoulder.
   b. An old sweater with a jelly stain on it.
   c. A pair of socks.
10. When you give a gift, you . . .
   a. try to wrap it beautifully — how a gift looks is as important as what's inside it.
   b. wrap it as best you can.
   c. skip the wrapping — it gets tossed out anyway.

SCORING

1. a10 b5 c0	4. a5 b3 c1	7. a1 b2 c4	9. a5 b2 c4
2. a5 b2 c1	5. a5 b3 c1	8. a2 b3 c5	10. a5 b3 c1
3. a3 b3 c6	6. a5 b2 c4		

HOW YOU RATE . . .

14-20 Gift goof. An old, deflated soccer ball for Uncle Ted? A brick for baby Sam? Perfume for your very allergic Auntie Sneza? Your gifts leave people dazed and confused. It's true — choosing the perfect gift for everyone on your list takes plenty of advance planning. It's hard work to get it right. No wonder you don't always nail it. Your pals don't mind. Too much.

21-35 Gift horse. Your gifts are solid and reliable. Everyone *does* need a new toothbrush. Or a box of gluten-free crackers. And yes, shoelaces are very useful. Next year, however, you might consider giving everyone different colours of dollar-store gloves so they feel a little, you know, special.

36-55 Gift genius. You are the Angel of Giftgiving. You spend weeks, no — months, studying your friends and family, trying to figure out what would be the exact right thing to make each person leap with joy. Then you save your pennies — or get to work crafting — to make their secret wish come true. And the wondrous things you can do with wrapping paper! Best of all — your joy in gift giving is contagious.

Credits